PERGAMON INTERNATIONAL LIBRARY
of Science, Technology, Engineering and Social Studies

*The 1000-volume original paperback library in aid of education,
industrial training and the enjoyment of leisure*

Publisher: Robert Maxwell, M.C.

Worked Examples in Engineering Field Theory

APPLIED ELECTRICITY AND ELECTRONICS DIVISION

General Editor: P. HAMMOND

Other Titles of Interest in the Pergamon International Library

ABRAHAMS and PRIDHAM
Semiconductor Circuits: Theory Design and Experiments

ABRAHAMS and PRIDHAM
Semiconductor Circuits: Worked Examples

BADEN FULLER
Microwaves

BADEN FULLER
Engineering Field Theory

BROOKES
Basic Electric Circuits, 2nd Edition

BROOKES
Instrumentation for Engineers and Physicists

BINNS and LAWRENSON
Analysis and Computation of Electric and Magnetic Field Problems, 2nd Edition

CHEN
Theory and Design of Broadband Matching Networks

COEKIN
High-Speed Pulse Techniques

CRANE
Electronics for Technicians

CRANE
Worked Examples in Basic Electronics

DUMMER and GRIFFIN
Electronic Reliability: Calculation and Design

FISHER and GATLAND
Electronics: From Theory into Practice, 2nd Edition

GATLAND
Electronic Engineering Applications of Two-Port Networks

HAMMOND
Applied Electromagnetism

HAMMOND
Electromagnetism for Engineers

HANCOCK
Matrix Analysis of Electrical Machinery, 2nd Edition

HARRIS and ROBSON
The Physical Basis of Electronics

HINDMARSH
Electrical Machines and their Applications, 3rd Edition

HOWSON
Mathematics for Electronic Technology, 2nd Edition

PRIDHAM
Solid State Circuits

THOMA
Introduction to Bond Graphs and their Applications

The terms of our inspection copy service apply to all the above books. Full details of all books listed will gladly be sent upon request.

Worked Examples
in
Engineering Field Theory

by

A. J. BADEN FULLER, M.A., C.ENG., M.I.E.E.

Lecturer, Department of Engineering, University of Leicester

PERGAMON PRESS

OXFORD · NEW YORK · TORONTO

SYDNEY · PARIS · FRANKFURT

U.K.	Pergamon Press Ltd., Headington Hill Hall, Oxford OX3 0BW, England
U.S.A.	Pergamon Press Inc., Maxwell House, Fairview Park, Elmsford, New York 10523, U.S.A.
CANADA	Pergamon of Canada Ltd., P.O. Box 9600, Don Mills M3C 2T9, Ontario, Canada
AUSTRALIA	Pergamon Press (Aust.) Pty. Ltd., 19a Boundary Street, Rushcutters Bay, N.S.W. 2011, Australia
FRANCE	Pergamon Press SARL, 24 rue des Ecoles, 75240 Paris, Cedex 05, France
WEST GERMANY	Pergamon Press GmbH, 6242 Kronberg-Taunus, Pferdstrasse 1, Frankfurt-am-Main, West Germany

First edition 1976

Library of Congress Cataloging in Publication Data

Baden Fuller, A   J
Worked examples in engineering field theory.
(Applied electricity and electronics)
Includes index.
1. Field theory (Physics)—Problems, exercises, etc.
I. Title. II. Title: Field theory.
QC661.B23   1976   530.1′42′076   75-38900
ISBN 0-08-018142-2
     0-08-018143-0 (H)

Printed in Great Britain by A. Wheaton & Co., Exeter

Contents

Preface

This book arises from a lecture course given by the author to first-year
students in the Department of Engineering in the University of Leicester. It
is written as a companion to the author's other book on the same subject.[†]
Most lecture courses are best supplemented by worked examples and by practice
in working problems. *Engineering Field Theory* gives a detailed exposition of
the subject for those who are completely new to the subject and is supplemented
by a few worked examples and some problems. This book provides a summary of
the same theory together with a large number of worked examples, worked
solutions to all the problems given in *Engineering Field Theory*, and an equal
number of completely new problems together with their worked solutions. It
will be found useful by those students who have difficulty with the problems
in *Engineering Field Theory* or who need further worked examples and more
problems to solve. It will also be found to give a valuable summary of the
theory for those who, attending a course of lectures, have a detailed set of
lecture notes.

The contents of the book fall naturally into two parts. The first
develops the concept of flux starting with electric flux and proceeding to
applications in gravitation, ideal fluid flow, and magnetism. The second
part introduces the concept of potential, again starting with electrical
potential, and proceeds to applications in gravitation, electric conduction,
fluid flow through permeable media, conductive heat transfer, ideal fluid
flow, and magnetism. Attention is confined to static fields, and although
there is a chapter on electromagnetic induction, time-varying fields are not
included. The book has 104 diagrams, 145 worked examples, and 244 problems
with their worked solutions.

It is difficult to thank the many people who have contributed to the
development of the Field Theory course at Leicester, and I hope that lack of
acknowledgement will not be taken to imply lack of gratitude. Many of the
problems given in the book have been used for many years in the form of
duplicated examples papers by our students. My thanks are due to my various
colleagues who have taught this course for their contributions over some
years to these problems.

Leicester   *A. J. Baden Fuller*

[†] A. J. Baden Fuller, *Engineering Field Theory*, Pergamon Press, Oxford, 1973.

Introduction

Field theory gives a unified mathematical theory which can be used in a number of different physical situations. This book provides a summary of the theory together with worked examples, problems, and their solutions in the context of all its many applications, gravitation, electrostatics, magnetism, electric current flow, conductive heat transfer, fluid flow, and seepage. Each chapter consists of the following sections:

> Theory
> Examples
> Problems
> Solutions
> Further Problems
> Further Solutions

except for Chapter 1 which is introductory and has no Further Problems. The first section consists of a summary of the theory needed to solve the problems covered in that chapter. Those students who require a detailed development of the theory are referred to the author's other book on the same subject[†] which provides just the detailed treatment needed. Then a section of Examples are given where each example is followed immediately by a detailed worked solution. The Problems are printed with their Solutions on a separate page so that the student may attempt the Problems before being confronted with the Solutions. For those students who require further exercise in problem-solving, there is then another set of Further Problems which cover again the complete subject matter of the chapter. The Further Solutions (to the Further Problems) have been shortened as far as is consistent with an understanding of the solution.

[†] A. J. Baden Fuller, *Engineering Field Theory*, Pergamon Press, Oxford, 1973.

Students who are already familiar with *Engineering Field Theory* will
find that all the Problems in that book are repeated in this book. They are
identified in Table I.1. There is no bibliography in this book. The reader
is referred to the bibliography in *Engineering Field Theory*.

TABLE I.1. TO IDENTIFY THE PROBLEMS TAKEN FROM *ENGINEERING FIELD THEORY*

Problem numbers in this book	Problem numbers in *Engineering Field Theory*
1.1-1.5	1.1-1.5
2.1-2.10	2.1-2.10
3.1-3.10	3.1-3.10
4.1-4.5	4.1-4.5
5.1-5.5	5.1-5.5
6.1-6.5	5.6-5.10
7.1-7.10	6.1-6.10
8.1-8.10	7.1-7.10
9.1-9.10	8.1-8.10
10.1-10.7	9.1-9.7
10.8-10.10	9.18-9.20
11.1-11.10	9.8-9.17
12.1-12.10	10.1-10.10
13.1-13.10	11.1-11.10
14.1-14.10	12.1-12.10

PART I

Flux

CHAPTER 1

Flux

THEORY

In the physical phenomena of electricity, magnetism, and gravitation,
one body exerts an influence on another body which is some distance away from
it. Both Newton's law (eqn. (5.1)) and Coulomb's law (eqns. (2.1) and (7.1))
describe the action of a force between two bodies without there being any
physical contact between these bodies to transmit the force. In order to be
able to understand the phenomenon so that we can make practical use of it, it
is necessary to postulate a method by which a force can be transmitted from
one body to another. We postulate that an imaginary fluid issues from a body
and exerts a force on any other body in its path. It is not necessary to
postulate a physical model that is *the truth;* it is quite sufficient to
suggest a system which fits all the observable facts. Field theory makes use
of the concept of the emission of an imaginary fluid called *flux* because it
is helpful in an understanding of field-effect phenomenon, because it fits
the observable facts, and because it enables one to predict what will happen
in other situations. The basis of field theory is the explanation of observed
facts in such a form that it may easily be used to predict results in design
situations. The explanation may be fictitious in real terms, but it is
acceptable and useful to engineers because it gives the correct answers. As
the theory is built round the flow of an imaginary fluid, it also becomes
applicable to systems of electric current flow, conductive heat transfer, and
certain systems of real fluid flow.

In electrostatics, the field concept of flux considers that the imaginary
fluid is relentlessly pouring out of the charged body and by virtue of its
motion exerts a force on any other charged body in its path. In one sense,
flux is an imaginary fluid because it is neither a liquid nor a gas and we
cannot detect it with our normal senses. However, electric flux is real in
another sense because an electric charge will always detect it. Similarly, a
magnet will always experience a force due to a magnetic flux, and any lump of
matter will experience a force due to gravitational flux.

This book uses the international system of units (SI units) which are
given in Table 1.1. The normally used multiple and submultiple prefixes are

TABLE 1.1. INTERNATIONAL SYSTEM OF UNITS
Basic Units

Quantity	Unit	Symbol
Length	metre	m
Mass	kilogram	kg
Time	second	s
Temperature	kelvin	K
Electric current	ampere	A

Supplementary Units

Quantity	Unit	Symbol
Plane angle	radian	rad
Solid angle	steradian	sr

Derived Units

Quantity	Unit	Symbol
Area	square metre	m^2
Volume	cubic metre	m^3
Frequency	hertz	Hz (c/s)
Density	kilogram per cubic metre	kg/m^3
Velocity	metre per second	m/s
Angular velocity	radian per second	rad/s
Acceleration	metre per second squared	m/s^2
Angular acceleration	radian per second squared	rad/s^2
Force	newton	N ($kg\ m/s^2$)
Pressure	newton per square metre	N/m^2
Energy, quantity of heat	joule	J (N m)
Power, rate of flow of heat	watt	W (J/s)
Thermal conductivity	watt per metre kelvin	W/m K
Electric charge, electric flux	coulomb	C (A s)
Electric potential difference, electromotive force	volt	V (W/A)
Electric intensity, field strength	volt per metre	V/m
Electric resistance	ohm	Ω (V/A)
Resistivity	ohm metre	Ω m
Conductance	siemens	S ($1/\Omega$)
Conductivity	siemens per metre	S/m
Capacitance	farad	F (A s/V)
Electric flux density	coulomb per square metre	C/m^2
Magnetic flux, pole strength	weber	Wb (V s)
Magnetic potential difference, magnetomotive force	ampere	A
Magnetic intensity, field strength	ampere per metre	A/m
Inductance	henry	H (V s/A)
Magnetic flux density	tesla	T (Wb/m^2)

given in Table 1.2.

TABLE 1.2. MULTIPLE AND SUBMULTIPLE PREFIXES

Multiple or submultiple	Prefix	Symbol	Pronunciation
10^{12}	tera	T	tĕr′å
10^{9}	giga	G	jĭ′gå
10^{6}	mega	M	mĕg′å
10^{3}	kilo	k	kĭl′ō
10^{-3}	milli	m	mĭl′ĭ
10^{-6}	micro	μ	mī′krō
10^{-9}	nano	n	năn′ō
10^{-12}	pico	p	pē′cō
10^{-15}	femto	f	fĕm′tō
10^{-18}	atto	a	ăt′tō

The unit of temperature difference is the kelvin; however, temperature is usually measured according to the Celsius scale, $^{\circ}$C, which has the same temperature increment as the kelvin but a different zero so that 0°C $= 273.15$ K. Methods of dimensions are sometimes used as a check on the validity of equations. In mechanical systems, the three basic dimensions are length L, mass M, and time T. The other basic dimension is the electric current A and it is also convenient to use the electric potential difference V.

A *scalar quantity* can be completely represented by its amplitude. A *vector quantity* needs to be described by a direction as well as its amplitude. Throughout this book, subscripts are used to denote the components of a vector, the subscripts x, y, z in rectangular coordinates and r, θ, z in cylindrical coordinates. To give directional information to expressions in vector equations which otherwise lack it, use is made of the unit vector **U** whose direction is noted by the subscript added. Some properties of a vector are:

Addition: if   **C** = **A** + **B**,

then $C_x = A_x + B_x, \quad C_y = A_y + B_y, \quad C_z = A_z + B_z.$

Multiplication by a scalar: if   **C** = k**A**,

then $C_x = kA_x, \quad C_y = kA_y, \quad C_z = kA_z.$

Scalar multiplication:

$$k = \mathbf{A} \cdot \mathbf{B} = A_x B_x + A_y B_y + A_z B_z.$$

$$\textit{Vector multiplication:} \quad \text{if} \quad \mathbf{C} = \mathbf{A} \times \mathbf{B},$$

then $\quad C_x = A_y B_z - A_z B_y, \quad C_y = A_z B_x - A_x B_z, \quad C_z = A_x B_y - A_y B_x.$

EXAMPLES

1.1. Find the dimensions of acceleration, force, and work or energy.

Answer. In mechanics, acceleration is defined by

$$\text{acceleration} = \frac{d^2 x}{dt^2}.$$

This is length divided by time squared. Therefore the dimensions of acceleration are $L\,T^{-2}$.

Force is defined as mass x acceleration. Therefore the dimensions of force are $M\,L\,T^{-2}$.

Work is defined as force x distance. Therefore the dimensions of work are $M\,L^2\,T^{-2}$. Alternatively, the dimensions of work or energy can be obtained from the formula for kinetic energy, energy = mass x (velocity)^2. In electrical terms, the energy is given by the product of potential difference, current, and time. Therefore, energy = V A T.

1.2. Find the dimensions and SI units of the permittivity constant ε_0 taken from Coulomb's law (eqn. (2.5)).

$$f = \frac{q_1 q_2}{4\pi \varepsilon_0 r^2}.$$

Answer. Rearranging the equation gives an expression for ε_0.

$$\varepsilon_0 = \frac{q_1 q_2}{4\pi r^2 f}.$$

Putting this expression into dimensional terms gives

$$(\varepsilon_0) \equiv \frac{(A\,T)^2}{L^2\,M\,L\,T^{-2}} = \frac{A^2\,T^2}{L\,M\,L^2\,T^{-2}}.$$

Putting energy into both its mechanical and electrical dimensions enables the denominator of the dimension to be changed into electrical dimensions.

$$(\text{energy}) \equiv M\,L^2\,T^{-2} = V\,A\,T.$$

Therefore $$(\varepsilon_0) \equiv \frac{A^2\,T^2}{L\,V\,A\,T} = \frac{A\,T}{V\,L}.$$

These are the dimensions of capacitance per unit length. Therefore, from Table 1.1, the SI units of permittivity are farads per metre.

1.3. State whether the following quantities are scalar or vector quantities, giving reasons: time, velocity, force, and height.

Answer. *Time* is a scalar quantity. Time only exists in one dimension and does not need directional information.

Velocity is a vector quantity. Movement can occur anywhere in three dimensions, and velocity needs to be specified by its direction as well as by its amplitude.

Force is a vector quantity. The direction of action of a force needs to be specified as well as the size of the force.

Height is a scalar quantity. The height of a mountain or hill above sea-level is independent of the path taken to reach the top of the mountain or hill. If the level of the sea drops by a uniform amount everywhere, the height of all the land above sea-level will increase by the same amount.

1.4. Show that the work done is the scalar product of the two vectors (force) and (distance moved by the point of application of that force).

Answer. Work is done by a force when the point of application of that force moves in the direction of the force. Where the direction of the force and the movement are not parallel, the work done is the product of the distance moved and that component of the force parallel to the direction of movement. If the distance moved $\mathbf{l}$ is resolved into three perpendicular components, l_x, l_y, and l_z, the total work done is the sum of the three products of each component of the movement with the parallel component of the force. Therefore,

$$W = F_x l_x + F_y l_y + F_z l_z = \mathbf{F} \cdot \mathbf{l}.$$

1.5. Show that the area of a parallelogram is the vector product of the vectors representing two adjacent sides of the parallelogram.

Answer. The area of a parallelogram is given by the product of the base and the height. Vector multiplication can be represented as the product of the amplitude of one vector and that component of the other vector which is perpendicular to the original vector. Therefore the area of a parallelogram, two adjacent sides of which are represented by the vectors $\mathbf{B}$ and $\mathbf{C}$, is given by

$$\mathbf{A} = \mathbf{B} \times \mathbf{C}.$$

The vector **A** is perpendicular to the plane of the parallelogram. It is a
useful convention that an area can be represented by a vector whose
amplitude is equal to the area and whose direction is perpendicular to the
plane of that area.

PROBLEMS

1.1. In terms of length L, mass M, and time T determine the dimensions of: area, volume, velocity, acceleration, force, density, pressure, stress, work, and power.

1.2. Give the SI units of each of the quantities listed in Problem 1.1.

1.3. Name several physical quantities that are scalars and several that are vectors taken from all branches of engineering and science.

1.4. The universal law of gravitation, Newton's law, states that the force of attraction between two spherical bodies of mass m_1 and m_2 a distance d apart is

$$\text{force} = \frac{Gm_1m_2}{d^2},$$

where G is the gravitational constant. Find the dimensions (L, M, T) of G. In SI units it has the numerical value of $6 \cdot 67 \times 10^{-11}$. In what units is this quantity measured?

1.5. In c.g.s. and e.s.u. units the force between two point charges a distance d apart in air is given by

$$\text{force (in dynes)} = \frac{q_1q_2}{d^2}.$$

The same relationship in SI units is

$$\text{force (in newtons)} = \frac{q_1q_2}{4\pi\varepsilon_0 d^2}.$$

If $3 \cdot 0 \times 10^9$ e.s.u. of current equals $1 \cdot 0$ A, derive the value of ε_0.

SOLUTIONS

1.1. The method of finding the dimensions of a quantity has been shown in the solution to Example 1.1. The dimensions are:

Area = L^2.

Volume = L^3.

Velocity = (distance)/(time) = $L\ T^{-1}$.

Acceleration = (velocity)/(time) = $L\ T^{-2}$.

Force = (mass) x (acceleration) = $M\ L\ T^{-2}$.

Density = (mass)/(volume) = $M\ L^{-3}$.

Pressure = (force)/(area) = $M\ L\ T^{-2}\ L^{-2}$ = $M\ L^{-1}\ T^{-2}$.

Stress = pressure = $M\ L^{-1}\ T^{-2}$.

Work = (force) x (distance) = $M\ L^2\ T^{-2}$.

Power = (work)/(time) = $M\ L^2\ T^{-3}$.

1.2. The SI units can be obtained directly from the dimensions of each quantity by substituting, M = kg, L = m, and T = s. Alternatively, the SI units of each quantity is given in Table 1.1. The required answers are:

Area: m^2.

Volume: m^3.

Velocity: m/s.

Acceleration: m/s^2.

Force: $kg\ m/s^2$ = N.

Density: kg/m^3.

Pressure = stress: $kg/m\ s^2$ = N/m^2.

Work: $kg\ m^2/s^2$ = N m = J.

Power: $kg\ m^2/s^3$ = J/s = W.

1.3. See the solution to Example 1.3 for further information.

Scalars: height or altitude, time, energy, volume, potential difference, temperature.

Vectors: force, velocity, acceleration, gradient, electric current.

1.4. Rearranging the equation in the question gives

$$G = \frac{F\ d^2}{m_1 m_2}.$$

Therefore the dimensions of G are given by

$$(G) \equiv \frac{M\ L\ L^2}{T^2\ M^2} = \frac{L^3}{M\ T^2}.$$

The units of G can be obtained from the components of the rearranged equation or from the reduced dimensions. They are

$$N\ m^2/kg^2 = m^3/kg\ s^2.$$

1.5. In the e.s.u. equation, q is measured in e.s.u. and d in cm. In the SI equation, q is measured in C and d in m. According to the SI equation, two charges of 1·0 C a distance 1·0 m apart exert a force of $1/4\pi\varepsilon_0$ N. Substituting these sizes into the e.s.u. equation gives a force of

$$\text{force} = \frac{9\cdot0 \times 10^{18}}{10^4} = 9\cdot0 \times 10^{14} \text{ dynes} = 9\cdot0 \times 10^9 \text{ N.}$$

Therefore

$$\frac{1}{4\pi\varepsilon_0} = 9\cdot0 \times 10^9, \quad \varepsilon_0 = \frac{1}{36\pi \times 10^9}.$$

CHAPTER 2

Electric Flux

THEORY

An imaginary field, called flux, is used to describe the effect of a
force acting at a distance. Flow of the fluid generates the force as a
result of interacting with the flow from another flux source. Flux is
incompressible and has no mass or density.

Coulomb's law describes the force between two point charges of
electricity. If d is the distance between point charges q_1 and q_2, the
force is given by

$$\mathbf{f} = 9 \times 10^9 \mathbf{U}_d \frac{q_1 q_2}{d^2} \tag{2.1}$$

where $\mathbf{U}_d$ is the unit vector in the direction of the dimension d and the
force is one of repulsion for two charges of the same polarity. The
numerical constant in Coulomb's law is a function of the system of units
used and can be considered to be an experimental constant. At a constant
distance from an isolated point charge, the force is independent of the
position at which it is measured. The field of force is generated by the
flow of flux from the point charge, and the flux flow will be uniform in
all directions around the charge. The total flux from the point charge
is Ψ and it is made equal to the strength of the charge and measured in
coulombs.

The *flux density* is defined as the flux per unit area. It is a
vectorial quantity with the vector direction parallel to the flux flow. The
flux density at a distance r from a point charge q, having an equivalent
flux source of strength Ψ, is given by

$$\mathbf{D} = \mathbf{U}_r \frac{q}{4\pi r^2} = \mathbf{U}_r \frac{\Psi}{4\pi r^2} \ \text{C/m}^2. \tag{2.2}$$

The formal definition of vector flux density is

$$\mathbf{D} = \lim_{\text{area} \to 0} \left(\frac{(\text{flux through the area})}{(\text{area})} \right) \tag{2.3}$$

acting in a direction perpendicular to the plane of the area. If $\mathbf{D}$ is the
vector flux density at the charge q_2 due to the charge q_1, substitution of
eqn. (2.2) into eqn. (2.1) when $d = r$ gives

$$\mathbf{f} = 36\pi \times 10^9 \; \mathbf{D}q_2. \tag{2.4}$$

The numerical constant in eqn. (2.4) has dimensions. Its reciprocal is defined as a fundamental electrical constant called the *permittivity constant*, with the symbol ε_0,

$$\varepsilon_0 = \frac{1}{36\pi} \times 10^{-9} \; \text{F/m}.$$

In terms of the permittivity constant, eqn. (2.1) becomes

$$\mathbf{f} = \mathbf{U}_d \frac{q_1 q_2}{4\pi\varepsilon_0 r^2}. \tag{2.5}$$

The definition of vector flux density may be written in terms of the vector representation of area. For a uniform flux density $\mathbf{D}$ flowing through a plane area represented by the vector $\mathbf{A}$ as shown in Fig. 2.1, the total flux is given by

$$\Psi = \mathbf{D} \cdot \mathbf{A}. \tag{2.6}$$

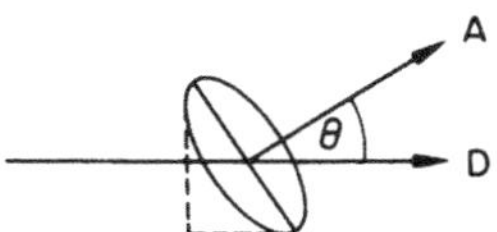

Fig. 2.1. Calculation of total flux with the
flux density oblique to a plane area.

For a non-uniform field, the flux density can still be considered to be uniform over a small element of area $d\mathbf{A}$, and the total flux can be obtained by integrating over the whole surface to be considered,

$$\Psi = \iint\limits_{\text{area}} \mathbf{D} \cdot d\mathbf{A}. \tag{2.7}$$

In terms of the permittivity constant, eqn. (2.4) becomes

$$\mathbf{f} = \frac{\mathbf{D}}{\varepsilon_0} q_2, \tag{2.8}$$

and the quantity $\mathbf{D}/\varepsilon_0$ is called the *electric field intensity*. It is defined by

$$\mathbf{E} = \lim_{\text{charge} \to 0} \left(\frac{(\text{force on the charge})}{(\text{charge})} \right) \tag{2.9}$$

The permittivity gives a direct relationship between the flux density and the field intensity,

$$\mathbf{D} = \varepsilon_{0}\mathbf{E}, \tag{2.10}$$

and eqn. (2.8) becomes

$$\mathbf{f} = \mathbf{E}q. \tag{2.11}$$

The principle of superposition states that for a number of charges the effect of each charge separately may be added vectorially to give the net force on another charge or the net field intensity or the net flux density due to any system of charges acting simultaneously.

Gauss's law states that the total flux out of any closed surface equals the algebraic sum of the total flux sources enclosed within that surface, or

FLUX OUT = CHARGE ENCLOSED

or, mathematically,

$$\iint_{\text{area}} \mathbf{D} \cdot d\mathbf{A} = \sum q_{\text{enclosed}} = \iiint_{\text{volume}} \rho \, dv, \tag{2.12}$$

where ρ is the charge density of a distributed charge.

Gauss's law is used to solve all flux problems involving symmetry, that is, those problems for which the flux density is uniform over the surface. The flux density due to a number of symmetrical charge distributions may be found by the application of Gauss's law. Some results are given here:

1.   At a distance r from the centre of a sphere radius R having a uniform charge of density ρ:

when $r > R$
$$\mathbf{D} = \mathbf{U}_{r} \frac{1}{3} \rho \frac{R^{3}}{r^{2}}; \tag{2.13}$$

when $r < R$
$$\mathbf{D} = \mathbf{U}_{r} \frac{1}{3} r\rho. \tag{2.14}$$

2.   At a distance r from a long, straight filament of line charge density q,

$$\mathbf{D} = \mathbf{U}_{r} \frac{q}{2\pi r}. \tag{2.15}$$

3.   At a distance d from a plane surface of charge density σ,

$$\mathbf{D} = \mathbf{U}_{d} \tfrac{1}{2}\sigma. \tag{2.16}$$

4.   The flux density due to two parallel surface charges of equal and opposite charge density is given in Fig. 2.2. The flux density is σ in the region between the charged surfaces and is zero outside.

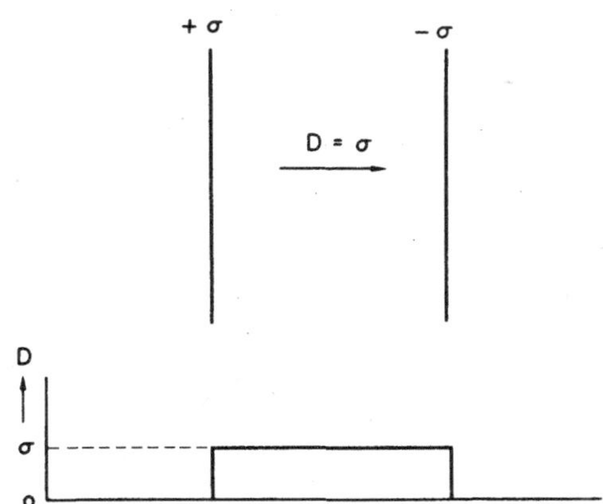

Fig. 2.2. The flux density in the region
between two parallel charged surfaces.

EXAMPLES

2.1.   What is the total flux out of an isolated point charge of 1·0 μC and what is its distribution in space?

Answer. The flux out of a point charge is equal to the strength of the point charge. Therefore the point charge is a flux source of strength

$$\Psi = q = 1 \cdot 0 \ \mu C.$$

The flux flows uniformly out of a point source of flux so that the flux distribution in space will be uniform on the surface of a sphere which is concentric with the point charge.

2.2.   What is the electric flux density at a distance of $0 \cdot 2$ m due to an isolated point charge of 20 nC?

Answer. As the question refers to a point charge in space, there is a uniform flux distribution on the surface of a sphere surrounding the point charge. The flux density is given by the total flux divided by the surface area of the enclosing sphere at the required radius. This is the same as the formula given in eqn. (2.2). Therefore

$$D = \frac{\psi}{A} = \frac{20 \times 10^{-9}}{4\pi \times (0.2)^2} = \frac{10^{-6}}{8\pi} = 3 \cdot 98 \times 10^{-8} \text{ C/m}^2 = 39 \cdot 8 \text{ nC/m}^2.$$

2.3.   What is the electric field intensity at a distance of $0 \cdot 2$ m from an isolated point charge of 20 nC?

Answer. In Example 2.2 we have already found the electric flux density at a distance of $0 \cdot 2$ m from a point charge of 20 nC. The field intensity is related to the flux density by the permittivity constant according to eqn. (2.10). Then, using the value for D from Example 2.2 gives

$$E = \frac{D}{\varepsilon_0} = \frac{36\pi \times 10^9}{8\pi \times 10^6} = 4 \cdot 5 \times 10^3 \text{ V/m}.$$

2.4.   What is the electric field intensity a distance of $1 \cdot 0$ m from a point charge of 10 nC?

Answer. This problem is very similar to Example 2.3. Instead of just substituting the numbers into the equations that were used in Examples 2.2 and 2.3, an alternative method will be used. The definition for field intensity, eqn. (2.9), indicates that the field intensity is numerically equal to the force exerted on a unit charge at the point in question. From Coulomb's law, eqn. (2.1), the force on a unit charge is given by

$$f = \frac{9 \times 10^9 \times 10^{-8} \times 1}{1 \cdot 0 \times 1 \cdot 0} = 90 \text{ N}.$$

Therefore the electric field intensity is 90 N/C. By dimensional analysis it can be shown that N/C is the same as V/m.

2.5.   Three point charges of 10 μC are located at the corners of a plane equilateral triangle of side $1 \cdot 0$ m. Find the force on any one charge due to the other two.

Answer. The charges and the forces due to these charges on one of the three are shown in Fig. 2.3. The forces f_1 and f_2 are due to the charges q_1 and q_2 respectively. From Coulomb's law, eqn. (2.5),

$$f_1 = f_2 = \frac{10^{-5} \times 10^{-5} \times 36\pi}{4\pi \times 1 \cdot 0 \times 1 \cdot 0 \times 10^{-9}} = 0 \cdot 9 \text{ N}.$$

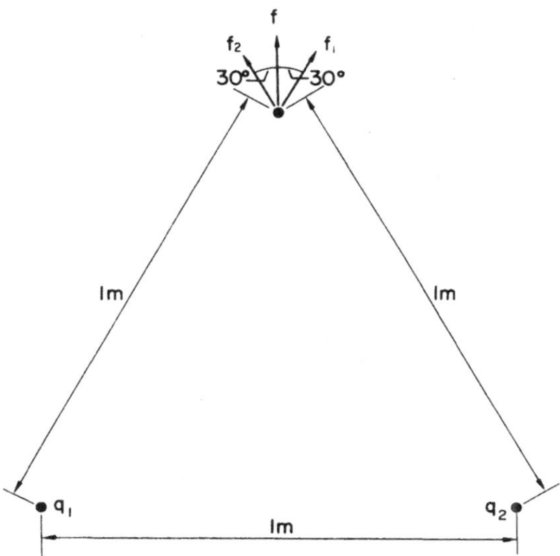

Fig. 2.3. Illustrating Example 2.5.

Using the principle of superposition, the forces due to the two charges
individually may be summed vectorially to give the net force due to both the
charges. From the diagram, the net force resolved in the horizontal
direction is zero and the total force acts in a vertical direction. The net
force is given by

$$f = (f_1 + f_2) \cos 30^0 = 1 \cdot 56 \text{ N}.$$

2.6.   A cube of side $1 \cdot 0$ mm is uniformly charged with a charge density of
10^{-6} C/m³. If the cube is enclosed inside a spherical shell of radius $1 \cdot 0$ m,
find the total electric flux flowing through the surface of the sphere. Can
anything be said about the flux density on the surface of the sphere?

Answer. The total charge in the body of the cube is given by the
product of the charge density and the volume. Therefore

$$Q = 10^{-6} \times (10^{-3})^3 = 10^{-15} \text{ C}.$$

From Gauss's law,

FLUX OUT = CHARGE ENCLOSED.

Therefore the total flux is given by

$$\Psi = Q = 10^{-15} \text{ C}.$$

The size of the enclosing sphere does not make any difference to the total
flux flowing out through its surface so long as the charged cube is inside
the surface of the enclosing sphere. As the cube is not spherically
symmetrical, the flux density on the surface of a concentric sphere is not
uniform. As the question does not state that the cube is at the centre of
the sphere, nothing can be said about the flux density on the surface of the
sphere except that it is not uniform. If the cube were at the centre of the
enclosing sphere, an *approximate* value of flux density at the surface of the
sphere may be obtained by assuming that the charge on the cube is replaced
by a point charge at the centre of the cube. Then the approximate value of
the flux density is given from eqn. (2.2) by

$$D = \frac{10^{-15}}{4\pi \times 1 \cdot 0} = 7 \cdot 96 \times 10^{-17} \text{ C/m}^2.$$

As the size of the cube is only 0·1 per cent of the radius of the sphere, it
may be expected that this is a good approximation.

2.7.   Find, from first principles, the flux density on the surface of a
sphere of radius 1·0 m having a uniform volume charge density of 1·0 uC/m³.

Answer. From Gauss's law,

FLUX OUT = CHARGE ENCLOSED.

Because the system is spherically symmetric, the flux density on the surface
of the sphere is uniform. The flux out of the surface of the sphere is equal to
the product of this flux density and the surface area of the sphere.
Therefore

$$\Psi = D \times 4\pi r^2.$$

Also the flux is equal to the total charge enclosed. Therefore

$$\Psi = \rho \times \frac{4}{3}\pi r^3.$$

Equating these two values for Ψ gives

$$D = \frac{1}{3}\,\rho r,$$

which is the same as eqn. (2.14). Substituting the values from the question
gives

$$D = \frac{1}{3} \times 10^{-6} \times 1 \cdot 0 = 3 \cdot 33 \times 10^{-7} \text{ C/m}^2.$$

2.8.   A charge of 10 μC is on a conducting sphere of diameter 1·0 m. Due to the mutual repulsion between each element of charge on the sphere, it may be assumed that the charge will distribute itself uniformly on the surface of the sphere. Find the electric field intensity at a distance of 2·0 m from the centre of the sphere.

Answer. As the charge distribution is uniform, the field will be spherically symmetrical and Gauss's law may be applied to the problem. Then

$$\text{Total flux} = \text{total charge} = 10^{-5} \text{ C.}$$

The flux density at a radius of 2·0 m is given by

$$D = \frac{(\text{charge})}{(\text{surface area})} = \frac{10^{-5}}{4\pi \times 4 \cdot 0}$$

and the electric field intensity is given by

$$E = \frac{D}{\varepsilon_0} = \frac{36\pi \times 10^9 \times 10^{-5}}{16\pi} = 2 \cdot 25 \times 10^4 \text{ V/m.}$$

2.9.   Find, from first principles, the flux density a distance of 100 mm away from a long, straight line filament of charge density 1·0 μC/m.

Answer. In order to determine the flux density it is necessary to determine the directions of symmetry if the problem is to be solved by Gauss's law. The field is circularly symmetric about the line of the filament in a plane perpendicular to the line of the filament. A Gaussian surface is taken at a radius of 100 mm from the filament. It is a cylindrical surface and is shown in Fig. 2.4. Due to symmetry, there will be no flux out of the ends of the cylinder. The charge enclosed is equal to the product of the charge density and the length of the filament enclosed by the cylinder. The flux out is equal to the product of the flux density and the surface area of the cylinder through which the flux flows. Equating these two quantities for the cylinder of dimensions shown in Fig. 2.4 gives

$$D = \frac{ql}{2\pi r l} = \frac{q}{2\pi r},$$

which is the same as eqn. (2.15). Substituting values from the question gives

$$D = \frac{10^{-6}}{2\pi \times 0 \cdot 1} = 1 \cdot 59 \times 10^{-6} \text{ C/m}^2.$$

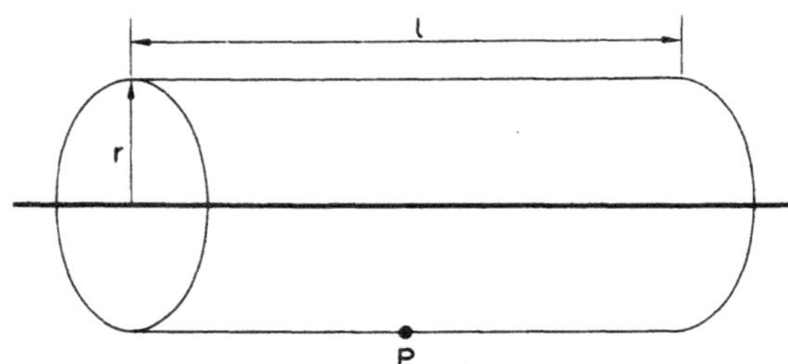

Fig. 2.4. A cylindrical Gaussian surface around
a filament of uniform charge density.

2.10.   Find the flux density due to an infinitely large thin sheet which
has a uniform surface charge density of 10 μC/m².

Answer.   The problem is symmetrical, so that the flux density will be
constant across any plane parallel to the charged plane. If a right circular
cylinder perpendicular to the plane as shown in Fig. 2.5 is chosen as the
Gaussian surface, the flux will flow through the ends of the cylinder and,
because of symmetry, no flux will flow through the curved sides of the
cylinder. If the cross-sectional area of the cylinder is A, the charge
enclosed inside the cylinder will be the product of the charge density on the
charge sheet and the cross-sectional area of the cylinder. Therefore

$$\Psi = 10^{-5}A \ \text{C.}$$

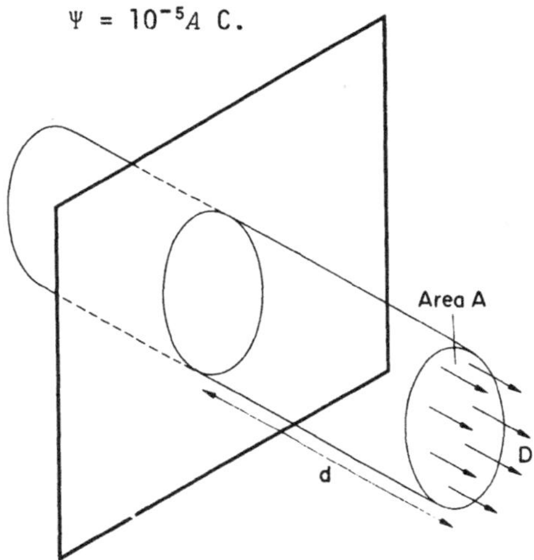

Fig. 2.5. A cylindrical Gaussian surface enclosing
part of a uniform surface charge density.

The total flux will be flowing out of both ends of the cylinder and, if the charge sheet bisects the cylinder as shown in Fig. 2.5, the flux density will be the same at each end of the cylinder. Therefore if the flux density is D,

$$\Psi = 2DA \ \text{C}.$$

Equating the two expressions for Ψ gives the value of the flux density,

$$D = 5 \cdot 0 \times 10^{-6} \ \text{C/m}^2.$$

It is seen that this expression for the flux density is independent of the distance from the charge sheet. This means that for a charge sheet of infinite extent the field is uniform on each side of the sheet with the flux density vector perpendicular to the plane of the sheet.

2.11. A charge of $9 \cdot 0 \ \mu\text{C}$ is deposited on to a flat sheet of metal of circular shape of diameter $0 \cdot 6$ m. Find the electric field intensity at a perpendicular distance of $1 \cdot 0$ mm and 10 km from the centre of the sheet.

Answer. (i) Very close to the centre of the sheet, the field will approximate to that of an infinite plane of charge. Assuming a uniform distribution of charge throughout the sheet, the surface charge density of the equivalent plane of charge is given by

$$\sigma = \frac{9 \times 10^{-6}}{\pi \times 9 \times 10^{-2}} = \frac{10^{-4}}{\pi} \ \text{C/m}^2.$$

The flux density is numerically equal to $\tfrac{1}{2}\sigma$ so that

$$E = \frac{36\pi \times 10^9}{2\pi \times 10^4} = 1 \cdot 8 \times 10^6 \ \text{V/m}.$$

(ii) A great distance away from the sheet, the charge will approximate to a point charge. The flux density due to a point charge is given by eqn. (2.2) so that the electric field intensity at the required point is given by

$$E = \frac{9 \times 10^{-6} \times 36\pi \times 10^9}{4\pi \times 10^8} = 8 \cdot 1 \times 10^{-4} \ \text{V/m}.$$

2.12. Two parallel metal sheets each having an area of $1 \cdot 0$ m^2 are situated $5 \cdot 0$ mm apart in air. If they are given equal and opposite charges of $1 \cdot 0$ μC, what is the electric field intensity in the space between the sheets? What happens when the sheets are both given equal charges of the same sign?

Answer. Due to the mutual attraction of the opposite charges on the two sheets, the charge on each will be distributed uniformly on the surface nearest to the other sheet. As the area of each sheet is much larger than the gap between them, they will approximate to infinite surface charges and the end effects may be neglected. The flux densities due to two equal and opposite charge densities are shown in Fig. 2.6. Using the principle of superposition, these flux densities may be combined to give the effect of two surface charges together. The combination is shown in Fig. 2.7, where it is seen that between the planes, the flux densities due to the two charges act together giving

$$D = D_1 + D_2 = \sigma.$$

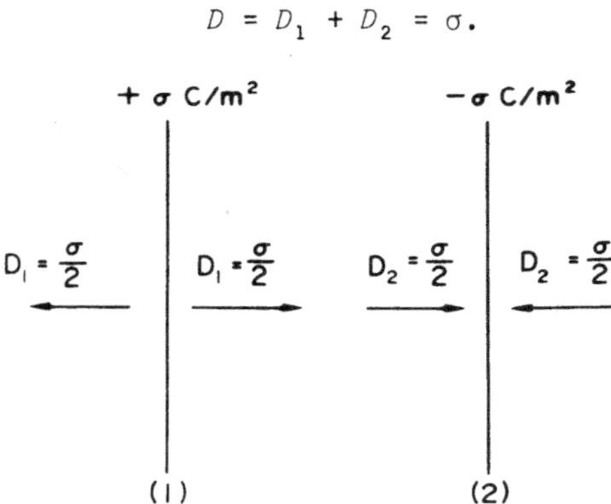

Fig. 2.6. Individual flux densities due to two equal and opposite surface charge densities.

Outside the two charged planes, the flux densities due to the two charges are oppositely directed and cancel one another out.

$$D = D_1 - D_2 = 0.$$

There is a uniform field between the two planes of charge and no field outside. The numerical value of the flux density is equal to the surface charge density. Therefore

$$D = 1 \cdot 0 \text{ μC/m}^2$$

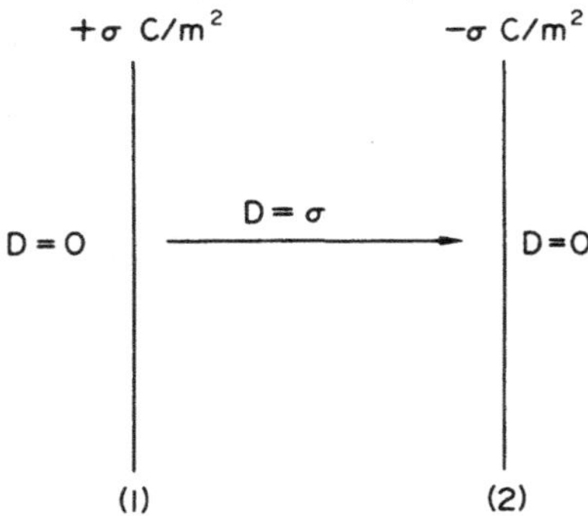

Fig. 2.7. Combined flux density due to two equal
and opposite surface charge densities.

in the region between the plates and zero outside this region.   Therefore
the field intensity between the sheets of charge is given by

$$E = \frac{D}{\varepsilon_o} = 1 \cdot 0 \times 10^{-6} \times 36\pi \times 10^9 = 1 \cdot 3 \times 10^5 \text{ V/m} = 113 \text{ kV/m}.$$

If the two sheets are given the same charge, the mutual repulsion
between all the charge will drive the charge to the outside surfaces of the
sheets. It is unlikely that the charge will distribute itself uniformly on
the outside of the sheets, and very little can be said about the field
around the sheets. However, if it could be assumed that the charge
distribution was uniform on the outside of each of the metal sheets, the
sign of the charge density on plane (2) in Figs. 2.6 and 2.7 will be changed
and so will the direction of the flux density D_2. This means that there
will be no field between the plates and a uniform field outside the plates
close to the plates. A long way away from the plates the field will be
similar to that due to a point charge.

PROBLEMS

2.1. What is the definition of solid angle? What is the maximum number of solid angle possible? Find the solid angle subtended at a point on the axis of a right circular cone of $8 \cdot 0$ m altitude and base of $6 \cdot 0$ m radius.

A point electric charge of $0 \cdot 01$ μC is mounted $8 \cdot 0$ mm above a circular disc of paper of diameter $12 \cdot 0$ mm. The perpendicular from the plane of the paper to the charge goes through the centre of the disc. Find the total electric flux passing through the disc. ($1 \cdot 26$ sr; $1 \cdot 0$ nC.)

2.2. Find the total flux flowing through any one face of a cube which has a point charge of 50 pC at its centre. ($8 \cdot 33$ pC.)

2.3. Two point charges of amplitude $+2 \cdot 0$ μC and $-1 \cdot 0$ μC are situated at two adjacent corners of a square of side $1 \cdot 0$ m. Find the force (in amplitude and direction) acting on a positive unit charge at either of the other two corners. ($15 \cdot 2$ kN at 78^{0} or 6.9 kN at $22 \cdot 5^{0}$.)

2.4. Working from first principles with Gauss's law, find the flux density at a distance of 10 mm away from a long, straight cylindrical conductor carrying a charge of 10 pC/m. ($0 \cdot 159$ μC/m .)

2.5. Point charges $q_1 = 10$ μC and $q_2 = 5$ μC are placed $0 \cdot 1$ m apart in air. Determine (a) the force between them, (b) the flux density at q_2 due to q_1, (c) the flux density at q_1 due to q_2, and (d) the flux density and field intensity at a point $0 \cdot 05$ m above the mid-point of the line joining the charges. (45 N; $79 \cdot 6$, $39 \cdot 8$, 178 μC/m²; $71 \cdot 5^{0}$, $20 \cdot 1$ MN/C.)

2.6. Six point charges of value $+1 \cdot 0$ μC, $+2 \cdot 0$ μC, $+3 \cdot 0$ μC, $+1 \cdot 0$ μC, $-1 \cdot 0$ μC, and $-2 \cdot 0$ μC are situated at the corners of a cube. If the cube is surrounded by an enclosing sphere, find the total electric flux flowing through the surface of the sphere. ($4 \cdot 0$ μC.)

2.7. Four equal point charges of 10 μC are situated at the corners of a square of side $1 \cdot 0$ m. Find the force acting on any one of the charges due to the other three by direct calculation from Coulomb's law and also by first finding the electric intensity. ($1 \cdot 72$ N.)

2.8. If air becomes ionized and conducts charge when the electric intensity exceeds $3 \cdot 0 \times 10^6$ V/m, find the largest charge that can be accommodated by a spherical metal shell $0 \cdot 5$ m diameter. (21 μC.)

2.9. The earth has a radius of $6 \cdot 37$ Mm. If the maximum permissible electric intensity on the surface is $3 \cdot 0$ MV/m, find the maximum possible charge that the earth could sustain. (13·5 GC.)

2.10. A thick spherical shell, internal radius 0·4 m and external radius 0·6 m, has a charge of 10 μC uniformly distributed throughout its material. Find the flux density at the surface of the shell and draw a graph showing the variation of electric field intensity with radial distance from the centre of the sphere to a radius of 2·0 m, indicating significant values. (2.2 μC/m².)

$$\text{SOLUTIONS}$$

2.1. Taking a circumscribed sphere, solid angle is defined by

$$\text{solid angle} = \frac{\text{area of surface of the sphere included in the angle}}{\text{square of the radius of the sphere}}.$$

$$\text{The maximum possible solid angle} = \frac{\text{area of surface of sphere}}{(\text{radius})^2} = 4\pi.$$

To find the value of an arbitrary solid angle it is necessary to find the surface area subtended by the angle on a circumscribing sphere of unit radius. In order to find the solid angle subtended by a cone, it is necessary to find the surface area of the equivalent circular segment of the sphere. The geometry is illustrated in Fig. 2.8. By integration,

$$\text{area} = \int_0^{\theta_1} 2\pi (r \sin \theta)\, r\, d\theta = 2\pi r^2 (1 - \cos \theta_1).$$

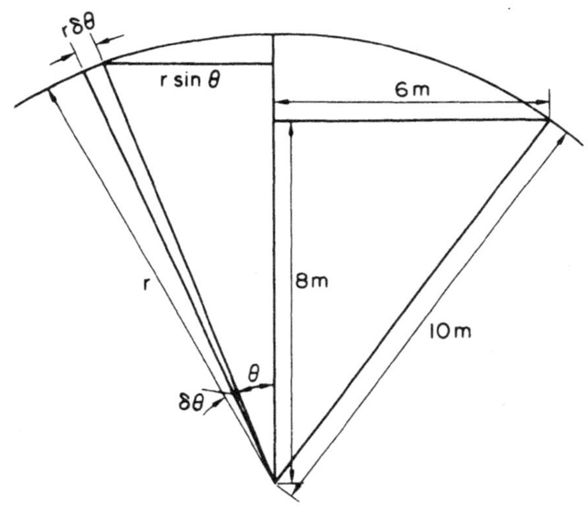

Fig. 2.8. Illustrating the solution to Problem 2.1, the calculation of solid angle.

For the dimensions of the cone given in the question, $\tan \theta_1 = \dfrac{6}{8}$.
Therefore

$$\text{area} = 2\pi r^2 \left(1 - \frac{8}{10}\right) = \frac{4}{10}\, \pi r^2.$$

Therefore

$$\text{solid angle} = \frac{\text{area}}{r^2} = \frac{4\pi}{10} = 1 \cdot 26 \text{ sr.}$$

For the second part of the question, the geometry is exactly similar to that of the first part. The solid angle subtended by the disc is one-tenth of the maximum possible so that, because the electric flux flows uniformly from a point charge, one-tenth of the total flux flows through the disc. Therefore

$$\text{flux} = \frac{0 \cdot 01 \times 10^{-6}}{10} = 1 \cdot 0 \times 10^{-9} \text{ C} = 1 \cdot 0 \text{ nC.}$$

2.2. Gauss's law states,

Flux out of a closed surface = charge enclosed.

Electric flux flows uniformly from a point charge so that, because of the symmetry of the cube, one-sixth of the total flux will flow through each of the six faces of the cube. Therefore

$$\text{flux} = \frac{50}{6} = 8 \cdot 33 \text{ pC.}$$

2.3. There is no symmetry in this problem and it has to be solved using Coulomb's law and the principle of superposition. The forces acting at the two corners are illustrated in Fig. 2.9. Then, applying Coulomb's law, the four forces are given by

$$F_1 = \frac{2 \times 10^{-6} \times 1 \times 36\pi \times 10^9}{4\pi \times 1^2} = 2 \times 9 \times 10^3 \text{ N,}$$

$$F_2 = \frac{1 \times 10^{-6} \times 1 \times 36\pi \times 10^9}{4\pi \times (\sqrt{2})^2} = \tfrac{1}{2} \times 9 \times 10^3 \text{ N,}$$

$$F_3 = \frac{1 \times 10^{-6} \times 1 \times 36\pi \times 10^9}{4\pi \times 1^2} = 1 \times 9 \times 10^3 \text{ N,}$$

$$F_4 = \frac{2 \times 10^{-6} \times 1 \times 36\pi \times 10^9}{4\pi \times (\sqrt{2})^2} = 1 \times 9 \times 10^3 \text{ N.}$$

At the point A, the vertical and horizontal components of the resultant force are given by

$$\text{vertical} = \left(2 - \frac{1}{2\sqrt{2}}\right) \times 9 \times 10^3 = 1 \cdot 646 \times 9 \times 10^3 \text{ N (downwards),}$$

$$\text{horizontal} = \frac{1}{2\sqrt{2}} \times 9 \times 10^3 = 0 \cdot 354 \times 9 \times 10^3 \text{ N (to the right).}$$

Therefore the resultant is given by

$$\text{force} = \sqrt{(2\cdot71 + 0\cdot125)} \times 9 \times 10^3 = 1\cdot68 \times 9 \times 10^3 \text{ N} = 15\cdot2 \text{ kN}$$

at an angle of $-\tan^{-1} \dfrac{1\cdot646}{0\cdot354} = -78^0$ with the horizontal.

At the point B, the vertical and horizontal components of the resultant force are given by

$$\text{vertical} = (1 - \frac{1}{\sqrt{2}}) \times 9 \times 10^3 = 0\cdot193 \times 9 \times 10^3 \text{ N (upwards)},$$

$$\text{horizontal} = \frac{1}{\sqrt{2}} \times 9 \times 10^3 = 0\cdot707 \times 9 \times 10^3 \text{ N (to the right)}.$$

Therefore the resultant is given by

$$\text{force} = \sqrt{(0\cdot500 + 0\cdot086)} \times 9 \times 10^3 = 0.765 \times 9 \times 10^3 \text{ N} = 6\cdot9 \text{ kN}$$

at an angle of $\tan^{-1} \dfrac{0\cdot293}{0\cdot707} = 22\cdot5^0$ with the horizontal.

Fig. 2.9.   Illustrating the solution to Problem 2.3.

2.4.   This problem is so similar to Example 2.9 that the derivation will not be repeated.   Inserting numbers into the relationship derived in the answer to that example gives

$$D = \frac{10 \times 10^{-9}}{2\pi \times 10^{-2}} = 1\cdot59 \times 10^{-7} \text{ C/m}^2 = 0\cdot159 \text{ μC/m}^2.$$

2.5.   This problem is another application of Coulomb's law and the principle of superposition.

(a)   The force is given by Coulomb's law:

$$\text{force} = \frac{10 \times 10^{-6} \times 5 \times 10^{-6} \times 36\pi \times 10^{9}}{4\pi \times 10^{-2}} = 9 \times 50 \times 10^{-1} = 45 \text{ N (repulsive)}.$$

(b)   The flux density due to a point charge is given by eqn. (2.2), therefore

$$D_1 = \frac{10 \times 10^{-6}}{4\pi \times 10^{-2}} = 0 \cdot 796 \times 10^{-4} \text{ C/m}^2 = 79 \cdot 6 \ \mu\text{C/m}^2.$$

(c)
$$D_2 = \frac{5 \times 10^{-6}}{4\pi \times 10^{-2}} = 39 \cdot 8 \ \mu\text{C/m}^2.$$

(d)   By the principle of superposition, the flux density due to each charge is calculated independently and they are summed vectorially. Therefore at the required point

$$D_1 = \frac{10 \times 10^{-6}}{4\pi \times \frac{1}{4} \times 10^{-2} \times (\sqrt{2})^2} = 1 \cdot 59 \times 10^{-4} \text{ C/m}^2,$$

$$D_2 = \frac{5 \times 10^{-6}}{4\pi \times \frac{1}{4} \times 10^{-2} \times (\sqrt{2})^2} = 0 \cdot 796 \times 10^{-4} \text{ C/m}^2,$$

and the vertical and horizontal components of the resultant flux density are given by

$$\text{vertical} = \frac{1}{\sqrt{2}} (1 \cdot 59 + 0 \cdot 796) \times 10^{-4} \text{ C/m}^2,$$

$$\text{horizontal} = \frac{1}{\sqrt{2}} (1 \cdot 59 - 0 \cdot 796) \times 10^{-4} \text{ C/m}^2.$$

Therefore the resultant is given by

$$D = \sqrt{\left(\frac{5 \cdot 96 + 0 \cdot 63}{2}\right)} \times 10^{-4} = 1 \cdot 78 \times 10^{-4} \text{ C/m}^2 = 178 \ \mu\text{C/m}^2$$

at an angle of $\tan^{-1} \dfrac{2 \cdot 386}{0 \cdot 794} = 71 \cdot 5^{0}$ with the horizontal.

The electric field intensity is given by

$$E = \frac{D}{\varepsilon_0} = 1 \cdot 78 \times 10^{-4} \times 36\pi \times 10^{9} = 2 \cdot 01 \times 10^{7} \text{ N/C} = 20 \cdot 1 \text{ MN/C or MV/m}.$$

2.6.   From Gauss's law, the total flux flowing through the enclosing sphere is the algebraic sum of the charges inside the sphere. Therefore

$$\text{flux} = 1 \cdot 0 + 2 \cdot 0 + 3 \cdot 0 + 1 \cdot 0 - 1 \cdot 0 - 2 \cdot 0 = 4 \cdot 0 \ \mu\text{C}.$$

2.7. The forces are shown in Fig. 2.10. *From Coulomb's law,*

$$F_1 = \frac{q^2}{4\pi\varepsilon_o d^2} = \frac{10^{-10} \times 36\pi \times 10^9}{4\pi \times (1\cdot0)^2} = 0\cdot90 \text{ N,}$$

$$F_2 = \frac{10^{-10} \times 36\pi \times 10^9}{4\pi \times (\sqrt{2})^2} = 0\cdot45 \text{ N.}$$

By vector addition, the resultant force is along the vector F_2, therefore

$$F = 0\cdot45 + 2 \times 0\cdot90 \times \frac{1}{\sqrt{2}} = 1\cdot72 \text{ N.}$$

By first finding the field intensity, for a point charge

$$E = \frac{q}{4\pi\varepsilon_o r^2}.$$

Therefore, continuing to use the notation of Fig. 2.10,

$$E_1 = \frac{10^{-5} \times 36\pi \times 10^9}{4\pi \times 1\cdot0} = 9\cdot0 \times 10^4 \text{ N/C;}$$

$$E_2 = \frac{10^{-5} \times 36\pi \times 10^9}{4\pi \times 2\cdot0} = 4\cdot5 \times 10^4 \text{ N/C.}$$

Therefore, by vector addition, the resultant field is given by

$$E = (4\cdot5 + 9\cdot0 \times \sqrt{2}) \times 10^4 = 1\cdot72 \times 10^5 \text{ N/C,}$$

and the force is given by substituting into eqn. (2.11), therefore

$$F = 1\cdot72 \times 10^5 \times 10^{-5} = 1\cdot72 \text{ N.}$$

Fig. 2.10. Illustrating the solution to Problem 2.7.

2.8. By Gauss's law, a charged conducting sphere behaves as if its total charge is concentrated at its centre. Therefore the field intensity on the surface of the sphere is given by the expression for a point charge,

$$E = \frac{q}{4\pi\varepsilon_0 r^2}.$$

When the electric field intensity exceeds $3 \cdot 0 \times 10^6$ V/m, some of the charge will leak away due to conduction in the air. Therefore the limiting condition is when $E = 3 \cdot 0 \times 10^6$ V/m. Putting numbers into the expression for E gives

$$3 \cdot 0 \times 10^6 = \frac{q \times 36\pi \times 10^9}{4\pi \times (0 \cdot 25)^2}.$$

Therefore

$$q = \frac{3 \cdot 0 \times 10^6}{9 \times 16 \times 10^9} = 2 \cdot 1 \times 10^{-5} \text{ C} = 21 \text{ }\mu\text{C}.$$

2.9. If we assume that the earth is a conducting sphere, this problem becomes the same as Problem 2.8 with different numbers. Therefore

$$q = \frac{3 \cdot 0 \times 10^6 \times 4\pi \times (6 \cdot 37 \times 10^6)^2}{36\pi \times 10^9} = 1 \cdot 35 \times 10^{10} \text{ C} = 13 \cdot 5 \text{ GC}.$$

2.10. At any radius r inside the spherical shell, the flux density D is given by Gauss's law. If the charge density is ρ C/m^3,

$$\text{enclosed charge} = \left(\tfrac{4}{3}\pi r^3 - \tfrac{4}{3}\pi(0 \cdot 4)^3\right)\rho,$$

$$\text{flux} = 4\pi r^2 D;$$

therefore

$$D = \left(\tfrac{1}{3}r - \tfrac{1}{3}\frac{0 \cdot 064}{r^2}\right)\rho.$$

The total charge on the shell is 10 μC, therefore

$$\text{total charge} = 10^{-5} = \tfrac{4}{3}\pi\left\{(0 \cdot 6)^3 - (0 \cdot 4)^3\right\}\rho.$$

Therefore

$$\rho = \frac{3 \times 10^{-5}}{4\pi \times 0 \cdot 152}.$$

Therefore D inside the shell is given by

$$D = \frac{10^{-5}}{4\pi \times 0 \cdot 152}\left(r - \frac{0 \cdot 064}{r^2}\right) \text{ C/m}^2$$

and outside the shell by

$$D = \frac{10^{-5}}{4\pi r^2} \text{ C/m}^2.$$

On the surface of the shell

$$D = \frac{10^{-5}}{4\pi \times 0 \cdot 36} = 2 \cdot 2 \times 10^{-6} \; C/m^2 = 2 \cdot 2 \; \mu C/m^2.$$

The electric intensity on the surface of the shell is given by

$$E = \frac{10^{-5} \times 36\pi \times 10^9}{4\pi \times 0 \cdot 36} = 2 \cdot 5 \times 10^5 \; V/m.$$

A graph of the variation of electric intensity with radial distance from the centre of the sphere is given in Fig. 2.11.

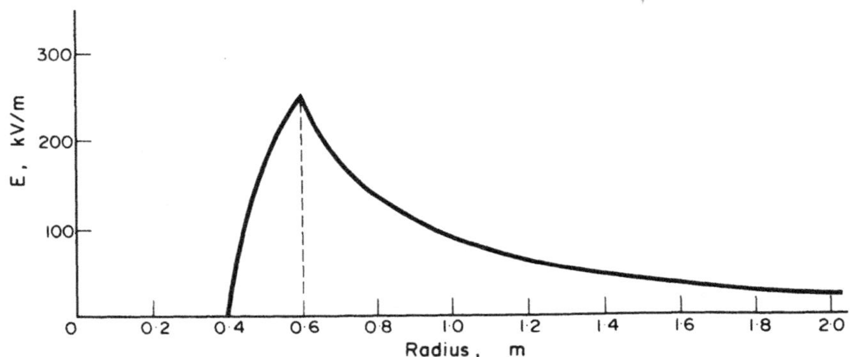

Fig. 2.11. The solution to Problem 2.10. The field
intensity due to a thick spherical shell of
uniform charge density.

FURTHER PROBLEMS

2.11. The circumference of a circle of radius $1 \cdot 0$ m is drawn on the surface of a sphere of radius $2 \cdot 0$ m. Find the solid angle subtended at its centre by the surface of the sphere enclosed by the circumference of the circle.

A point electric charge of $1 \cdot 0$ μC is situated at the centre of an insulating sphere of $1 \cdot 30$ m diameter. The circular rim of a bicycle wheel of $0 \cdot 65$ m diameter is placed on the surface of the sphere; find the total electric flux passing through the surface enclosed by the wheel.

$$(0 \cdot 84 \; sr; \quad 67 \; nC.)$$

2.12. A point charge of 10 nC is 10 mm outside the surface of an insulating sphere of 1·0 m diameter. Find the net flux flowing into the surface of the sphere.

2.13. For the charge and sphere as described in Problem 2.12, find the maximum and minimum values of the electric field intensity at the surface of the sphere.   (900 kV/m;   88 V/m.)

2.14. Three point electric charges are situated in a straight line 0·1 m apart. They have charges, 2·0 μC, -1·0 μC, and 2·0 μC, with the negative charge in the centre. Find the forces on each charge due to the other two. (0·9 N attractive;   0 on centre charge.)

2.15. A line charge filament of 10 nC/m passes perpendicularly through the centre of an insulating circular disc of 1·0 m diameter. Two equal point charges of 0·1 μC are placed on the circumference of the disc diametrically opposite one another. Find the force on these point charges.
(0·126 mN repulsive.)

2.16. A solid sphere of 1·0 m diameter has a uniform charge density of 10 nC/m^3. Find the electric flux density at distances of 0·25 m and 2·0 m from the centre of the sphere.   (834, 104 pC/m^2.)

2.17. Find the electric field intensity at a distance of 10 m away from a long, straight conductor which carries an electric charge density of 22·2 μC/m.   (40 kV/m.)

2.18. Find a general expression for the flux density inside and outside an infinitely long straight tube having an inner radius a and outer radius b whose material has a uniform charge density ρ. If a = 6 mm, b = 10 mm, and ρ = 0·1 μC/m^3, find the flux density at a distance of 8 mm and 100 mm from the centre of the tube.   (175, 32 pC/m^2.)

2.19. An infinite plane sheet has a uniform charge density of 10 μC/m^2. What is the flux density at a distance of 50 mm from the plate?
(5·0 μC/m^2.)

2.20. A parallel plate capacitor consists of two metal plates of area 1·0 m^2, 1·0 mm apart. Find the electric field intensity in the gap between the plates when they have equal and opposite charges of 0·1 μC. Assume a uniform field between the plates.   (11·3 kV/m.)

FURTHER SOLUTIONS

2.11.   The area of the surface of the sphere is obtained by integration of the element shown in Fig. 2.8 (page 26).   Therefore

$$\text{area} = 2\pi r^2 \int_0^{\sin^{-1}\frac{1}{2}} \sin\theta\, d\theta = 2\pi r^2 \left(-\frac{\sqrt{3}}{2} + 1\right) = 0\cdot134 \times 2\pi r^2.$$

Therefore $$\text{solid angle} = 0\cdot134 \times 2\pi = 0\cdot844 \text{ sr},$$

$$\text{flux} = \frac{10^{-6} \times 0\cdot134 \times 2\pi}{4\pi} = 6\cdot7 \times 10^{-8} \text{ C} = 67 \text{ nC}.$$

2.12.   From Gauss's law, as there is no charge enclosed within the insulating sphere, there is no net flux flowing into or out of the surface of the sphere.   The answer is zero.

2.13. $$E_1 = \frac{10^{-8} \times 36\pi \times 10^9}{4\pi \times 10^{-4}} = 9\cdot0 \times 10^5 \text{ V/m} = 900 \text{ kV/m},$$

$$E_2 = \frac{10^{-8} \times 36\pi \times 10^9}{4\pi \times 1\cdot02} = 88 \text{ V/m}.$$

2.14.   The force on one end charge is given from Coulomb's law,

$$F = \frac{2\cdot0 \times 10^{-6} \times 1\cdot0 \times 10^{16} \times 36\pi \times 10^9}{4\pi \times 10^{-2}} + \frac{2\cdot0 \times 10^{-6} \times 2\cdot0 \times 10^{-6} \times 36\pi \times 10^9}{4\pi \times 4 \times 10^{-2}}$$

$$= 9 \times 10^{-1} (-2\cdot0 + 1\cdot0) = -9\cdot0 \times 10^{-1} \text{ N} = -0\cdot9 \text{ N}.$$

There will be no net force on the centre charge as the problem is symmetrical and the centre charge will experience equal and opposite forces due to each end charge.

2.15.   As there is a line charge as well as a point charge, it is convenient to find the force through the medium of the field intensity rather than through Coulomb's law.   Therefore

$$E = \frac{10^{-8} \times 36\pi \times 10^9}{2\pi \times 0\cdot5} + \frac{10^{-7} \times 36\pi \times 10^9}{4\pi \times 1\cdot0} = 1260 \text{ N/C}.$$

By symmetry, the electric field at each point charge due to the other charges is the same.   Therefore the force on each is the same:

$$F = 1260 \times 10^{-7} \text{ N} = 0\cdot126 \text{ mN}.$$

As all charges are positive, the force is repulsive.

2.16. Applying eqns. (2.13) and (2.14),

$$D = \frac{1}{3}\rho r = \frac{1}{3} \times 10^{-8} \times 0 \cdot 25 = 8 \cdot 34 \times 10^{-10} \text{ C/m}^2 = 834 \text{ pC/m}^2,$$

$$D = \frac{1}{3}\rho \frac{R^3}{r^2} = \frac{1}{3} \times 10^{-8} \times \frac{0 \cdot 125}{4 \cdot 0} = 1 \cdot 04 \times 10^{-10} \text{ C/m}^2 = 104 \text{ pC/m}^2,$$

2.17. $E = \dfrac{22 \cdot 2 \times 10^{-6} \times 36\pi \times 10^9}{2\pi \times 10} = 4 \cdot 0 \times 10^4 \text{ V/m} = 40 \text{ kV/m}.$

2.18. Applying Gauss's law to a cylindrical surface concentric with the axis of the tube gives, at any radius r, where $r < b$ for a unit length,

$$\text{enclosed charge} = \pi\rho(r^2 - a^2) = 2\pi r D.$$

Therefore $D = \tfrac{1}{2}\rho(r - a^2/r).$

When $r > b$,
$$\text{enclosed charge} = \pi\rho(b^2 - a^2) = 2\pi r D.$$

Therefore $D = \tfrac{1}{2}\rho(b^2 - a^2)/r.$

When $r = 8$ mm,

$$D = \tfrac{1}{2} \times 10^{-7} (8 \times 10^{-3} - 36 \times 10^{-3}/8) = 1 \cdot 75 \times 10^{-10} \text{ C/m}^2 = 175 \text{ pC/m}^2.$$

When $r = 100$ mm,

$$D = \tfrac{1}{2} \times 10^{-7}(100 - 36) \times 10^{-3}/10^2 = 3 \cdot 2 \times 10^{-11} \text{ C/m}^2 = 32 \text{ pC/m}^2.$$

2.19. The infinite plane sheet of charge gives rise to a uniform field that is independent of distance,

$$D = \tfrac{1}{2}\sigma = 5 \cdot 0 \text{ }\mu\text{C/m} .$$

2.20. The two parallel charged plates will behave as if they were two parallel planes of uniform charge density. The field between the planes is uniform,

$$\sigma = 10^{-7} \text{ C/m}^2, \quad D = \sigma = 10^{-7} \text{ C/m}^2.$$

Therefore $E = 10^{-7} \times 36\pi \times 10^9 = 1 \cdot 13 \times 10^4 \text{ V/m} = 11 \cdot 3 \text{ kV/m}.$

CHAPTER 3

Flux Function

THEORY

We imagine the flux to be carried through the field in tubes, each tube carrying the same amount of flux. For a two-dimensional field the tubes can be represented by lines with the same amount of flux between each pair of lines; the density of the lines is proportional to the magnitude of the flux density vector at every point. No flux crosses any line so that the flux is parallel to a tangent to the flux lines at any point.

Definition of flux function: arbitrarily designate one line of flux as having the flux function value zero ($\psi = 0$); the flux function at any point in the field is then defined as the amount of flux encompassed in going from the line of zero flux function to the point of interest.

Sign convention: the flux function is defined as increasing to the left when looking in the direction of flux flow, that is in the direction of the flux density vector; *alternatively:* the flux function increases along any path which is crossed by flux from left to right.

It follows that lines of flux are also lines of constant flux function. From Fig. 3.1, the difference in flux function value between the points A and B is given by

$$\Psi = \psi_B - \psi_A = \int_A^B \mathbf{D} \cdot d\mathbf{A}. \tag{3.1}$$

If eqn. (3.1) is put into its differential form,

$$D_x = \frac{\partial \psi}{\partial y} \tag{3.2}$$

and

$$D_y = -\frac{\partial \psi}{\partial x}, \tag{3.3}$$

and in cylindrical polar coordinates,

$$D_r = \frac{1}{r}\frac{\partial \psi}{\partial \theta} \tag{3.4}$$

and

$$D_\theta = -\frac{\partial \psi}{\partial r}. \tag{3.5}$$

In the most general case, the flux density is the vector sum of the two components of the flux density.

36

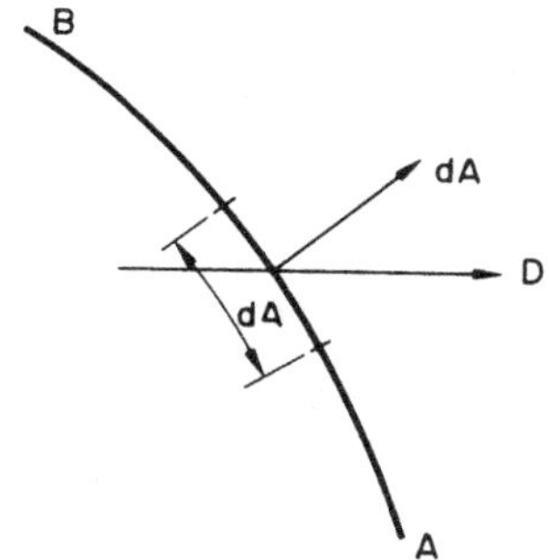

Fig. 3.1. Flux density in a two-dimensional field.

Expressions for the flux function of a number of different fields due to different combinations of charges are now given.

For the uniform flux density D directed in the x-direction,

$$\psi = Dy, \tag{3.6}$$

$$\psi = Dr \sin\theta. \tag{3.7}$$

For a line charge density q or line source of flux,

$$\psi = \frac{q\theta}{2\pi}. \tag{3.8}$$

For a line sink of flux,

$$\psi = -\frac{q\theta}{2\pi}. \tag{3.9}$$

The flux function lines and values for a line source are shown in Fig. 3.2;

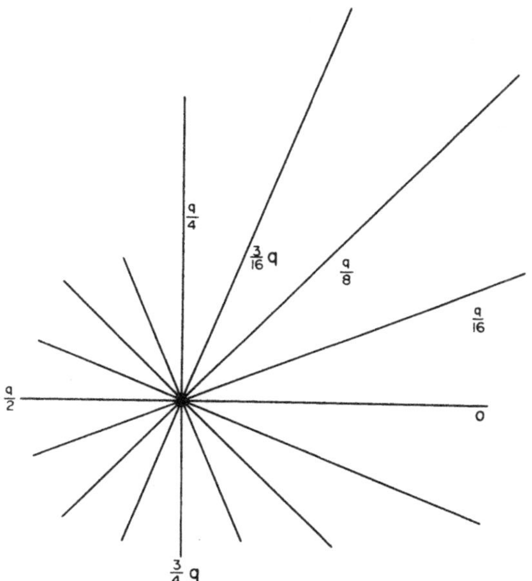

Fig. 3.2. The flux function values of a line source of strength q C/m.

those of a line sink will be the negative of those for a line source.

For a line source parallel to but not coincident with an equal line sink, the flux function lines are segments of circles. They are shown in Fig. 3.3. The geometry for obtaining the flux function of the combination is shown in Fig. 3.4, where the flux function is seen to be

$$\psi = \psi_1 + \psi_2 = \frac{q}{2\pi}(\theta_1 - \theta_2) = -\frac{q\alpha}{2\pi}. \tag{3.10}$$

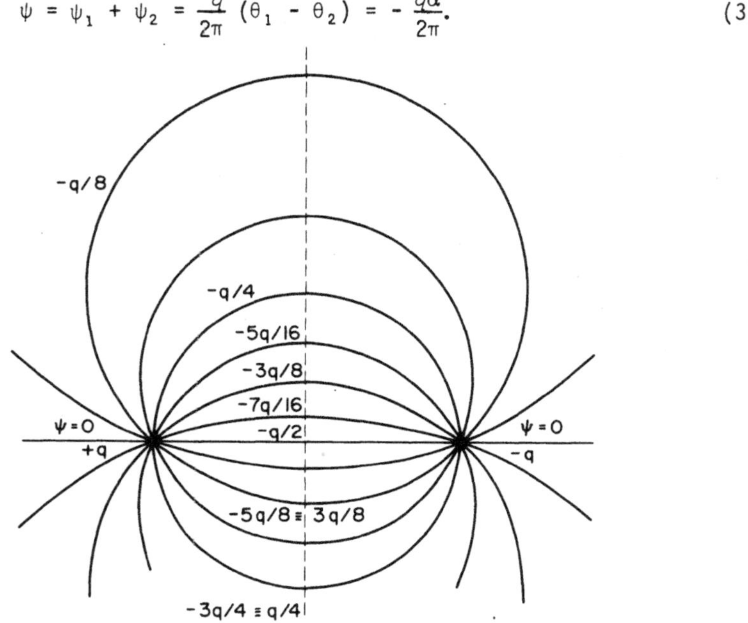

Fig. 3.3. The flux function values due to the combined field of a line source and a line sink, of strength $\pm q$ C/m.

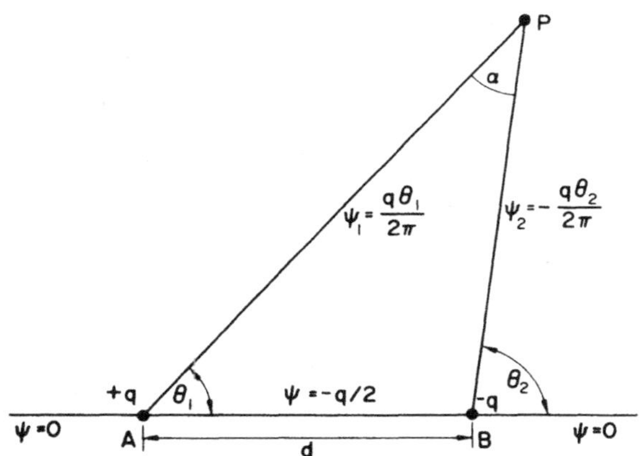

Fig. 3.4. Illustrating a mathematical expression for the flux function values due to the combined field of a line source and a line sink.

The locus of the points having the angle α constant will be the segment of a
circle passing through the points A and B.

The doublet consists of a line source which is coincident with a
parallel line sink. The strength of the doublet is given by

$$c = qd, \tag{3.11}$$

where q is the strength of the line source or the line sink and d is the
distance between them. As they are coincident, d is infinitely small and
consequently q is infinitely large so that their product remains finite.
The flux function is given by

$$\psi = - \frac{c \sin \theta}{2\pi r}. \tag{3.12}$$

The flux function lines are still circles, and are shown in Fig. 3.5. If the
radius of a line of constant flux function is a, the flux function value is
given by

$$\psi = - \frac{c}{4\pi a}. \tag{3.13}$$

Other field patterns may be generated by suitable combinations of sources
and sinks of flux.

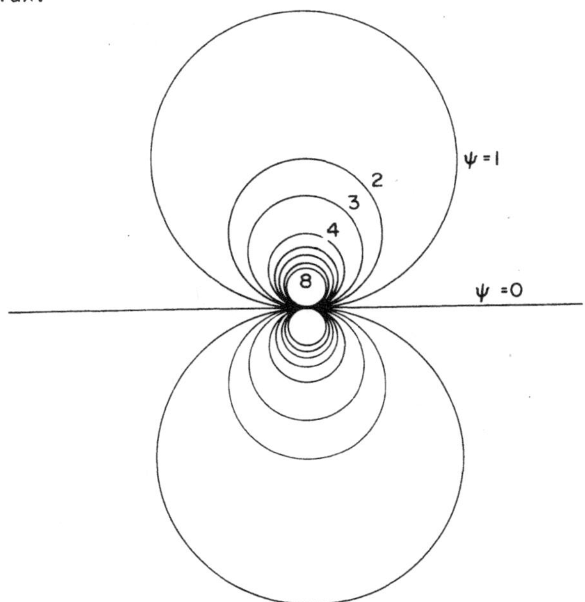

Fig. 3.5. The flux function values of a doublet.

EXAMPLES

3.1. A uniform field has a flux density of 3·0 nC/m²; find the shape and spacing of the flux function lines. Take the increment between the flux function lines as 10^{-12} C/m.

Answer. A uniform depth of field is assumed in determining flux function lines. Let the depth be 1·0 m, i.e. unit depth. The shape of the flux function lines in a uniform field is straight and parallel to one another and to the direction of the flux density. The spacing of the flux function lines is given by substituting numbers into eqn. (3.6). Therefore

$$y = \frac{10^{-12}}{3 \times 10^{-9}} = \frac{1}{3} \times 10^{-3} \text{ m} = 0\cdot333 \text{ mm.}$$

3.2. Derive, from first principles, an expression for the flux function in the field due to a line filament of charge of 6·28 µC/m.

Answer. Due to symmetry, the flux will flow radially out from the line source and will give a two-dimensional field. In order to find the flux function values in the field, it is necessary to integrate along a path perpendicular to the flux lines. This path of integration is circular and is shown in Fig. 3.6. The flux density at any distance r from the line source is given by eqn. (2.15), therefore

$$D = \frac{6\cdot28 \times 10^{-6}}{2\pi r} = \frac{10^{-6}}{r}.$$

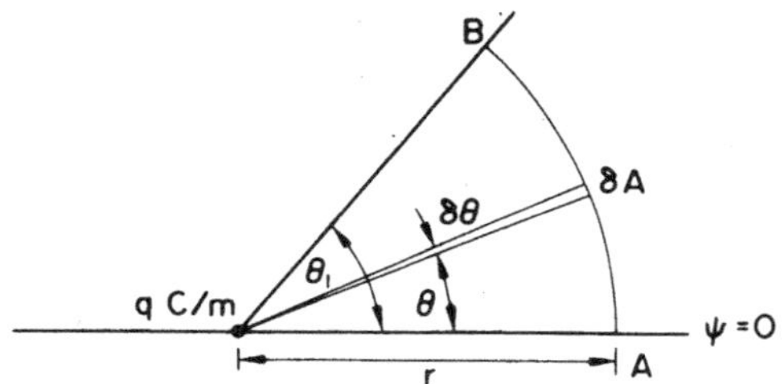

Fig. 3.6. The flux function of a line filament of charge.

Assuming a unit depth of field and following a circular path, the elemental area is given by

$$dA = r \, d\theta.$$

The flux function is obtained by substituting into eqn. (3.1)

$$\psi_1 = \frac{10^{-6}}{r} \int_0^{\theta_1} r \, d\theta = 10^{-6} \, \theta_1.$$

3.3. Find an expression for the flux function in the field due to a line sink having a numerical strength of 10 μC/m.

Answer. It is only necessary to remember that a line sink is the negative of a line source. The flux function can be found by substitution into eqn. (3.9) which is derived in the same way as in the solution to the last example. Therefore

$$\psi = - \frac{10^{-5} \, \theta}{2\pi} = - 1 \cdot 59 \times 10^{-6} \, \theta.$$

3.4. A line source of strength q C/m is situated at the point (1, 0) m and a line sink of strength q C/m is situated at the point (-1, 0) m. Find the value of the flux function in the field at the points (-1, 1), (0, 1·62), and (1, 1) m.

Answer. The value of the flux function is obtained by substituting numbers into eqn. (3.10) as obtained from Fig. 3.4.

At the point (-1, 1), $\theta_1 = \tfrac{1}{2}\pi$, $\theta_2 = \pi - \tan^{-1}\tfrac{1}{2} = (\pi - 0\cdot148\pi)$.

Therefore $\psi = \dfrac{q}{2\pi} (\theta_1 - \theta_2) = \tfrac{1}{2}q(0\cdot500 - 1\cdot000 + 0\cdot148) = - 0\cdot176q$ C/m.

At the point (0, 1·62), $\theta_1 = \tan^{-1} 1\cdot62$, $\theta_2 = \pi - \tan^{-1} 1\cdot62 = (\pi - 0\cdot324\pi)$.

Therefore $\psi = \tfrac{1}{2}q(0\cdot324 - 1\cdot000 + 0\cdot324) = - 0\cdot176q$ C/m.

At the point (1, 1), $\theta_1 = \tan^{-1} \tfrac{1}{2} = 0\cdot148\pi$, $\theta_2 = \tfrac{1}{2}\pi$.

Therefore $\psi = \tfrac{1}{2}q(0\cdot148 - 0\cdot500) = - 0\cdot176q$ C/m.

3.5. Find the maximum value of flux function 10 m away from two parallel charged wires 3·0 mm apart. The wires, which may be assumed to be thin, have equal and opposite charges of 10 μC/m.

Answer. It is assumed that the distance of 10 m is measured from the mid-point between the wires although the scale of the problem is such that no error is introduced by assuming that the distance is measured from either of the two wires. From the results of Example 3.4 and by considering Fig. 3.4, it is seen that the maximum value of flux function is on a plane perpendicular to the line joining the two wires.

The formula for the flux function due to two parallel line charges is given in eqn. (3.10). The angle α is equal to the angle subtended by an arc of 3·0 mm at a radius of 10 m. Therefore

$$\alpha = 3 \times 10^{-3}/10 = 3 \times 10^{-4} \text{ rad}$$

and
$$\psi = \frac{3 \times 10^{-4} \times 10^{-5}}{2\pi} = 4 \cdot 77 \times 10^{-10} \text{ C/m}.$$

It is possible that the two lines close together approximate to a doublet so we will also calculate the value of the flux function by substitution into the formula for the doublet, eqn. (3.12). The strength of the equivalent doublet is given by

$$c = qd = 10^{-5} \times 3 \times 10^{-3} = 3 \times 10^{-8}.$$

Therefore the flux function is given by

$$\psi = \frac{3 \times 10^{-8}}{2\pi \times 10} = 4 \cdot 77 \times 10^{-10} \text{ C/m}.$$

This result shows that there are practical situations which approximate to a doublet.

3.6. Take a uniform flux density of $4 \cdot 0$ μC/m and combine it with a line source of $0 \cdot 8$ μC/m. Plot a combined flux line pattern by summing the flux function values of the two fields and calculate the position of the stagnation point.

Answer. If we take flux function lines for every $0 \cdot 10$ μC/m, the flux line spacing for the uniform field will be 25 mm. These will be parallel straight lines in the direction of the flux density vector. For the line source, the flux lines will be radial straight lines from the line filament. To give the same scale of flux function, there will be eight lines radiating from the line source. These two flux patterns are drawn in Fig. 3.7 together with the flux line pattern of the combined field. We can also find an analytic expression for the flux function of the combined field. The flux function of a uniform field is given by eqn. (3.7) and that for a line source by eqn. (3.8). Therefore the flux function of the combination is given by

$$\psi = Dr \sin \theta + \frac{q\theta}{2\pi},$$

where the line of the source is taken as the origin of the coordinate system. At the stagnation point, the electric field vector is zero. From eqns. (3.4) and (3.5) the electric field components are given by

$$D_\theta = -\frac{\partial \psi}{\partial r} = -D \sin \theta,$$

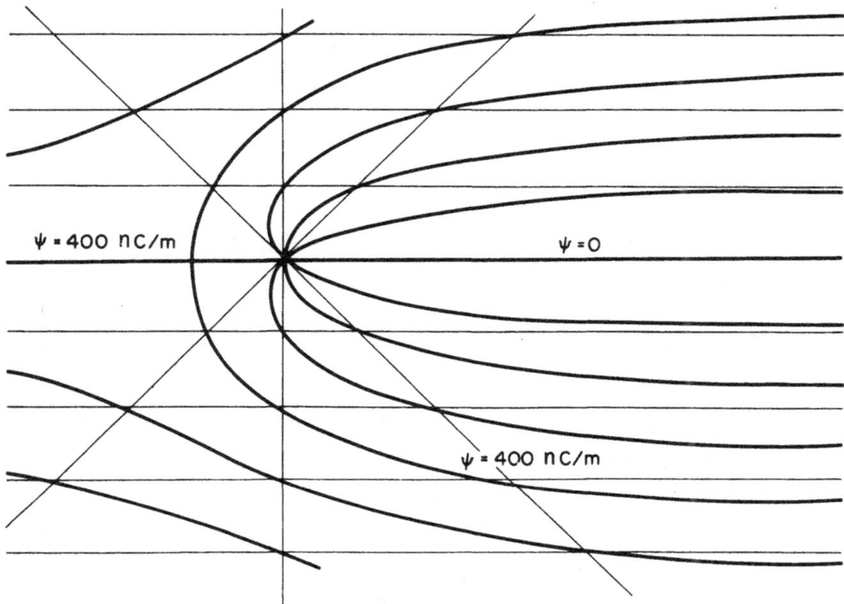

Fig. 3.7. The field pattern of the combination of a line
source with a uniform field.

$$D_r = \frac{1}{r}\frac{\partial \psi}{\partial \theta} = \frac{1}{r}Dr\cos\theta + \frac{q}{2\pi r}.$$

At $\theta = \pi$, D_θ is identically equal to zero. However, D_r is only equal to zero
when

$$D = \frac{q}{2\pi r}.$$

That is, when
$$r = \frac{q}{2\pi D}.$$

Inserting the numbers from the problem gives

$$r = \frac{0.8}{2\pi \times 4 \cdot 0} = 0 \cdot 0318 \text{ m} = 31 \cdot 8 \text{ mm}.$$

Therefore the stagnation point is at a distance of 31·8 mm from the line of
the line source.

3.7. Find the strength of the line source required to provide a
stagnation point in a uniform field of 20 V/m at a distance of 1·0 m from the
line source.

Answer. The flux pattern will be similar to that shown in Fig. 3.7 of the combination of a line source with a uniform field. The theoretical analysis of the combination is given in the solution to Example 3.6, and the relationship of the various parameters for the position of the stagnation point is given by

$$q = 2\pi r D.$$

The uniform flux density D is given in terms of the electric field intensity and is related to it by $D = \epsilon_o E$. Therefore, inserting numbers from the question,

$$q = 2\pi \times 1{\cdot}0 \times 20\epsilon_o = \frac{40\pi}{36\pi \times 10^9} = 1{\cdot}11 \times 10^{-9} \text{ C/m} = 1{\cdot}11 \text{ nC/m}.$$

3.8. Construct a flux pattern for the combination of a doublet of strength 628 nC and a uniform flux density of 40 μC/m^2, such that the line joining the elements of the doublet is parallel to the uniform field.

Answer. The first decision to be made is the flux difference between flux function lines. In this case, take one flux function line for every $1{\cdot}0$ μC/m. Then, in the uniform field, the flux function line spacing will be given by

$$y = \frac{10^{-6}}{40 \times 10^{-6}} = 0{\cdot}025 \text{ m} = 25 \text{ mm}.$$

The radius of the flux function line circles for the doublet are given by eqn (3.13), whence

$$a = \frac{c}{4\pi\psi}.$$

When $\psi = 1{\cdot}0$ μC/m, the radius is given by

$$a = \frac{628 \times 10^{-9}}{4\pi \times 10^{-6}} = 0{\cdot}050 \text{ m} = 50 \text{ mm}.$$

Therefore it is quite easy to construct full size a diagram similar to Fig. 3.5 for the doublet and to add to it the paralled lines of the uniform field. Omitting the construction flux function lines, the flux function pattern for the combination is shown in Fig. 3.8. It is seen that there is one circular flux function line and that this line separates the flux circulation through the doublet from the flux due to the uniform field.

An analytic expression for the flux function of the combination is given by substitution from eqns. (3.7) and (3.12), therefore

$$\psi = D r \sin \theta - \frac{c \sin \theta}{2\pi r}.$$

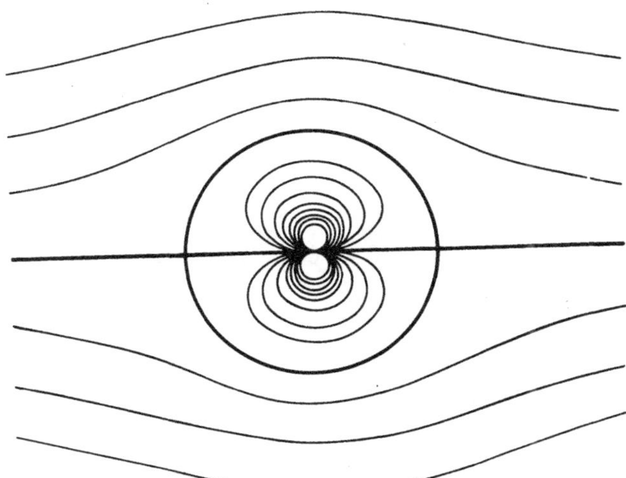

Fig. 3.8. The field pattern of the combination of a doublet
with a uniform field.

This expression is identically equal to zero independently of θ when

$$r^2 = \frac{c}{2\pi D}.$$

This expression gives the radius of the circular flux function line. Let
this value of radius be a, then the expression for the flux function becomes

$$\psi = Dr \sin \theta \left(1 - \frac{a^2}{r^2}\right).$$

For the conditions of this example, the radius of the circular flux function
line is given by

$$a = \sqrt{\left(\frac{c}{2\pi D}\right)} = \sqrt{\left(\frac{628 \times 10^{-9}}{2\pi \times 40 \times 10^{-6}}\right)} = \sqrt{(0 \cdot 0025)} = 0 \cdot 050 \text{ m} = 50 \text{ mm}.$$

This circular flux function line can form a flux boundary in the field.

3.9. Find the strength of the equivalent doublet needed in combination
with a uniform field of 100 V/m to represent a circular flux boundary of
6·0 m diameter in the uniform field.

Answer. The theory of the combination of a doublet field with a uniform
field is given in the solution to Example 3.8, where the radius of the
circular flux boundary is given by

$$r^2 = \frac{c}{2\pi D}.$$

Therefore

$$c = 2\pi D r^2.$$

Putting numbers into this equation gives the strength of the equivalent doublet. Because

$$D = \varepsilon_o E = \frac{100}{36\pi \times 10^9} = C/m^2,$$

$$c = \frac{2\pi \times 100 \times 9}{36\pi \times 10^9} = 50 \times 10^{-9} \, C = 50 \text{ nC.}$$

3.10. Construct a flux pattern for the combination of a line-source line-sink pair with a uniform field. The strength of the line source is 80 nC/m and it is parallel to and distant 40 mm from an equal line sink of strength -80 nC/m. The line joining the source and sink is parallel to the direction of the uniform field of strength $1 \cdot 0 \, \mu C/m^2$.

Answer. The flux pattern for the line-source line-sink pair is shown in Fig. 3.3, where flux function lines are drawn for every $q/16$. From the values given in our problem, this is for every $5 \cdot 0$ nC/m. Therefore, to give the same flux function difference for the uniform field lines, their spacing is given by

$$y = \frac{\psi}{D} = \frac{5 \times 10^{-9}}{10^{-6}} = 5 \cdot 0 \times 10^{-3} \, m = 5 \cdot 0 \text{ mm.}$$

The complete pattern is given in Fig. 3.9, where it is seen that there is an elliptical flux function line. Therefore this combination can be used to represent an elliptical flux boundary.

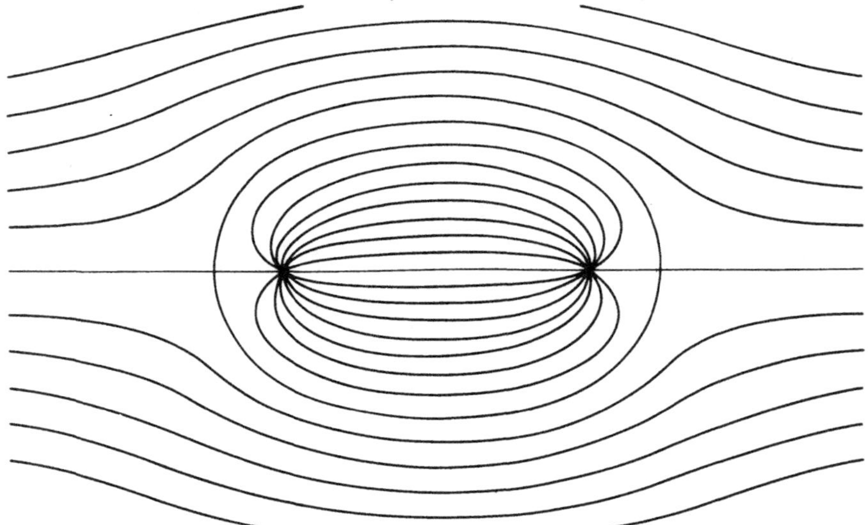

Fig. 3.9. The field pattern of the combination of a line-source line-sink pair with a uniform field.

PROBLEMS

3.1. Two hollow conducting metal spheres are concentric. The outer one
has a diameter of $0 \cdot 2$ m and is charged initially with $+0 \cdot 10 \mu C$. The inner
one has a diameter of $0 \cdot 1$ m and is charged at some time later with $-0 \cdot 04 \ \mu C$.
The charge on the inner sphere must induce an equal but opposite charge on
the inside surface of the outer sphere. By considering the charge
distribution on the system and sketching the flux patterns, determine (a) the
flux density midway between the two spheres, and (b) the electric intensity
in air at $0 \cdot 2$ m from the centre of the system. Repeat the calculation using
Gauss's law. If the inner sphere is moved to an eccentric position without
touching the outer sphere, are these results changed, and how?

$(-566 \ nC/m^2; \ 13 \cdot 5 \ kV/m.)$

3.2. Parallel line charges are located $+5$ nC/m at $(0, 0)$ m, $+4$ nC/m at
$(3, 0)$ m, -6 nC/m at $(0, 4)$ m. Find **E** and **D** at the position $(3, 4)$ m, (a)
by vector addition of the field quantities, and (b) by plotting the flux
function in the region of the point. $(41 \ V/m; \ 360 \ pC/m^2.)$

3.3. By plotting the flux function, find the flux distribution due to
three equal parallel line charges all the same distance apart so that, in
section, the charges are at the corners of an equilateral triangle.

3.4. Verify the truth of Fig. 3.3 by summing the flux function values
of a line source and line sink.

3.5. Take a uniform flux density of 10 nC/m^2 and combine it with a
line source of $2 \cdot 0$ nC/m. Sum the flux function values and compare the
resulting flux pattern with that given in Fig. 3.7.

3.6. Calculate the spacing of the flux function lines at the point
$(1, 0)$ m for the combination used in Problem 3.5. What is the spacing at an
infinite distance along the same axis? Hence, at an infinite distance away,
find the width of the boundary formed by the flux line passing through the
stagnation point. $(97 \times 10^6 \ m^2/C; \ 10^8 \ m^2/C; \ 0 \cdot 20 \ m.)$

3.7. Calculate the radius of the circular flux line which occurs with
a combination of a uniform flux density of 100 nC/m^2 and a doublet of
strength $7 \cdot 0$ nC. $(0 \cdot 106 \ m.)$

3.8. By adding flux function values, construct the field pattern of
the combination given in Problem 3.7 and compare it with Fig. 3.8.

3.9. Certain materials have the property of almost preventing flux from entering them so that the surface of the material becomes a flux boundary. Calculate the strength of the doublet required to give a flux pattern equivalent to that around a cylinder of such a material of 10 mm diameter mounted perpendicular to a uniform flux density of 16 nC/m^2. (2·5 pC.)

3.10. A line-source line-sink pair is mounted in a uniform field with the distance between them at 45^0 to the direction of the uniform field. Construct the flux pattern of the combination by (a) adding the flux function of the pair (given in Fig. 3.3) to the flux function of the uniform field, and (b) adding the flux function pattern of a flux line source and uniform field (given in Fig. 3.7) to the flux function pattern of a flux line sink and uniform field (Fig. 3.7 reversed). (Use the results to Problems 3.4 and 3.5.)

SOLUTIONS

3.1. When the inner sphere is charged with -0·04 µC, an equal and opposite charge will collect on the inside surface of the outer sphere. This means that the charge on the outer sphere will be divided between its inner and outer surfaces. Therefore,

$$\text{charge on inner surface} = + 0\cdot04 \text{ µC,}$$
$$\text{charge on outer surface} = + 0\cdot06 \text{ µC.}$$

By symmetry all these charges will be distributed uniformly around the surface of the spherical shells as shown in Fig. 3.10.

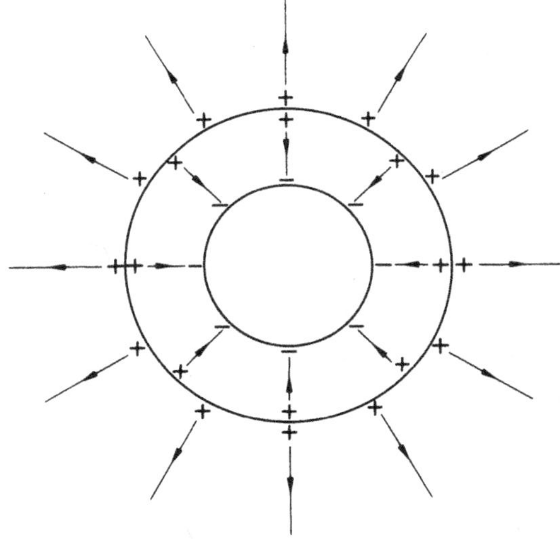

Fig. 3.10. Illustrating the solution to Problem 3.1.

(a) Midway between the spheres the total flux is -0·04 µC. This is uniformly distributed so that the flux density at a radius of 0·075 m is given by

$$D = \frac{-0\cdot04 \times 10^{-6}}{4\pi \times (0\cdot075)^2} = -5\cdot66 \times 10^{-7} \text{ C/m}^2 = -566 \text{ nC/m}^2.$$

(b) External to the outer sphere, the total flux is +0·06 µC. This is also uniformly distributed so that the flux density at radius 0·2 is given by

$$D = \frac{0 \cdot 06 \times 10^{-6}}{4\pi \times (0 \cdot 2)^2} \text{ C/m}^2.$$

Therefore

$$E = \frac{0 \cdot 06 \times 10^{-6} \times 36\pi \times 10^9}{4\pi \times 0 \cdot 04} = 1 \cdot 35 \times 10^4 \text{ V/m} = 13 \cdot 5 \text{ kV/m}.$$

By Gauss's law:

    (a) Enclosed charge = $-0 \cdot 04$ μC,

         flux out   = $D \times 4\pi \times (0 \cdot 075)^2$.

Therefore $D = -566 \text{ nC/m}^2$.

    (b) Enclosed charge = $(-0 \cdot 04 + 0 \cdot 10)$ μC,

         flux out   = $D \times 4\pi \times (0 \cdot 20)^2$

and $E = D/\varepsilon_0.$

Therefore $E = \dfrac{0 \cdot 06 \times 10^{-6} \times 36\pi \times 10^9}{4\pi \times (0 \cdot 2)^2} = 13 \cdot 5 \times 10^3 \text{ V/m}.$

So it is seen that the use of Gauss's law enables the results to be obtained with much less thought and trouble.

If the inner sphere is moved:

    (a) The distribution is no longer uniform between the two spherical shells since the charges will tend to concentrate where the shells are closest. The flux density midway between the two spheres will change.

    (b) Since each tube of flux inside the spherical shell terminates in a charge inside the spherical shell, the charge distribution on the outside of the spherical shell is unaffected and the flux density outside the shell remains symmetrical with the same value as before. The field outside a spherical conducting shell is unaffected by any change in distribution of the charges inside the spherical shell.

    3.2. The flux density due to a line charge is given by eqn. (2.15). The geometry of the problem is illustrated in Fig. 3.11. The flux densities at the point D due to the line charges at the other points are:

$$D_A = \frac{-6}{2\pi \times 3} = -\frac{1}{\pi} \text{ nC/m}^2,$$

$$D_B = \frac{5}{2\pi \times 5} = \frac{1}{2\pi} \text{ nC/m}^2,$$

$$D_C = \frac{4}{2\pi \times 4} = \frac{1}{2\pi} \text{ nC/m}^2.$$

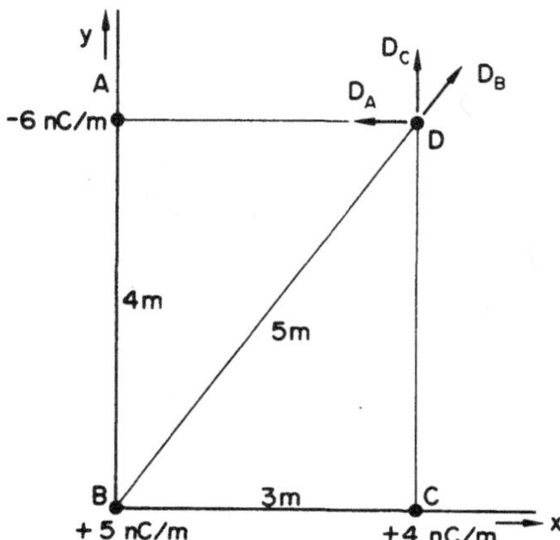

Fig. 3.11.   Illustrating the geometry of Problem 3.2.

By vector addition of these flux densities, the horizontal and vertical
components of the resultant flux density are given by

$$D_x = -\frac{1}{\pi} + \frac{3}{5}\frac{1}{2\pi} = -\frac{1}{\pi}(0 \cdot 7)\ \text{nC/m}^2$$

and

$$D_y = \frac{1}{2\pi} + \frac{4}{5}\frac{1}{2\pi} = +\frac{1}{\pi}(0 \cdot 9)\ \text{nC/m}^2.$$

Therefore the resultant flux density is given by

$$D = \frac{1}{\pi}\sqrt{(0 \cdot 49 + 0 \cdot 81)} = \frac{1 \cdot 14}{\pi} = 0 \cdot 36\ \text{nC/m}^2$$

and

$$E = 36\pi \times 10^9 \times \frac{1 \cdot 14}{\pi} \times 10^{-9} = 41\ \text{V/m}.$$

The angular direction of the flux density vector is given by

$$\text{angle} = \tan^{-1}\frac{0 \cdot 7}{0 \cdot 9} = 38^\circ.$$

But the angle is obtuse if the horizontal is taken as the direction of zero
angle. Therefore the direction of the flux density vector is at 128° to the
horizontal.

Plotting the flux function in the region of the point D is shown in
Fig. 3.12. Three flux function lines have been drawn for each flux source,
and their values marked against them. As it is difficult to determine the
flux function values for the addition of the three at once, the dotted lines

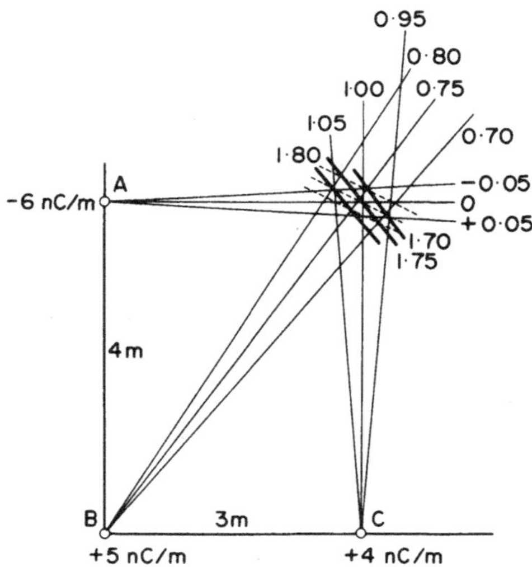

Fig. 3.12.   The flux function plot in the solution to
Problem 3.2.   The dotted lines are flux function lines
for the combination flux function of the source at *A*
and the source at *C*.

are flux function lines for the combination of the source at *A* with the
source at *C*. The other labelled lines give the flux function values of the
finally combined field. It is seen that the flux function line passing
through the point of interest has the value 1·75 nC/m. From this diagram,
the spacing between the flux function lines is 0·15 m and 0·13 m. This
gives an average spacing of 0·14 m. Then the flux density is given by

$$D = \frac{\delta\psi}{\delta l} = \frac{0·05}{0·14} = 0·36 \text{ nC/m},$$

which is in good agreement with the calculated value. The measured slope of
the flux line is 128°.

3.3. The technique of flux plotting is to draw the flux function lines for
each line charge and to add the effect of all three. This is easiest to
perform by drawing the flux function due to two positive line charges using
a technique similar to that used to draw Fig. 3.14. Such a diagram is shown
in Fig. 6.6c (p.109). Afterwards the effect of the third line charge is
added to the other two. All the construction lines for a similar problem
are shown in Fig. 3.12. The final solution for this problem is shown in

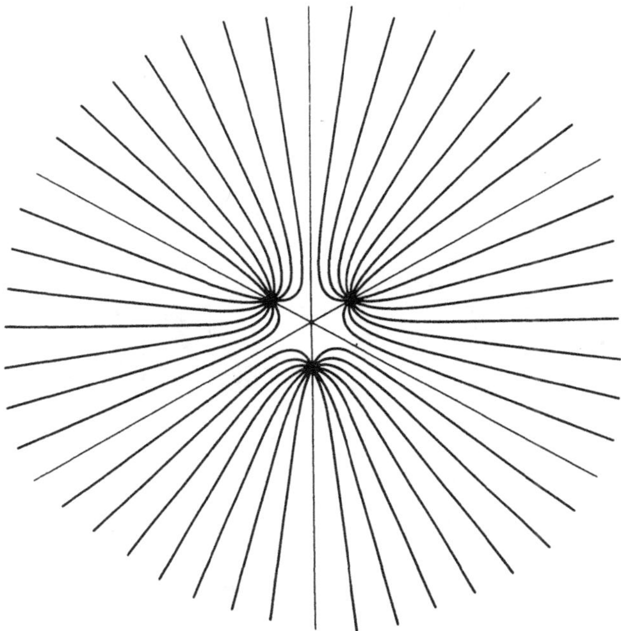

Fig. 3.13. The flux function lines for three equal line
sources. The solution to Problem 3.3.

Fig. 3.13. A large distance away, the flux pattern will be radial, the same
as that due to a single line charge of three times the strength.

3.4. The flux function lines for a line source and a line sink are
drawn in Fig. 3.14. Four segments of circles which are the flux function
lines of the combination are also shown. The complete solution will be the
same as Fig. 3.3.

3.5. The flux lines out of a line source are radial. Take sixteen
lines around the point, then the flux increment between lines is 0·125 nĊ/m.
For the same flux increment between lines, the line spacing in the uniform
field is 0·0125 m or 12·5 mm. Taking these values, the diagram ought to be
the same as that given in Fig. 3.7.

3.6. An expression for the flux function of the combination is given
in the solution to Example 3.6. It is

$$\psi = Dr \sin \theta + \frac{q\theta}{2\pi}.$$

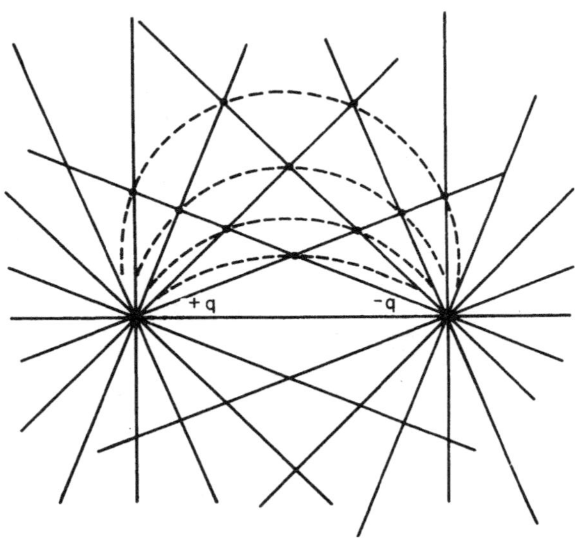

Fig. 3.14. The flux function lines due to a line source
and a line sink. The solution to Problem 3.4.

From Fig. 3.7 it can be deduced that the flux function lines in the region
of the point (1, 0) are parallel to the direction of uniform flow. The
spacing is the inverse of the density of flux function lines. In cylindrical
polar coordinates, this is given by

$$\frac{1}{D_r} = 1 \bigg/ \left(\frac{1}{r}\frac{\partial\psi}{\partial\theta}\right).$$

But

$$\frac{1}{r}\frac{\partial\psi}{\partial\theta} = D\cos\theta + \frac{q}{2\pi r}.$$

At the point (1, 0), $\theta = 0$ and $r = 1$; therefore, inserting numbers from the
question,

$$\frac{1}{r}\frac{\partial\psi}{\partial\theta} = 10 + \frac{2}{2\pi} = 10\cdot32 \text{ nC/m}^2$$

and the spacing is the inverse of this. Therefore the spacing = 97×10^6
m^2/C. At an infinite distance along the same axis,

$$\frac{1}{r}\frac{\partial\psi}{\partial\theta} = D = 10 \text{ nC/m}^2.$$

Therefore the spacing = 10^8 m^2/C.

The value of the flux function line passing through the stagnation point
is $\frac{1}{2}q = 1\cdot0$ nC/m. The spacing of the two branches of this stagnation line

an infinite distance away is given by twice the flux function value of the line multiplied by the line spacing at infinity. Therefore the required spacing is given by

$$l = 2 \times \tfrac{1}{2}q \times (\text{spacing}) = 2 \times 10^{-9} \times 10^{8} = 0\cdot20 \text{ m}.$$

3.7. In the solution to Example 3.8, the radius of the circular flux function line is found to be given by

$$r = \sqrt{\frac{c}{2\pi D}}.$$

Inserting numbers from the problem into this equation gives

$$r = \sqrt{\left(\frac{7 \times 10^{-9}}{2\pi \times 100 \times 10^{-9}}\right)} = 0\cdot106 \text{ m}.$$

3.8. For simplicity in the construction of the flux function pattern, we will make the increment of the flux function lines such that the spacing of the lines in the uniform field is a simple multiple of the radius of the circular flux line. Let the spacing be half the radius that has been determined in the solution to Problem 3.7. Therefore the increment between the flux function lines is given by

$$\delta\psi = 100 \times 0\cdot053 = 5\cdot3 \text{ nC/m}.$$

The radius of the flux function line circles for the doublet are given by eqn. (3.13). When $\psi = 5\cdot3$ nC/m, the radius is given by

$$a = \frac{c}{4\pi\psi} = \frac{7 \times 10^{-9}}{4\pi \times 5\cdot3 \times 10^{-9}} = 0\cdot106 \text{ m}.$$

The other flux function circles for the doublet will have radii which are simple fractions of this radius. When the combined field is drawn with the flux function increment we have chosen, the diagram will be similar to Fig. 3.8.

3.9. For the combination of a doublet and a uniform field, the radius of the circular flux boundary is given by

$$r^2 = \frac{c}{2\pi D}.$$

In this problem, we are required to find the strength of the equivalent doublet to give a particular radius of circular flux line. Therefore

$$c = 2\pi D r^2 = 2\pi \times 16 \times 10^{-9} \times (5 \times 10^{-3})^2 = 2\cdot5 \times 10^{-12} \text{ C} = 2\cdot5 \text{ pC}.$$

3.10. The diagram of the combined field due to a line source and a line sink is given in Fig. 3.3. This diagram is then combined with the parallel flux function lines of a uniform field but with the direction of the uniform field at 45° to the line joining the line source and the line sink. The actual result obtained will be a function of the relative strength of the line-source line-sink pair and the uniform field. One possible answer to this problem is given in Fig. 3.15.

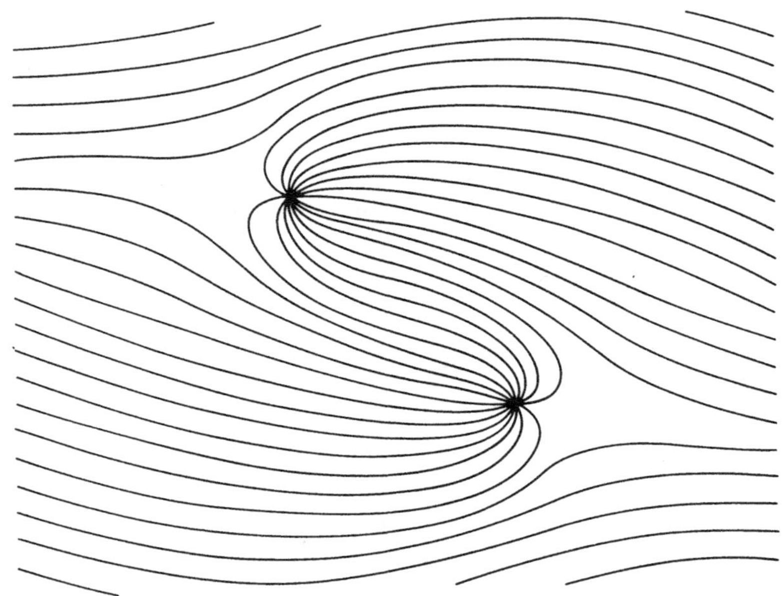

Fig. 3.15. The flux function lines due to a line source and a line sink set at 45° to a uniform field. The solution to Problem 3.10.

FURTHER PROBLEMS

3.11. Find the shape and spacing of the flux function lines which have a flux function increment of $1 \cdot 0$ pC/m (a) for a uniform field of 20 V/m, and (b) for a line filament charge of $0 \cdot 032$ nC/m. ($5 \cdot 66$ mm; $11\frac{1}{4}^{\circ}$.)

3.12. Line charges are located at the following points where the charge density is measured in μC/m and the distance coordinates in m: $+ 2 \cdot 0$ at $(-2, 0)$, $+ 3 \cdot 0$ at $(0, -1)$, and $- 2 \cdot 0$ at $(2, 0)$. Find the strength of the resultant flux density at the point, $(0, 2)$. ($0 \cdot 225$ μC/m^2.)

3.13. By plotting the flux function, find the flux distribution due to a line charge filament of $2q$ C/m parallel to but not coincident with a line charge filament of $-q$ C/m.

3.14. A line source of 1·0 µC/m is situated at the point (1, 0) m and a line sink of strength -1·0 µC/m is situated at the point (-1, 0) m. Find the value of the electric flux density due to the combined field at the points (-1, 1), (0, 1) and (1, 1) m.   (0·142, 0·159, 0·142 µC/m².)

3.15. Find the maximum value of flux function 1·0 m away from the midpoint of two parallel charged wires 0·1 m apart. The wires, which may be assumed to be thin, have equal and opposite charges of 1·0 µC/m. (16·1 nC/m.)

3.16. A line source of strength 10 µC/m is combined with a uniform field of strength 1·0 µC/m². Find the distance between the stagnation point and the line of the line source.   (1·59 m.)

3.17. A flux pattern is similar to that generated by the combination of a line source with a uniform field. A long way from the line source, the separation of the flux function lines is 20 mm and the total distance between the two halves of the line through the stagnation point is 0·1 m. Suggest a suitable combination of line source with uniform field to produce a model of this field pattern.   (1·0 C/m²; 0·1 C/m.)

3.18. A doublet of strength 6·3 µC is combined with a uniform field of 1·0 µC/m². Find the shape and position of the stagnation flux line. (1·0 m.)

3.19. Find the strength of the equivalent doublet needed to model a circular flux boundary of 0·1 m radius in an otherwise uniform field of 9·0 kV/m.   (5·0 nC.)

3.20. A line-source line-sink pair, of strength ±1·0 µC/m, distant 1·0 m apart, is combined with a uniform field of 0·2 µC/m², such that the line joining the line source and line sink is parallel to the uniform field. Find the width at its perpendicular axis of the elliptical stagnation flux line.   (1·7 m.)

FURTHER SOLUTIONS

3.11. The flux density of the uniform field is given by

$$D = \frac{20}{36\pi \times 10^9}.$$

The spacing of the flux function lines in the uniform field is given by

$$\delta l = \frac{\delta\psi}{D} = \frac{10^{-12} \times 36\pi \times 10^9}{20} = 5\cdot66 \times 10^{-3}\ \text{m} = 5\cdot66\ \text{mm}.$$

For the *uniform field*, the flux function lines are parallel equally spaced lines which are 5·66 mm apart.

For the *line filament of charge*, the flux function lines are equally spaced radial lines having the line of charge as origin. There will be thirty-two of these radial lines from the line charge so that each line will be at an angle of $11\frac{1}{4}^{\circ}$ to its neighbours.

3.12. The student is invited to draw his own diagram for this problem. The flux density due to a line charge is given by eqn. (2.15). The horizontal and vertical components of the flux density at the point (0, 2) are given by

$$D_h = \frac{2\cdot0 \times 10^{-6}}{2\pi \times 2\sqrt2} \times \frac{1}{\sqrt2} + \frac{2\cdot0 \times 10^{-6}}{2\pi \times 2\sqrt2} \times \frac{1}{\sqrt2} + 0 = \frac{10^{-6}}{2\pi},$$

$$D_v = \frac{2\cdot0 \times 10^{-6}}{2\pi \times 2\sqrt2} \times \frac{1}{\sqrt2} - \frac{2\cdot0 \times 10^{-6}}{2\pi \times 2\sqrt2} \times \frac{1}{\sqrt2} + \frac{3\cdot0 \times 10^{-6}}{2\pi \times 3} = \frac{10^{-6}}{2\pi}$$

Therefore, $\qquad D = \dfrac{10^{-6}}{2\pi} \times \sqrt2 = 0\cdot225 \times 10^{-6}\ \text{C/m}^2 = 0\cdot225\ \mu\text{C/m}^2.$

3.13. The required flux function plot is shown in Fig. 3.16.

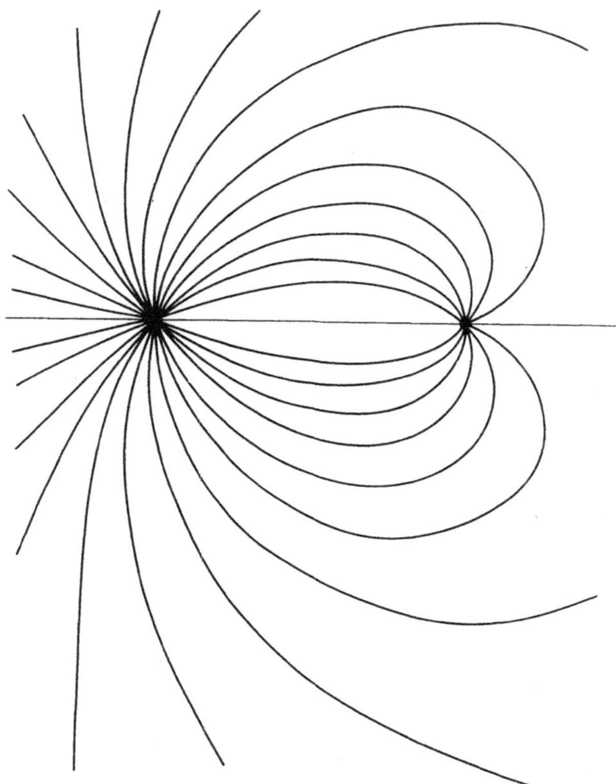

Fig. 3.16. The flux function pattern due to a line source
of +2q C/m parallel to a line sink of -q C/m.   The
solution to Problem 3.13.

3.14. The student is invited to obtain the solution to this problem
graphically. This may be done by measuring the spacing of the flux function
lines from Fig. 3.3. The increment between the flux function lines is
$\frac{1}{16}q = \frac{1}{16}$ μC/m. At the point (0, 1) the spacing is $\delta l = \frac{1}{2}(0\cdot50 + 0\cdot30) =$
0·40 m. Therefore the flux density is given by

$$D = \frac{\delta\psi}{\delta l} = \frac{1}{16 \times 0\cdot40} = 0\cdot157 \text{ μC/m}^2 .$$

At the point $(1, 1)$, the spacing is $\delta l = \frac{1}{2}(0\cdot 50 + 0\cdot 35) = 0\cdot 425$ m.
Therefore the flux density is given by

$$D = \frac{\delta \psi}{\delta l} = \frac{1}{16 \times 0\cdot 425} = 0\cdot 147 \ \mu C/m^2.$$

By calculation the flux density is obtained by resolving components of
the field at each point and adding them together. The field at the point
$(0, 1)$ is given by

$$D = \frac{1}{2\pi \times \sqrt{2}} \times \frac{1}{\sqrt{2}} \times 2 = 0\cdot 159 \ \mu C/m^2.$$

The field at the point $(1, 1)$ is given by

$$D_v = \frac{1}{2\pi} - \frac{1}{2\pi \times \sqrt{5}} \times \frac{1}{\sqrt{5}} = \frac{2}{5\pi},$$

$$D_h = \frac{1}{2\pi \times \sqrt{5}} \times \frac{2}{\sqrt{5}} = \frac{1}{5\pi}.$$

Therefore

$$D = \frac{1}{5\pi} \sqrt{(1 + 4)} = 0\cdot 142 \ \mu C/m^2.$$

It is seen that the graphical solution is within about 3 per cent of the
true value. The accuracy of the graphical solution could have been improved
by increasing the number of flux function lines at the points of interest.

3.15. Using eqn. (3.10), the flux function for a line-source line-sink
pair,

$$\alpha = 2 \ \tan^{-1} \frac{0\cdot 05}{1\cdot 0} = \frac{2 \times 2\cdot 9 \times 2\pi}{360} \ \text{rad}.$$

Therefore

$$\psi = \frac{10^{-6}}{2\pi} \times \frac{2 \times 2\cdot 9 \times 2\pi}{360} = 1\cdot 61 \times 10^{-8} \ C/m = 16\cdot 1 \ nC/m.$$

Considering the line-source line-sink pair as approximating to a doublet
gives

$$\psi = \frac{10^{-6} \times 0\cdot 1}{2\pi \times 1\cdot 0} = 1\cdot 59 \times 10^{-8} \ C/m = 15\cdot 9 \ nC/m.$$

Here it is seen that there is about 1 per cent error in considering the line-
source line-sink pair as a doublet.

3.16. The distance between the line source and the stagnation point has been derived in the solution to Example 3.6. Inserting numbers from this problem into the relationship given there gives

$$r = \frac{q}{2\pi D} = \frac{10 \times 10^{-6}}{2\pi \times 10^{-6}} = 1\cdot59 \text{ m.}$$

3.17. The separation of the flux function lines in the uniform field is 20 mm = δy. From the separation of the stagnation flux function line, we know that 100 mm = $5\delta y$; therefore, $q = 5\delta\psi$. Let the uniform field be $1\cdot0$ C/m^2. Then

$$\delta\psi = 1\cdot0 \times 0\cdot02 = 0\cdot02 \text{ C/m.}$$

Therefore
$$q = 5 \times 0\cdot02 = 0\cdot10 \text{ C/m.}$$

The general relationship is that $D = 10q$.

3.18. The combination of a doublet with a uniform field gives a circular stagnation flux function line. The relationship for the radius of this circular line is given in the solution to Example 3.8. Inserting numbers into this relationship gives

$$r = \sqrt{\left(\frac{c}{2\pi D}\right)} = \sqrt{\left(\frac{6\cdot3 \times 10^{-6}}{2\pi \times 10^{-6}}\right)} = 1\cdot0 \text{ m.}$$

3.19. The flux density of the uniform field is given by

$$D = \frac{9 \times 10^3}{36\pi \times 10^9} = \frac{1}{4\pi \times 10^6} \text{ C/m}^2$$

and the strength of the equivalent doublet by

$$c = 2\pi D r^2 = \frac{2\pi \times 10^{-2}}{4\pi \times 10^6} = 0\cdot5 \times 10^{-8} \text{ C} = 5\cdot0 \text{ nC.}$$

3.20. A mathematical expression for the flux function of the combination of a uniform field with a line-source line-sink pair is obtained from a combination of eqns. (3.6) and (3.10). It is

$$\psi = Dy - \frac{q\alpha}{2\pi}.$$

A typical flux function pattern is shown in Fig. 3.9. The value of the stagnation flux function line is usually taken to be zero when the origin of the coordinate system defining the uniform field passes through the line joining the line source and the line-sink. Then we have the condition

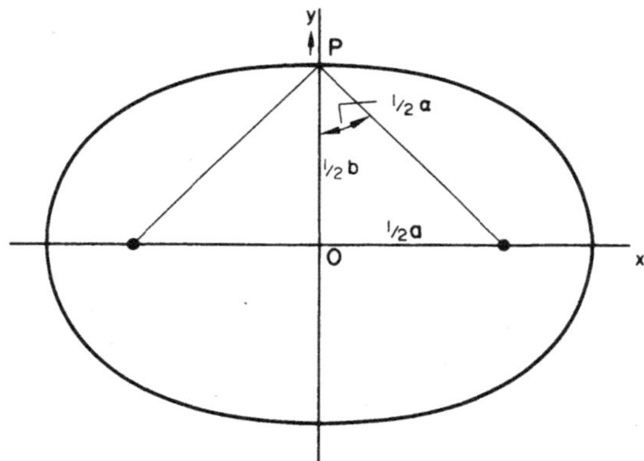

Fig. 3.17.   Illustrating the geometry of Problem 3.20.

$$Dy = \frac{q\alpha}{2\pi}.$$

The geometry of the problem is illustrated in Fig. 3.17. The required width will be the dimension b shown in Fig. 3.17; α is given by

$$\alpha = 2 \tan^{-1} \frac{a}{b}$$

Therefore, because the flux function at the point P is zero,

$$D \times \tfrac{1}{2}b = \frac{q}{2\pi} \times 2 \tan^{-1} \frac{a}{b}.$$

or

$$\frac{a}{b} = \tan \frac{\pi Db}{2q}.$$

Inserting numbers gives the relationship

$$\frac{1}{b} = \tan \frac{\pi \times 0 \cdot 2b}{2 \times 1 \cdot 0} = \tan \frac{\pi b}{10}.$$

This needs to be solved numerically and a computer may be used; alternatively, a plot of each side of the equation is made against b and the solution is that value of b at which the two curves intersect. This occurs when $b = 1 \cdot 7$ m.

CHAPTER 4

Electrical Materials

THEORY

Atoms consist of a positively charged nucleus surrounded by a cloud of negatively charged electrons. In an insulator, all the electrons are rigidly attached to each atom and there is no freedom of movement of electrons through the material. Under the influence of an electric field, each atom becomes polarized, giving rise to a non-uniform charge distribution in the atom. In a conductor, some electrons in each atom are completely free to move around in the solid. Usually the electrons transfer charge to the surface of the conductor so as to cancel the electric flux which gives rise to the original electric field intensity. Therefore there can be no electric field intensity inside a conductor.

If an electric charge of one sign is placed inside the volume of a conducting body, it will all flow out to the surface under the mutually repulsive Coulomb forces. Experimental evidence shows that the surface density increases rapidly where the convex curvature is sharpest and decreases with concave curvature. The flux density immediately above the surface of a conductor is equal to the surface charge density on the conductor at that point. The direction of the flux density is perpendicular to the surface. Therefore the field intensity just outside a charged conducting surface is given by

$$E_O = \frac{\sigma}{\epsilon_O} \text{ V/m,} \tag{4.1}$$

where σ is the charge density on the surface. Inside the surface of the conducting body, the electric intensity is always zero, therefore

$$E_i = 0. \tag{4.2}$$

However, at the position of the charge itself, the field varies across the charge from E_O outside the conductor to zero inside the conductor, so that the average intensity is $\frac{1}{2}E_O$. Therefore the electric intensity in the surface is given by

$$E_s = \frac{\sigma}{2\epsilon_O} \text{ V/m.} \tag{4.3}$$

As there is an electric field intensity in the region of a charge, there will also be a force acting on the charge. Therefore there will be a force on a charged conducting surface. In fact, there will be a force on any charged surface. However, for the charged conducting surface, the force per unit area of surface, which is also the pressure on the surface due to the field, is given by

$$p = \frac{\sigma^2}{2\varepsilon_o} \ \text{N/m}^2 .$$
(4.4)

In terms of the adjacent field quantities, the force is given by

$$\text{force} = \tfrac{1}{2}DE \ \text{N/m}^2 .$$
(4.5)

When an insulating medium is placed in an electric field, the atoms of the insulator become polarized so as to reduce the overall electric field intensity inside the insulator. Without the insulator present, the electric field intensity is given by

$$\mathbf{E} = \mathbf{D}/\varepsilon_o .$$
(4.6)

Provided that the electric flux lines are perpendicular to the boundary of the insulator, the field intensity inside the insulator is given by

$$\mathbf{E} = \mathbf{D}/\varepsilon ,$$
(4.7)

where ε is the permittivity. It has two components: the permittivity constant ε_o which is a dimensional constant - a fundamental electric constant which is solely a function of the system of units used, and the relative permittivity ε_r which is a dimensionless constant and is a property of the material in which the field exists. Therefore

$$\varepsilon = \varepsilon_o \varepsilon_r ,$$
(4.8)

and eqn. (4.7) becomes

$$\mathbf{E} = \mathbf{D}/\varepsilon_o \varepsilon_r .$$

EXAMPLES

4.1. A uniformly charged plane surface having a charge density of $1 \cdot 0$ $\mu C/m^2$ is situated $0 \cdot 1$ m distant from a parallel charged plane surface having a charge density of $-1 \cdot 0$ $\mu C/m^2$. A flat metal plate 10 mm thick is inserted into the space between the two charged surfaces parallel to the surfaces. Find the flux density and field intensity in each of the regions between the two charged planes.

Answer. The flux density due to two parallel charges of equal and opposite charge density is given in Fig. 2.2 (p. 15).   When the metal plate is inserted into this field, equal and opposite charges are induced on to the surfaces of the conducting plate so that the field intensity inside the metal of the plate is zero. Therefore in the region between the charged surface and the metal plate,

$$D = \sigma = 1 \cdot 0 \ \mu C/m^2,$$

$$E = D/\varepsilon_o = 10^{-6} \times 36\pi \times 10^9 = 113 \times 10^3 \ V/m = 113 \ kV/m.$$

Inside the conducting metal plate, the induced charge is such as to cancel any existing field intensity so that $D = 0$ and $E = 0$.

4.2. A solid spherical conductor of diameter 10 mm is charged with 314 nC. Find the flux density on the surface of the conductor.

Answer. The problem will be solved in two ways.

(i) By using Gauss's law. The surface of the sphere will be the Gaussian surface. Therefore the total charge $= D \times 4\pi r^2$. Therefore

$$D = \frac{314 \times 10^{-9}}{4\pi \times (0 \cdot 005)^2} = 10^{-3} \ C/m^2 = 1 \cdot 0 \ mC/m^2.$$

(ii) By finding the flux density at the surface of a conductor. Because the sphere is symmetrical, the charge distribution will be symmetrical and the surface charge density is equal to the total charge divided by the surface area of the sphere. Numerically it is the same as the expression given for D above, therefore

$$\sigma = 1 \cdot 0 \ mC/m^2.$$

The surface flux density equals the surface charge density, therefore

$$D = 1 \cdot 0 \ mC/m^2.$$

4.3. Find an expression for the field between two parallel plane conducting plates of area A with equal and opposite charges of Q.

Answer. Due to the mutual attraction of the equal and opposite charges, the charge on each plate will collect on the inside surfaces and there will be no charge on the outside. The charge distribution is shown in Fig. 4.1.

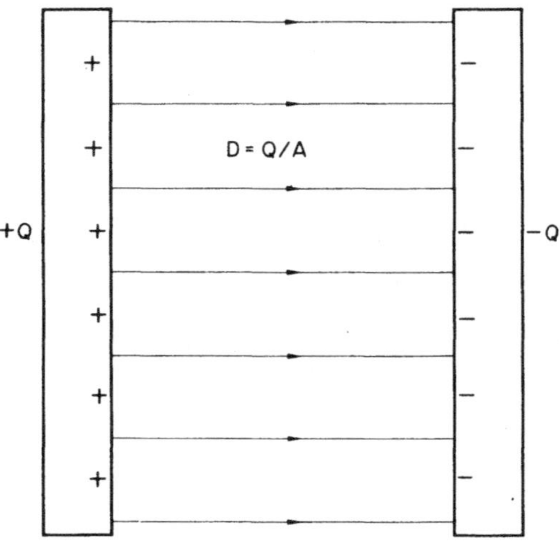

Fig. 4.1. Charge and flux distribution between two parallel
charged conducting plates.

The charge density on each plate is Q/A C/m^2. The flux density close to the inside surface is given by $D = Q/A$. If the effects at the edges of the plates are neglected, the flux density will be uniform between the plates and will be zero outside the plates. The field will be similar to that due to two equal and opposite surface charge densities. The field will also be zero inside the metal of each of the conducting plates. Therefore the field between the two conducting plates will be uniform and equal to

$$D = Q/A \ \text{C/m}^2.$$

4.4. A conducting metallic sphere of 10 mm radius is charged with 1·0 pC. Find the electric field intensity inside, on, and just outside the surface.

Answer. The charge will be uniformly distributed on the surface of the conducting sphere. Therefore

$$\sigma = \frac{10^{-12}}{4\pi \times 10^{-4}}.$$

The electric intensities are given by eqns. (4.1)–(4.3); therefore

$$E_o = \frac{\sigma}{\varepsilon_o} = \frac{36\pi \times 10^9}{4\pi \times 10^8} = 90 \text{ V/m},$$

$$E_s = 45 \text{ V/m},$$

$$E_i = 0.$$

4.5. Derive an expression for the pressure on a charged conducting surface due to the charge on the surface.

Answer. Consider a small element of surface as shown in Fig. 4.2.

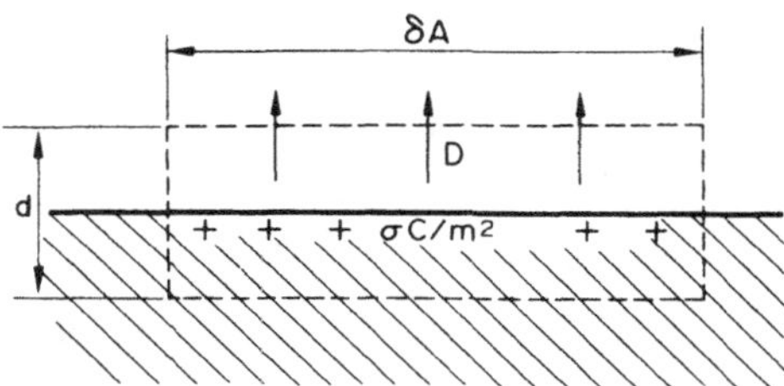

Fig. 4.2. A small element of surface of a charged conductor.

The surface has a charge density of σ C/m^2 and the element is of area δA. The force on the element of surface is given by the product of the charge and the field intensity. The field intensity in the surface is given by eqn. (4.3) so that

$$\delta f = E_s \delta q = E_s \sigma \delta A \text{ N}.$$

The force will be parallel to the field intensity E_s which is perpendicular to the surface. Therefore the pressure equals the force per unit area,

$$p = \frac{\delta f}{\delta A} = E_s \sigma = \frac{\sigma^2}{2\varepsilon_o} \text{ N/m}^2.$$

4.6. Find the bursting pressure exerted by a charge of 100 pC on a uniform conducting sphere of 1·0 mm diameter.

Answer. The charge density on the surface of the sphere is given by

$$\sigma = \frac{10^{-10}}{4\pi \times \frac{1}{4} \times 10^{-6}} \text{ C/m}^2$$

and the pressure by

$$p = \frac{\sigma^2}{2\varepsilon_o} = \frac{36\pi \times 10^9}{2 \times \pi^2 \times 10^8} = 57 \text{ N/m}^2.$$

4.7. Two electrically insulated parallel metal plates of area $1\cdot0$ m^2 are $3\cdot0$ mm apart and have equal and opposite charges of 10 µC. Neglecting any fringing effects of the field at the edges of the plates, find the force of attraction between the plates due to their electrical charge when the space between the plates is filled with air.

Answer. The two parallel charged plates form a parallel plate capacitor which is a very good approximation to the system of two parallel charge planes of uniform charge density. It must be assumed that the charge on the metal plates is uniformly distributed on the inside face of each plate. Then the field between the plates is uniform and is due to two parallel equal and opposite planes of uniform surface charge density. It is given by

$$D = \sigma = Q/(\text{area}) = 10 \text{ µC/m}^2.$$

The force per unit area is given by eqns. (4.4) and (4.5). Therefore

$$f = \tfrac{1}{2}DE = \tfrac{1}{2}D^2/\varepsilon_o = \tfrac{1}{2} \times 10^{-5} \times 10^{-5} \times 36\pi \times 10^9 = 1\cdot8\pi \text{ N/m}^2.$$
$$\text{Total force} = f \text{ (area)} = 5\cdot65 \text{ N.}$$

4.8. An insulating medium is situated between two parallel planes of charge. By considering the polarization of a representative number of atoms in the insulator, obtain an expression for the relative permittivity of the insulator in terms of the induced charges on the surface of the insulator.

Answer. The insulator with a large number of polarized atoms is shown in Fig. 4.3 between two parallel lines of charge. In so far as the field outside the insulator is concerned, the polarized atoms may be replaced by an induced charge on the surface of the insulator. The system of induced charges is shown in Fig. 4.4 together with the flux densities due to both the applied charges and the induced charges. The total flux density in the insulating medium is given by

$$D_t = D - D_i = \sigma - \sigma'.$$

Then inside the insulator the electric field intensity is given by

$$E = \frac{D_t}{\varepsilon_o} = \frac{\sigma - \sigma'}{\varepsilon_o} = \frac{D}{\varepsilon_o}\left(\frac{\sigma - \sigma'}{\sigma}\right).$$

But in terms of the relative permittivity and the externally applied flux density, the field intensity is given by $E = \dfrac{D}{\varepsilon_o \varepsilon_r}$.

Therefore ε_r is given by $\varepsilon_r = \left(\dfrac{\sigma}{\sigma - \sigma'}\right).$

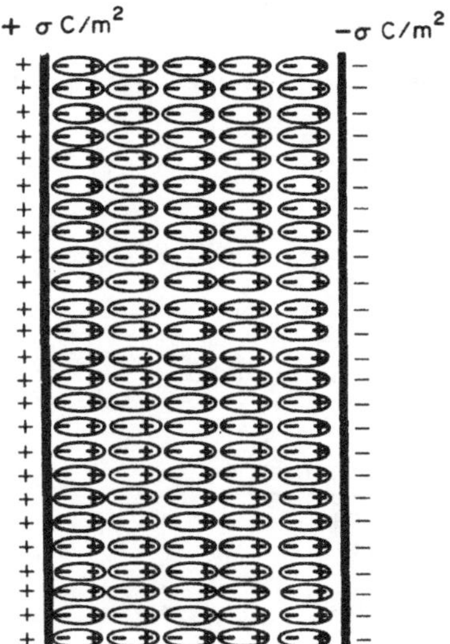

Fig. 4.3. Polarization of the atoms of a dielectric when
between two charge sheets.

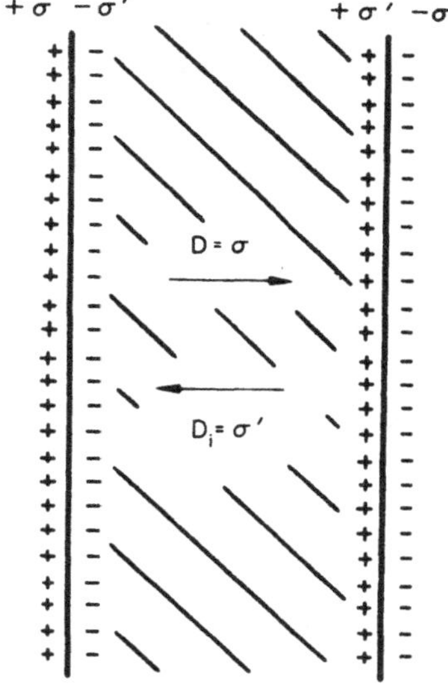

Fig. 4.4. Resultant effect of the polarization of the atoms
of the dielectric between two charge sheets.

4.9. A conducting metallic sphere of diameter 50 mm is charged with 10 nC and is surrounded by an insulating medium of relative permittivity 3·0. Find the surface charge density, the flux density, and the field intensity at a radius of 26 mm and 100 mm.

Answer. The surface charge density is given by

$$\sigma = \frac{Q}{4\pi r^2} = \frac{10^{-8}}{4\pi \times 2\cdot5^2 \times 10^{-4}} = \frac{10^{-4}}{25\pi} = 1\cdot27 \times 10^{-6} \ \text{C/m}^2.$$

At a radius of 26 mm,

$$D = \frac{Q}{4\pi r^2} = \frac{10^{-8}}{4\pi \times 676 \times 10^{-6}} = 1\cdot17 \times 10^{-6} \ \text{C/m}^2,$$

$$E = \frac{D}{\varepsilon} = \frac{10^{-8} \times 36\pi \times 10^{9}}{4\pi \times 676 \times 10^{-6} \times 3} = 4\cdot44 \times 10^{4} \ \text{V/m} = 44\cdot4 \ \text{kV/m}.$$

At a radius of 100 mm,

$$D = \frac{10^{-8}}{4\pi \times 10^{-2}} = 7\cdot96 \times 10^{-8} \ \text{C/m}^2,$$

$$E = \frac{10^{-6} \times 36\pi \times 10^{9}}{4\pi \times 3} = 3\cdot0 \times 10^{3} \ \text{V/m} = 3\cdot0 \ \text{kV/m}.$$

4.10. A long straight wire has a uniform charge of 11·1 pC/m and is insulated with a material having a relative permittivity of 2·5. The outside diameter of the insulation on the wire is 5·0 mm. Find the electric field intensity on each side of the surface of the insulator.

Answer. A long straight uniformly charged wire is approximately equivalent to a uniform charge filament. The flux density at any radius is given by eqn. (2.15); therefore, in the air space surrounding the insulation,

$$E = \frac{11\cdot1 \times 10^{-12} \times 36\pi \times 10^{9}}{2\pi \times 2\cdot5 \times 10^{-3}} = 80 \ \text{V/m}.$$

In the insulation itself,

$$E = \frac{80}{2\cdot5} = 32 \ \text{V/m}.$$

PROBLEMS

4.1. A conducting metallic sphere of diameter 0·1 m is charged with
0·1 μC and is surrounded by an insulating medium of relative permittivity 4.
Find the surface charge density, the electric field intensities inside, on,
and just outside the surface, the flux density at 0·2 m radius, and the
mechanical pressure on the surface. $(3·2 \ \mu C/m^2; \ 0, \ 180, \ 90 \ kV/m; \ 0·2 \ \mu C/m^2;$
$0·58 \ N/m^2.)$

4.2. A point charge of 10 pC is surrounded by a sphere of material of
relative permittivity 90 and 0·2 m diameter which in turn is surrounded by
a spherical shell of material of relative permittivity 5 and 2·0 m outside
diameter which is surrounded by vacuum. Calculate the electric intensity on
each side of each of the boundaries of the spherical shell.
$(0·1, \ 1·8, \ 0·018, \ 0·09 \ V/m.)$

4.3. If the point charge in Problem 4.2 was uniformly distributed on
the surface of a small conducting sphere at the centre of the material of
high relative permittivity, calculate the radius of the sphere so that the
electric field intensity does not exceed 15 V/m. Plot the variation of
electric field intensity with radius. $(8·2 \ mm.)$

4.4. A thick spherical shell, internal radius 0·4 m and external radius
0·6 m, has a charge of 10 μC uniformly distributed throughout its material.
The material has a relative permittivity of 3 and is surrounded by air. Draw
a graph showing the variation of electric field intensity with radial
distance from the centre of the sphere to a point 1·5 m from the surface of
the shell indicating significant values.

4.5. The base of a thunder-cloud may be regarded as a uniformly
charged plane area and the electric flux flow perpendicular to that plane.
One raindrop will act as a 1·0 mm diameter circular element of the charged
plane area. If the electric intensity immediately adjacent to the base of
the cloud is 10^5 V/m, find the electric charge density in the base of the
cloud and the force on one raindrop due to electric charge.
$(885 \ nC/m^2; \ 34·8 \ nN.)$

SOLUTIONS

4.1. It is assumed that the charge distributes itself uniformly on the surface of the sphere. Then the charge density is given by

$$\sigma = \frac{Q}{4\pi r^2} = \frac{0\cdot 1 \times 10^{-6}}{4\pi \times (0\cdot 05)^2} = 3\cdot 2 \times 10^{-6} \ C/m^2 = 3\cdot 2 \ \mu C/m^2 .$$

The electric intensity inside the conducting sphere is zero,

$$E_i = 0.$$

The electric intensity on the charged surface,

$$E_s = \frac{1}{2} \frac{\sigma}{\varepsilon_o} = \frac{10^{-5} \times 36\pi \times 10^9}{2\pi} = 18 \times 10^4 \ V/m = 180 \ kV/m.$$

The electric intensity just outside the charged surface,

$$E_o = \frac{D}{\varepsilon_o \varepsilon_r} = \frac{\sigma}{\varepsilon_o \varepsilon_r} = \tfrac{1}{4} \times 360 = 90 \ kV/m.$$

The flux density at a radius of $0\cdot 2$ m is given by application of Gauss's law,

$$D = \frac{10^{-7}}{4\pi \times (0\cdot 2)^2} = 2\cdot 0 \times 10^{-7} \ C/m^2 = 0\cdot 2 \ \mu C/m^2 .$$

The mechanical pressure is given by eqn. (4.4), therefore

$$p = \frac{1}{2} \frac{\sigma^2}{\varepsilon_o} = \tfrac{1}{2} \times 3\cdot 2 \times 3\cdot 2 \times 10^{-12} \times 36\pi \times 10^9 = 0\cdot 58 \ N/m^2 .$$

4.2. The electric field intensity at any radius is given by

$$E = \frac{Q}{4\pi \varepsilon_o \varepsilon_r r^2}.$$

Certain parts of this expression are independent of radius or material. Therefore, inserting some numbers,

$$E = \frac{10 \times 10^{-12} \times 36\pi \times 10^9}{4\pi \varepsilon_r r^2} = \frac{90 \times 10^{-3}}{\varepsilon_r r^2} \ V/m.$$

Inside the inner boundary of the spherical shell, $r = 0\cdot 1$ m and $\varepsilon_r = 90$, therefore

$$E = \frac{90 \times 10^{-3}}{90 \times 10^{-2}} = 0\cdot 1 \ V/m.$$

Outside the inner boundary of the spherical shell, $r = 0\cdot 1$ m and $\varepsilon_r = 5$, therefore

$$E = \frac{90 \times 10^{-3}}{5 \times 10^{-2}} = 1\cdot 8 \ V/m.$$

Inside the outer boundary of the spherical shell, $r = 1 \cdot 0$ m and $\varepsilon_r = 5$, therefore

$$E = \frac{90 \times 10^{-3}}{5 \times 1 \cdot 0} = 0 \cdot 018 \text{ V/m}.$$

Outside the outer boundary of the spherical shell, $r = 1 \cdot 0$ m and $\varepsilon_r = 1$, therefore

$$E = \frac{90 \times 10^{-3}}{1 \times 1 \cdot 0} = 0 \cdot 09 \text{ V/m}.$$

4.3. The problem is to find the radius for a given value of field intensity. Using the formula obtained in the solution to Problem 4.2,

$$r^2 = \frac{90 \times 10^{-3}}{\varepsilon_r \times E} = \frac{90 \times 10^{-3}}{90 \times 15} = 6 \cdot 68 \times 10^{-5},$$

therefore $r = 8 \cdot 2$ mm.

A plot of the variation of the electric field intensity with radius is given in Fig. 4.5.

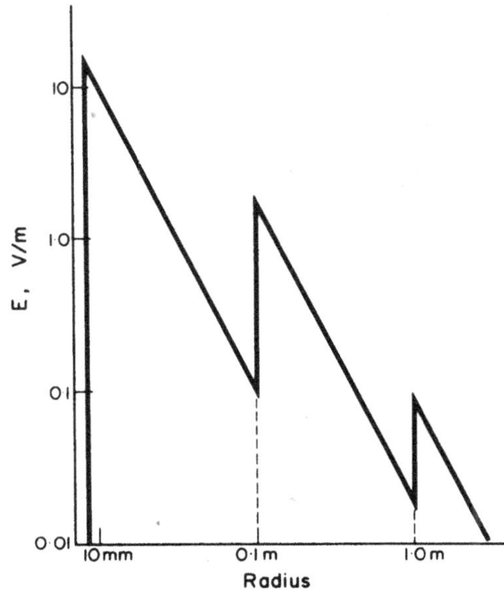

Fig. 4.5. The solution to Problem 4.3. Note the logarithmic scales for the graph giving a useful plot for a wide variation of size of the two variables.

4.4. This problem is the same as Problem 2.10 except that the relative permittivity of the charged material is $3 \cdot 0$. Working from the solution to

Problem 2.10 and Fig. 2.11 (page 32), the solution to this problem is given in Fig. 4.6.

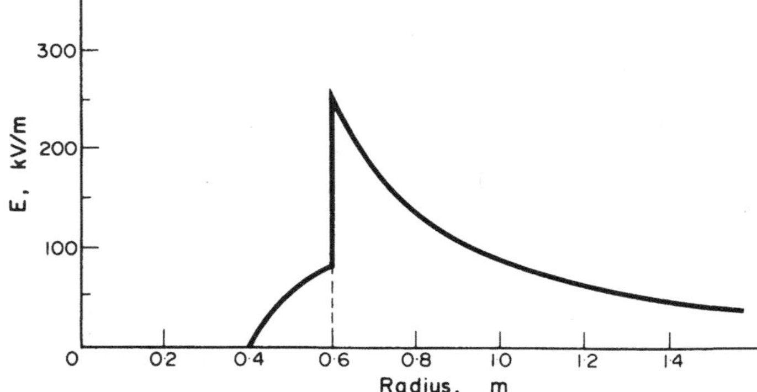

Fig. 4.6. The solution to Problem 4.4. The field intensity due to a thick spherical shell of uniform charge density.

4.5. From the electric field intensity given,

$$D = \frac{10^5}{36\pi \times 10^9} = 8\cdot85 \times 10^{-7} \ C/m^2 = 885 \ nC/m^2.$$

The flux density adjacent to the charged surface is the same as the charge density on the surface, therefore

$$\sigma = 885 \ nC/m^2.$$

One raindrop consists of a circular element of this charged surface of area

$$A = \pi \ (10^{-3} \times 0\cdot5)^2 = 25\pi \times 10^{-8} \ m^2.$$

The force per unit area is given by eqn. (4.5). Therefore the force on one raindrop is given by

$$F = \tfrac{1}{2} \times 25\pi \times 10^{-8} \times 10^5 \times 885 \times 10^{-6} = 3\cdot48 \times 10^{-8} \ N = 34\cdot8 \ nN.$$

FURTHER PROBLEMS

4.6. A long flat metal plate, 20 mm thick, is placed perpendicularly to a uniform electric field of 30 kV/m. Find the induced charge density on the plate and the field intensity inside the plate.   (265 nC/m^2; 0 V/m.)

4.7. A long flat insulating plate, 20 mm thick, is placed perpendicularly to a uniform electric field of 30 kV/m. The material is of relative permittivity 3·0. Find the induced charge density on the plate and the field intensity inside the plate.   (0 C/m^2; 10 kV/m.)

4.8. A solid spherical conductor of diameter 26 mm is charged with 1·09 pC. Find the flux density on the surface of the conductor and the field intensity at a radius of 14 mm.   (513 pC/m^2; 50 V/m.)

4.9. A thick spherical shell of inside diameter 20 mm and outside diameter 30 mm is made of an insulating material of relative permittivity 2·5. Its inside surface is coated with a conducting metallic film which is given an electric charge of 10 nC. Find the maximum electric field intensity due to the charge.   (400 kV/m.)

4.10. A conducting metallic sphere of 50 mm radius is given a charge of 333 pC. Find the electric field intensity inside, on, and just outside the surface of the sphere and the mechanical pressure on the surface due to the charge.   (0, 0·6, 1·2 kV/m; 6·37 µN/m^2.)

4.11. A metallised balloon of 10 m radius gains a static charge of 50 µC. Find the apparent increase in pressure in the balloon due to the electrostatic charge.   (89·5 µN/m^2.)

4.12. A long straight wire of diameter 2·0 mm has insulation covering it of outside diameter 10 mm. The wire has a line charge density of 10 nC/m and the relative permittivity of the insulation material is 9·0. Find the maximum electric field intensity due to the charge.   (36 kV/m.)

4.13. If the maximum electric field intensity in air is 3·0 MV/m and that in the insulation is 10 MV/m, find the maximum charge density that could be given to the wire of Problem 4.12 without the maximum electric field intensity being exceeded.   (834 nC/m.)

FURTHER SOLUTIONS

4.6. Inside a conductor, $E = 0$. The induced charge density is equal to the adjacent flux density, therefore

$$\sigma = D = \frac{30 \times 10^3}{36\pi \times 10^9} = 2 \cdot 65 \times 10^{-7} \text{ C/m}^2 = 265 \text{ nC/m}^2.$$

4.7. There is no induced charge density on an insulating material, therefore $\sigma = 0$.

$$E_{\text{inside}} = \frac{D}{\varepsilon_0 \varepsilon_r} = \frac{E_{\text{outside}}}{\varepsilon_r} = 10 \text{ kV/m}.$$

4.8. From Gauss's law,

$$D = \frac{1 \cdot 09 \times 10^{-12}}{4\pi \times 169 \times 10^{-6}} = 5 \cdot 13 \times 10^{-10} \text{ C/m}^2 = 513 \text{ pC/m}^2,$$

$$E = \frac{1 \cdot 09 \times 10^{-12} \times 36\pi \times 10^9}{4\pi \times 196 \times 10^{-6}} = 50 \text{ V/m}.$$

4.9. At 10 mm radius (within the insulating material),

$$E = \frac{10^{-8} \times 36\pi \times 10^9}{4\pi \times 10^{-4} \times 2 \cdot 5} = 3 \cdot 6 \times 10^5 \text{ V/m} = 360 \text{ kV/m}$$

at 15 mm radius (outside the insulating material),

$$E = \frac{10^{-8} \times 36\pi \times 10^9}{4\pi \times 10^{-4} \times 2 \cdot 25} = 400 \text{ kV/m}.$$

4.10. $E_{\text{inside}} = 0,$

$$E_{\text{outside}} = \frac{333 \times 10^{-12} \times 36\pi \times 10^9}{4\pi \times 25 \times 10^{-4}} = 1 \cdot 2 \times 10^3 \text{ V/m} = 1 \cdot 2 \text{ kV/m},$$

$$E_{\text{on}} = \tfrac{1}{2} E_{\text{outside}} = 600 \text{ V/m}.$$

$$\text{Pressure} = \tfrac{1}{2} \varepsilon_0 E^2_{\text{outside}} = \frac{1}{2} \times \frac{1200 \times 1200}{36\pi \times 10^9} = 6 \cdot 37 \times 10^{-6} \text{ N/m}^2 = 6 \cdot 37 \ \mu\text{N/m}^2.$$

4.11. Charge density $= \sigma = \dfrac{50 \times 10^{-6}}{4\pi \times 100} = \dfrac{10^{-6}}{8\pi}.$

$$\text{Pressure} = \frac{1}{2} \frac{\sigma^2}{\varepsilon_0} = \frac{10^{-12} \times 36\pi \times 10^9}{2 \times 64 \times \pi^2} = 8 \cdot 95 \times 10^{-5} \text{ N/m}^2 = 89 \cdot 5 \ \mu\text{N/m}^2.$$

4.12. At $1 \cdot 0$ mm radius,

$$E = \frac{q}{2\pi r \varepsilon} = \frac{10^{-8} \times 36\pi \times 10^9}{2\pi \times 10^{-3} \times 9 \cdot 0} = 2 \cdot 0 \times 10^4 \text{ V/m} = 20 \text{ kV/m}.$$

At 5·0 mm radius,

$$E = \frac{10^{-8} \times 36\pi \times 10^9}{2\pi \times 5 \times 10^{-3}} = 36 \times 10^3 \text{ V/m} = 36 \text{ kV/m}.$$

Therefore the maximum value of field intensity occurs at the surface of the insulation.

4.13. In air,

$$E_{max} = 3 \cdot 0 \times 10^6 \text{ V/m}.$$

Therefore, using the limiting value at 5·0 mm radius,

$$q = \frac{3 \cdot 0 \times 10^6 \times 2\pi \times 5 \times 10^{-3}}{36\pi \times 10^9} = 8 \cdot 34 \times 10^{-7} \text{ C/m}^2 = 834 \text{ nC/m}^2.$$

In the insulation,

$$E_{max} = 10 \times 10^6 \text{ V/m}.$$

Therefore, using the limiting value at 1·0 mm radius,

$$q = \frac{10^7 \times 2\pi \times 10^{-3} \times 9 \cdot 0}{36\pi \times 10^9} = 5 \cdot 0 \times 10^{-6} \text{ C/m}^2 = 5 \cdot 0 \text{ } \mu\text{C/m}^2.$$

Therefore the limiting condition is governed by the field intensity outside the surface of the insulation.

CHAPTER 5

Gravitational Flux

THEORY

Newton's law describes the force of attraction between any two masses. If d is the distance between the point masses m_1 and m_2, the force is given by

$$\mathbf{f} = - G \, \mathbf{U}_d \, \frac{m_1 m_2}{d^2}, \tag{5.1}$$

where G is the gravitational constant. Because there is an inverse square law of attraction, we postulate a gravitational flux similar to the electric flux. The total flux from a point mass is Ψ; it is equal to the mass and measured in kilograms. A negative sign occurs in Newton's law in comparison with Coulomb's law because there is a force of *attraction* between two masses (which must both be considered to have positive mass) whereas there is a force of *repulsion* between two electric charges of the same sign. The flux density at a distance r from a point mass m, having an equivalent flux source of strength Ψ, is given by

$$\mathbf{D}_g = \mathbf{U}_r \, \frac{m}{4\pi r^2} = \mathbf{U}_r \, \frac{\Psi}{4\pi r^2} \; \text{kg/m}^2 , \tag{5.2}$$

and there will be gravitational equivalents of eqns. (2.3), (2.6), and (2.7). The force exerted on a point mass in a gravitational flux density is given by

$$\mathbf{f} = - 4\pi G \, \mathbf{D}_g m. \tag{5.3}$$

The gravitational permittivity constant is given by

$$k = - \frac{1}{4\pi G}, \tag{5.4}$$

$$k = - 1 \cdot 2 \times 10^9 \; \text{kg s}^2/\text{m}^3 .$$

The *gravitational field intensity* is defined by

$$\mathbf{g} = \lim_{\text{mass} \to 0} \left\{ \frac{(\text{force on mass})}{(\text{mass})} \right\}. \tag{5.5}$$

It is the acceleration due to gravity that is experienced on the surface of any planet. Then

$$\mathbf{g} = - 4\pi G \, \mathbf{D}_g \tag{5.6}$$

or

$$\mathbf{D}_g = k\mathbf{g}, \tag{5.7}$$

78

and the force on a point mass is given by

$$\mathbf{f} = \mathbf{g}m. \tag{5.8}$$

Some of the analogous relationships between electric fields and gravitational
fields are given in Table 5.1.

TABLE 5.1. ANALOGOUS RELATIONSHIPS IN ELECTRIC AND GRAVITATIONAL FIELDS

Electric Fields	Gravitational Fields
$\mathbf{f} = \mathbf{U}_d \dfrac{q_1 q_2}{4\pi\epsilon_o d^2}$	$\mathbf{f} = \mathbf{U}_d \dfrac{m_1 m_2}{4\pi k d^2}$
$\epsilon_o = \dfrac{1}{4\pi \times 9 \times 10^9}$	$k = -\dfrac{1}{4\pi G}$
$\mathbf{D} = \epsilon_o \mathbf{E}$	$\mathbf{D}_g = k\mathbf{g}$
$\mathbf{f} = \mathbf{E}q$	$\mathbf{f} = \mathbf{g}m$

EXAMPLES

5.1. Find, from first principles, the gravitational field intensity below and above the surface of a planet of uniform density ρ and of radius R.

Answer. The problem is illustrated in Fig. 5.1, where r is the distance of the point at which the field intensity is required from the centre of the spherical planet. As shown in the diagram, $r > R$. A Gaussian surface is taken through the point P at a constant radius r from the centre of the planet. Then, from Gauss's law, the total flux out of the surface equals the mass enclosed in the surface. Therefore

$$\Psi = \frac{4}{3}\pi R^3 \rho,$$

and the total flux out of the surface

$$\Psi = D_g\, 4\pi r^2.$$

Therefore
$$D_g = \frac{\frac{4}{3}\pi R^3 \rho}{4\pi r^2}.$$

By symmetry, the vector flux density is directed along a radial line from the centre of the planet, therefore the gravitational field intensity is given by

$$\mathbf{g} = \frac{\mathbf{D}_g}{k} = 4\pi G\,\frac{1}{3}\,\rho\,\frac{R^3}{r^2}\,\mathbf{U}_r.$$

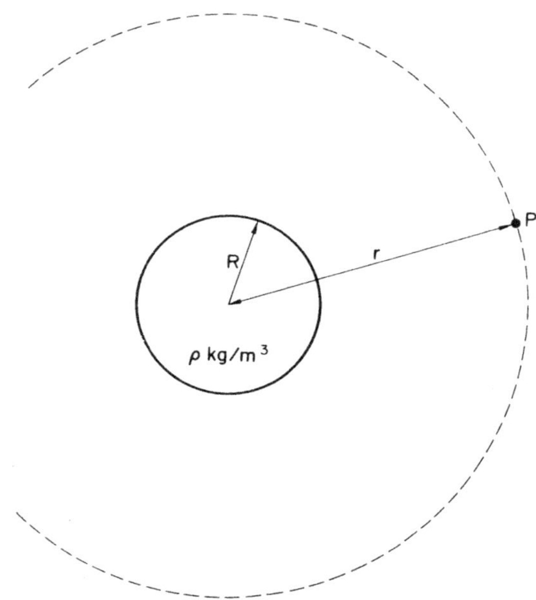

Fig. 5.1.   A spherical Gaussian surface taken through a point
P outside a sphere of uniform density.

Inside the surface of the planet, where $r < R$, only the mass inside the Gaussian surface through the point P is effective in contributing to the flux density at P. As before, the Gaussian surface through P will be a sphere of radius r. Therefore the flux out is given by

$$\Psi = \frac{4}{3}\pi r^3 \rho,$$

and the flux density at radius r is given by

$$\mathbf{D}_g = \frac{\frac{4}{3}\pi r^3 \rho}{4\pi r^2}\mathbf{U}_r.$$

Therefore the gravitational field intensity is given by

$$\mathbf{g} = \frac{\mathbf{D}_g}{k} = -4\pi G\frac{1}{3}\rho r\,\mathbf{U}_r.$$

5.2. Knowing the value of g on the surface of the earth, calculate the mean density of the earth.

Answer. An expression for the gravitational field intensity inside or on the surface of a uniform spherical body is given as the last equation in the solution to the last example,

$$g = -4\pi G\,\frac{1}{3}\rho r.$$

Various data for the earth is given in Table 5.2. Substituting values from that table for G, g, and the radius of the earth, gives

$$\rho = \frac{3 \times 9{\cdot}81}{4\pi \times 6{\cdot}67 \times 10^{-11} \times 6{\cdot}37 \times 10^6} = 5{\cdot}5 \times 10^3 \text{ kg/m}^3.$$

TABLE 5.2.   GRAVITATIONAL DATA

Density of water is 1000 kg/m³
$G = 6{\cdot}67 \times 10^{-11}$ m³/kg s²
g on the surface of the earth is 9·81 m/s²
Radius of the earth is 6·37 Mm
Radius of the orbit of the earth is 150 Gm
Radius of the moon is 1·74 Mm
Radius of the orbit of the moon is 380 Mm

5.3. Calculate the speed needed to maintain a satellite at a height of 400 km above the surface of the earth.

Answer. By an application of Gauss's law, the gravitational field intensity at a distance r from the centre of the earth is given by

$$g = - 4\pi G \frac{1}{3}\rho \frac{R^3}{r^2}.$$

The force acting on a satellite of mass m is directed towards the centre of the earth and is given by

$$\text{force} = mg.$$

The centripetal force acting on the satellite by virtue of its circular motion at a speed v is given by

$$\text{force} = \frac{mv^2}{r}.$$

These two forces are equal and opposite so that the satellite remains in a stable orbit. Therefore

$$v^2 = \frac{4\pi G}{3} \frac{R^3 \rho}{r}.$$

The density of the earth has been calculated and is given as the solution to the last example. The value of G and the radius of the earth are given in Table 5.2. The distance r is given by

$$r = R + 400 \text{ km} = 6\cdot77 \text{ Mm}.$$

Inserting the numbers into the equation gives

$$v^2 = \frac{4\pi \times 6\cdot67 \times 10^{-11} \times (6\cdot37)^3 \times 10^{18} \times 5\cdot5 \times 10^3}{3 \times 6\cdot77 \times 10^6} = 5\cdot87 \times 10^7.$$

Therefore
$$v = 7\cdot66 \times 10^3 \text{ m/s} = 7\cdot66 \text{ km/s}.$$

5.4. Calculate the height needed to maintain a satellite in a stationary orbit above the equator.

Answer. For the satellite to remain in a stationary orbit above the equator, the satellite must orbit the earth at the same speed as the earth is rotating on its axis. The frequency of rotation of the satellite is given by

$$\omega = \frac{2\pi}{24 \times 3600} = 7\cdot27 \times 10^{-5} \text{ rad/s}.$$

The gravitational force at any radius r on a satellite mass m was given in the solution to the last example, it is

$$\text{force} = m \, 4\pi G \frac{1}{3}\rho \frac{R^3}{r^2}.$$

The centripetal force acting on the satellite by virtue of its circular motion is given in terms of its angular velocity by

$$\text{force} = m\omega^2 r.$$

Equating these two forces gives an expression for the radius,

$$\left(\frac{r}{R}\right)^3 = \frac{4\pi G\rho}{3\omega^2}$$

Inserting numbers gives

$$\left(\frac{r}{R}\right)^3 = \frac{4\pi \times 6\cdot67 \times 10^{-11} \times 5\cdot5 \times 10^3}{3 \times (7\cdot27)^2 \times 10^{-10}} = 290\cdot7.$$

Therefore $r = 6\cdot62\ R.$

The height of a satellite is usually given as the dimension above the surface of the earth. Therefore

$$\text{height} = r - R = 5\cdot62\ R = 35\cdot8\ \text{Mm}.$$

5.5. A satellite orbits the moon close to its surface. If the mean density of the moon is $3\cdot3 \times 10^3$ kg/m^3, calculate the time of revolution of the satellite.

Answer. Equating the gravitational force to the centripetal force as in the last example, gives

$$m\omega^2 r = m 4\pi G \frac{1}{3}\rho \frac{R^3}{r^2}.$$

When the satellite is orbiting at the surface of the planet, the two radii are equal and the equation becomes

$$\omega^2 = \frac{4}{3}\pi G\rho.$$

Inserting numbers gives

$$\omega^2 = \frac{4}{3}\pi \times 6\cdot67 \times 10^{-11} \times 3\cdot3 \times 10^3 = 9\cdot22 \times 10^{-7}.$$

Therefore   . $\omega = 9\cdot60 \times 10^{-4}$ rad/s.

$$\text{Time of revolution} = \frac{2\pi}{\omega} = \frac{2\pi}{9\cdot60 \times 10^{-4}} = 6\cdot55 \times 10^3\,\text{s}.$$

This is the same as a time of $1\cdot82$ hours.

5.6. Calculate the escape velocity for a projectile from a planet. Compare this with the escape velocity of a satellite already in orbit.

Answer. The escape velocity is calculated by equating the kinetic energy given to the projectile at launch with the energy needed to overcome the total gravitational force as the projectile travels an infinite distance away from the planet. This assumes that there are no drag losses as the projectile leaves the atmosphere of the planet. The energy needed to overcome gravitational forces is equal to the product of the force and the distance moved by the projectile in the direction of the force. Therefore the result

will be independent of the path of the projectile. The gravitational field
intensity at any radius r external to a planet of radius R and uniform density
ρ has been derived in the solution to Example 5.1 and is given by

$$g = \frac{4}{3}\pi G\rho \,\frac{R^3}{r^2}.$$

The total energy expended in leaving the planet is given by

$$W = \int_{R}^{\infty} mg\,dr,$$

where m is the mass of the projectile. Therefore

$$W = m \int_{R}^{\infty} \frac{4}{3}\pi G\rho \,\frac{R^3}{r^2}\,dr = \frac{4}{3}\pi G\rho m R^3 \left(\frac{1}{R} - \frac{1}{\infty}\right) = \frac{4}{3}\pi G\rho m R^2 .$$

This energy expended in overcoming gravitational forces is equal to the
kinetic energy imparted to the projectile at launch. Therefore

$$W = \tfrac{1}{2} m v^2 .$$

Therefore the escape velocity of the projectile from the surface of the planet
is given by

$$v_1^2 = \frac{8}{3}\pi G\rho R^2 .$$

If the projectile is already in orbit around the planet at a radius r_1, the
energy required to overcome gravitation is given by

$$W = \int_{r_1}^{\infty} mg\,dr = \frac{4}{3}\pi G\rho m \,\frac{R^3}{r_1}.$$

Therefore the escape velocity needed to be given to the projectile in orbit
is given by

$$v_2^2 = \frac{8}{3}\pi G\rho \,\frac{R^3}{r_1}.$$

However, in order to remain in orbit to begin with, the projectile will
already have a speed which has been derived in the solution to Example 5.3.
It is given by

$$v_3^2 = \frac{4}{3}\pi G\rho \,\frac{R^3}{r_1}.$$

Therefore its existing speed needs to be increased by a factor $\sqrt{2}$, so further
reducing the energy needed to launch a projectile from a satellite already in
orbit.

5.7. A planet is thought to consist of a core of material of greater
density than the material comprising its surface. Estimate the average

density of the core. The density of the surface material is 3000 kg/m^3 and
its thickness is 500 km. The radius of the planet is 3·0 Mm and the
gravitational field intensity at its surface is 3·0 m/s^2.

Answer. The planet is assumed to consist of two uniform parts - an inner
core of density ρ_1 and radius r_1, and an outer shell of density ρ_2 and outer
radius r_2. Therefore the mass of the planet is given by

$$M = \frac{4}{3}\pi \left\{ r_1^3 \rho_1 + (r_2^3 - r_1^3)\rho_2 \right\}$$

and the gravitational intensity on its surface is given by

$$g = \frac{4\pi GM}{4\pi r_2^2}.$$

Therefore

$$\rho_1 = \frac{3 g r_2^2}{4\pi G r_1^3} - \left(\frac{r_2^3}{r_1^3} - 1 \right) \rho_2.$$

Inserting numbers from the question gives

$$\rho_1 = \frac{3 \times 3 \cdot 0 \times 3 \cdot 0^2 \times 10^{12}}{4\pi \times 6 \cdot 67 \times 10^{-11} \times 2 \cdot 5^3 \times 10^{18}} - \left(\frac{3 \cdot 0^3}{2 \cdot 5^3} - 1 \right) 3 \cdot 0 \times 10^3$$

$$= 6 \cdot 18 \times 10^3 - 2 \cdot 18 \times 10^3 = 4 \cdot 0 \times 10^3 \text{ kg/m}^3.$$

5.8. Find an expression for the gravitational forces acting on a
projectile which went straight from the surface of the earth to the surface of
the moon. Hence or otherwise find the distance from the surface of the earth
at which such a projectile becomes weightless.

Answer. Let r_e and ρ_e be the radius and density of the earth respectively
and r_m and ρ_m be the radius and density of the moon. Then the total
gravitational force acting on a projectile of mass m when it is at a distance
d from the surface of the earth on the line joining the centres of the earth
and moon is given by

$$\text{force} = \frac{m 4\pi G \frac{4}{3}\pi \rho_e r_e^3}{4\pi (r_e + d)^2} - \frac{m 4\pi G \frac{4}{3}\pi \rho_m r_m^3}{4\pi (r_o - r_e - d)^2},$$

where r_o is the radius of the orbit of the moon around the earth. Simplifying
this expression and taking the various dimensions from Table 5.2 and taking
the densities of the earth and the moon from Examples 5.2 and 5.5 respectively,
gives,

$$\text{force} = m\frac{4}{3}\pi \times 6\cdot 67 \times 10^{-11} \times 10^3 \times 10^6 \left(\frac{5\cdot 5 \times 6\cdot 37^3}{(6\cdot 37 + d)^2} - \frac{3\cdot 3 \times 1\cdot 74^3}{(380 - 6\cdot 37 - d)^2} \right)$$

$$= 0\cdot 28m \left(\frac{1422}{(6\cdot 37 + d)^2} - \frac{17\cdot 38}{(373\cdot 6 - d)^2} \right) \text{N},$$

where d is measured in Mm.

The weightless condition is given when the force is zero. Therefore

$$\frac{1422}{(6\cdot 37 + d)^2} = \frac{17\cdot 38}{(373\cdot 6 - d)^2}.$$

Inverting both sides of the equation and taking the positive square root of each side gives

$$\frac{6\cdot 37 + d}{37\cdot 71} = \frac{373\cdot 6 - d}{4\cdot 169}.$$

Therefore
$$d = 335\cdot 8 \text{ Mm}.$$

PROBLEMS

5.1. Calculate the mass of the earth knowing the acceleration due to gravity at its surface. $(5\cdot97 \times 10^{24}$ kg.)

5.2. If the sun's attraction on the earth produces the centripetal acceleration corresponding approximately to the motion of the earth round the sun in a circular orbit, find the mass of the sun and compare it with the earth's mass. $(2\cdot01 \times 10^{30}$; 10^6:3.)

5.3. What equal and opposite electrical charges, placed on the sun and the earth, would produce an attraction equal to the gravitational attraction between these bodies. $(2\cdot98 \times 10^{17}$ C.)

5.4. Draw a graph showing the variation of the acceleration due to gravity along a line from the centre of the earth to a point 20 Mm from the surface of the earth.

5.5. Calculate the force acting on $1\cdot0$ kg mass located midway between the centres of the earth and moon.   (11 mN.)

SOLUTIONS

5.1. Applying Gauss's law and eqn. (5.6) to a solid sphere gives

$$g = - \frac{4\pi GM}{4\pi r^2}.$$

Therefore, ignoring the negative sign,

$$M = \frac{gr^2}{G} = \frac{9\cdot81 \times (6\cdot37)^2 \times 10^{12}}{6\cdot67 \times 10^{-11}} = 5\cdot97 \times 10^{24} \text{ kg.}$$

5.2. From Newton's law, the force of attraction is given by

$$f = \frac{GMM_s}{r^2}$$

and the centripetal force is given by

$$f = Mr\omega^2.$$

Equating these two expressions for the force gives

$$M_s = \frac{r^3\omega^2}{G}.$$

Inserting the dimensions from Table 5.2 and taking the period of orbit of the earth as 365 days, gives

$$M_s = \frac{(150 \times 10^9)^3 \times (2\pi)^2}{6\cdot67 \times 10^{-11} \times (3600 \times 24 \times 365)^2} = 2\cdot01 \times 10^{30} \text{ kg.}$$

Using the solution to Problem 5.1, the ratio is given by

$$\text{mass sun : mass earth} = 10^6 : 3.$$

5.3. From Newton's law, the force is given by

$$f = \frac{GMM_s}{r^2}.$$

From Coulomb's law, the force due to the equal and opposite charges is given by

$$f = \frac{Q^2}{4\pi\varepsilon_0 r^2} = 9 \times 10^9 \frac{Q^2}{r^2}.$$

Equating these two relationships for the force gives an expression for the charge,

$$Q^2 = \frac{GMM_s}{9 \times 10^9}.$$

Inserting numbers into this expression gives

$$Q^2 = \frac{6 \cdot 67 \times 10^{-11} \times 5 \cdot 97 \times 10^{24} \times 2 \cdot 01 \times 10^{30}}{9 \times 10^9} = 8 \cdot 893 \times 10^{34}.$$

Therefore $\qquad Q = 2 \cdot 98 \times 10^{17}$ C.

5.4. By application of Gauss's law it has been shown in the solution to Example 5.1 that the flux density inside and outside a sphere of radius R is given by

$$D_{g(\text{inside})} = \tfrac{1}{3} r\rho.$$

$$D_{g(\text{outside})} = \tfrac{1}{3} \frac{R^3}{r^2} \rho.$$

The maximum value of the acceleration due to gravity is at the surface of the earth and is given by

$$g = -4\pi G \tfrac{1}{3} R\rho = -9 \cdot 81 \text{ m/s}.$$

Above the surface of the earth,

$$g = -\frac{4}{3}\pi G R\rho \left(\frac{R}{r}\right)^2 = -9 \cdot 81 \left(\frac{R}{r}\right)^2.$$

The graph of the variation of the gravitational acceleration from the centre of the earth is shown in Fig. 5.2.

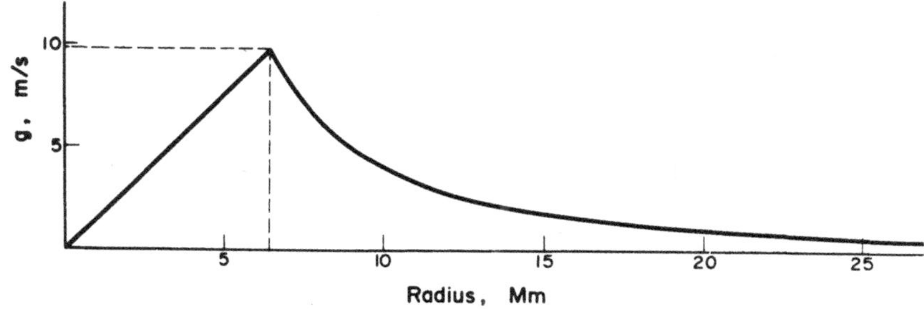

Fig. 5.2. The solution to Problem 5.4. The variation of gravitational intensity along a radial line from the centre of the earth.

5.5. The force may be calculated from Newton's law using the Principle of Superposition:

$$\text{force due to the earth} = \frac{GM_e m}{d^2},$$

$$\text{force due to the moon} = \frac{GM_m m}{d^2}.$$

Therefore the resultant force is given by

$$f = \frac{Gm}{d^2}(M_e - M_m).$$

The dimensions of the earth and the moon are given in Table 5.2. The mean densities of the earth and moon are given in Examples 5.2 and 5.5 respectively. Therefore, inserting numbers into this relationship gives

$$f = \frac{4\pi \times 6{\cdot}67 \times 10^{-11} \times 1{\cdot}0}{3 \times (190 \times 10^6)^2}\{5{\cdot}5 \times 10^3 \times (6{\cdot}37 \times 10^6)^3 - 3{\cdot}3 \times 10^3 \times$$

$$\times (1{\cdot}76 \times 10^6)^3\} = 10{\cdot}9 \times 10^{-3} \text{ N} = 10{\cdot}9 \text{ mN}.$$

FURTHER PROBLEMS

5.6. Three small balls, each of mass 1·0 kg, are situated 1·0 m apart in one plane at the apexes of an equilateral triangle. Find the gravitational force on any one due to the other two.   (115 pN.)

5.7. Find the gravitational force of attraction between three parallel long straight wires, each of mass 1·0 kg/m, which are situated 1·0 m apart at the apexes of an equilateral triangle.   (230 pN/m.)

5.8. Plot the variation of gravitational flux density both inside and outside a uniform solid cylinder of infinite length and density ρ.

5.9. Gravitational acceleration on the surface of the moon is found to be 1·62 m/s. Find the mass and density of the moon.   ($7\cdot35 \times 10^{22}$ kg; $3\cdot34 \times 10^{3}$ kg/m^3.)

5.10. Calculate the speed needed to maintain a satellite in an orbit at a height of 500 km above the surface of the moon.   (1·5 km/s.)

5.11. A satellite orbits the moon in 130 min; calculate the radius of its orbit.   (1·96 Mm.)

5.12. Assuming that Mars orbits the sun in an approximately circular path with a period of 687 days, find the radius of Mars's orbit round the sun. Take the mass of the sun to be $2\cdot0 \times 10^{30}$ kg.   (228 Gm.)

5.13. Calculate the escape velocity of an inert projectile from the earth neglecting any air friction effects. (11·2 km/s.)

FURTHER SOLUTIONS

5.6. A diagram of the forces acting on one mass will be similar to Fig. 2.3 (page 17), with the forces oppositely directed. From Newton's law, the force due to one other mass is given by

$$f_1 = f_2 = \frac{GM_1 M_1}{d^2} = \frac{6 \cdot 67 \times 10^{-11} \times 1 \cdot 0 \times 1 \cdot 0}{1 \cdot 0} = 6 \cdot 67 \times 10^{-11} \text{ N.}$$

By the principle of superposition, the force due to two masses is given by

$$f = 2 f_1 \cos 30^0 = 2 \times 6 \cdot 67 \times 10^{-11} \times 0 \cdot 866 = 1 \cdot 15 \times 10^{-10} \text{ N} = 115 \text{ pN.}$$

5.7. The geometry of this problem is similar to that shown in Fig. 2.3 (page 17), with the forces oppositely directed. For one long straight wire, by Gauss's law,

$$D_g = \frac{m}{2\pi r}.$$

Therefore
$$g = -4\pi G D_g = \frac{2Gm}{r}.$$

The force on a small element of another similar wire

$$\delta f = g m \delta x = \frac{2Gm^2 \delta x}{r}.$$

Therefore the force per unit length due to one other wire is given by

$$f_1 = \frac{2Gm^2}{r} = \frac{2 \times 6 \cdot 67 \times 10^{-11} \times 1 \cdot 0 \times 1 \cdot 0}{1 \cdot 0}.$$

Therefore the resultant force due to two wires is given by

$$f = 2 f_1 \cos 30^0 = 2 \times 6 \cdot 67 \times 10^{-11} \times 2 \times 0 \cdot 866 = 2 \cdot 30 \times 10^{-10} \text{ N/m} = 230 \text{ pN/m.}$$

5.8. By the application of Gauss's law, for a point inside the body of the cylinder,

$$\pi r^2 \rho = D_g 2\pi r.$$

Therefore
$$D_g = \tfrac{1}{2}\rho r.$$

For a point outside the body of the cylinder, where the radius of the cylinder is R,

$$\pi R^2 \rho = D_g 2\pi r.$$

Therefore
$$D_g = \tfrac{1}{2}\rho \frac{R^2}{r}.$$

The required graph is plotted in Fig. 5.3.

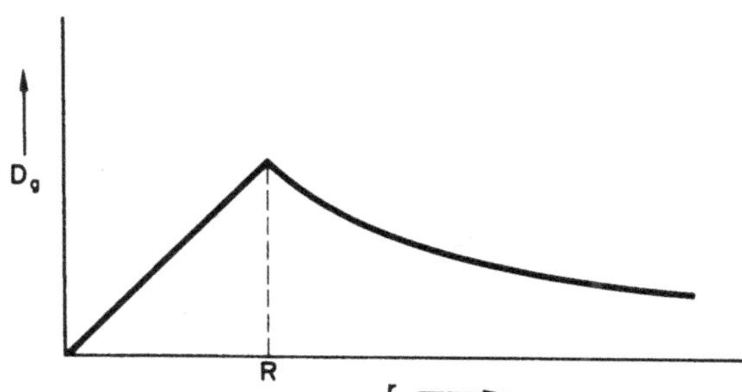

Fig. 5.3. The solution to Problem 5.8. The variation of flux
density inside and outside a solid cylinder of uniform
density.

5.9. For any solid sphere, on its surface

$$g = \frac{GM}{r^2} = 4\pi G \tfrac{1}{3}\rho r.$$

Therefore $M = \dfrac{gr^2}{G} = \dfrac{1\cdot62 \times (1\cdot74 \times 10^6)^2}{6\cdot67 \times 10^{-11}} = 7\cdot36 \times 10^{22}$ kg

and $\rho = \dfrac{3g}{4\pi Gr} = \dfrac{3 \times 1\cdot62}{4\pi \times 6\cdot67 \times 10^{-11} \times 1\cdot74 \times 10^6} = 3\cdot34 \times 10^3$ kg/m^3.

5.10. Gravitational intensity at any radius r is given by

$$g = \frac{GM}{r^2}.$$

Centrifugal acceleration at the same radius is given by

$$a = \frac{v^2}{r}.$$

Therefore $v^2 = \dfrac{GM}{r} = \dfrac{6\cdot67 \times 10^{-11} \times 7\cdot36 \times 10^{22}}{2\cdot24 \times 10^6} = 2\cdot19 \times 10^6.$

Therefore $v = 1\cdot48 \times 10^3$ m/s $= 1\cdot48$ km/s.

5.11. Using the relationships obtained in the solution to Example 5.4,

$$\omega = \frac{2\pi}{130 \times 60} = 8\cdot06 \times 10^{-4} \text{ rad/s}$$

and $\left(\dfrac{r}{R}\right)^3 = \dfrac{4\pi G\rho}{3\omega^2} = \dfrac{4\pi \times 6\cdot67 \times 10^{-11} \times 3\cdot33 \times 10^3}{3 \times (8\cdot06)^2 \times 10^{-8}} = 1\cdot43.$

Therefore $$r = 1 \cdot 12R = 1 \cdot 96 \text{ Mm}.$$

This radius is the same as a height of 220 km above the surface.

5.12. Using the relationship derived in the solution to Example 5.4,

$$\omega = \frac{2\pi}{687 \times 24 \times 3600} = 1 \cdot 06 \times 10^{-7} \text{ rad/s,}$$

$$r^3 = \frac{GM}{\omega^2} = \frac{6 \cdot 67 \times 10^{-11} \times 2 \cdot 0 \times 10^{30}}{(1 \cdot 06)^2 \times 10^{-14}} = 11 \cdot 87 \times 10^{33}.$$

Therefore $$r = 2 \cdot 28 \times 10^{11} \text{ m} = 228 \text{ Gm}.$$

5.13. Using the relationship derived in the solution to Example 5.6,

$$v^2 = \frac{8}{3}\pi G\rho R^2 = \frac{8}{3}\pi \times 6 \cdot 67 \times 10^{-11} \times 5 \cdot 5 \times 10^3 \times (6 \cdot 37 \times 10^6)^2 = 1 \cdot 25 \times 10^8.$$

Therefore $$v = 1 \cdot 12 \times 10^4 \text{ m/s} = 11 \cdot 2 \text{ km/s}.$$

CHAPTER 6

Fluid Flow Field

THEORY

Flux, having the properties of a fluid, can be used to describe certain conditions of real fluid flow. If the fluid flows steadily, flux lines can be drawn with tangents parallel to the fluid velocity. In the fluid field they are called *stream lines*, but are the same as the flux lines of field theory. The volume flow is called the discharge or flux of the fluid. It originates from a source and terminates in a sink of fluid, and the strength of the source and sink is equal in magnitude to the flux of the fluid.

The relationship between pressure and velocity and density in a fluid is given by *Bernoulli's equation*,

$$P + \tfrac{1}{2}\rho v^2 = \text{constant} \tag{6.1}$$

provided that the fluid is incompressible. The fluid that can be completely described in terms of field theory is called an *ideal fluid*. It must be incompressible, i.e. of constant density, and inviscid, i.e. no viscous forces acting on the fluid. In a real fluid, viscous forces often cause a departure from ideal fluid flow only in the boundary layer where there is a velocity gradient from the boundary to the position at which the full stream speed is reached. Therefore, a real fluid can still be modelled by taking a false boundary for the model outside the boundary layer parallel to the direction of flow.

Gauss's law for the fluid field states that the total fluid flow out of any closed surface equals the algebraic sum of the strength of the fluid sources enclosed within that surface. The strength of any source is measured in units of volume flow, i.e. m^3/s. Then the units of flux density become $(m^3/s)/m^2$, which is the same as velocity. Therefore

flux = discharge of fluid in m^3/s,

v = flux density = velocity of the fluid in m/s.

All the other results of Gauss's law to two-dimensional and one-dimensional fields can be applied to two-dimensional and one-dimensional fluid flow. The source is equivalent to a positive charge in electrostatics, and the sink is equivalent to a negative charge in electrostatics.

95

The *stream function* in fluid flow is equivalent to the flux function in field theory. Therefore the discharge between any two stream lines A and B, as shown in Fig. 6.1, is given by

$$Q = \Psi = \psi_B - \psi_A = \int_A^B \mathbf{v} \cdot d\mathbf{A}. \tag{6.2}$$

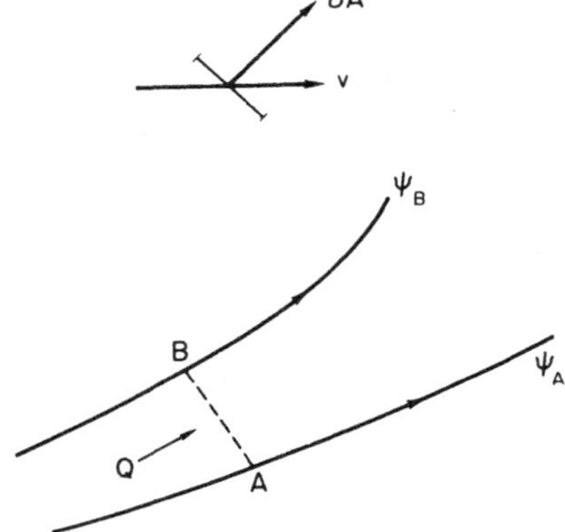

Fig. 6.1. Fluid flow through an element of area $\delta\mathbf{A}$ or between two stream lines.

In an ideal fluid, each stream line or flux function line can be replaced by a boundary since the fluid flow is parallel to the boundary. The velocity can be obtained from the flux function, so that, similar to eqns. (3.2) and (3.3)(page 36),

$$v_x = \frac{\partial \psi}{\partial y}, \qquad v_y = - \frac{\partial \psi}{\partial x}; \tag{6.3}$$

and, similar to eqns. (3.4) and (3.5)(page 36),

$$v_r = \frac{1}{r}\frac{\partial \psi}{\partial \theta}, \qquad v_\theta = - \frac{\partial \psi}{\partial r}. \tag{6.4}$$

In an ideal fluid, the pressure is solely dependent on the velocity. Therefore, from Bernoulli's equation,

$$P = P_0 - \tfrac{1}{2}\rho v^2, \tag{6.5}$$

where P_0 is the pressure of any stationary fluid in the same system.

EXAMPLES

6.1. A venturi meter operates by reducing the area of flow and measuring the pressure difference between the fluid in the reduced section and the unrestricted section of fluid flow. Such a venturi meter in a circular pipe consists of a smooth taper to a section of half the original diameter. If the fluid is water having a density of 1000 kg/m^3 and the pressure difference is measured to be 60 mm head of water, find the velocity of the water flow.

Answer. This problem involves the application of Bernoulli's equation. Assuming that the water flow is horizontal and behaves as an ideal fluid, then if P_1 and v_1 are the pressure and velocity in the undisturbed flow, and P_2 and v_2 are the same quantities at the restricted section,

$$P_1 + \tfrac{1}{2}\rho v_1^2 = P_2 + \tfrac{1}{2}\rho v_2^2.$$

Therefore

$$P_1 - P_2 = \tfrac{1}{2}\rho(v_2^2 - v_1^2).$$

It must be assumed that there is uniform volume flow through the pipe, so that the ratio of the velocities is proportional to the cross-sectional area of flow. Therefore

$$v_2 = 4v_1.$$

The pressure difference is given in head of water, therefore

$$P_1 - P_2 = \rho g h = \tfrac{1}{2}\rho(v_2^2 - v_1^2) = \tfrac{1}{2}\rho v_1^2(16 - 1).$$

Therefore

$$v_1^2 = \frac{2gh}{15} = \frac{2 \times 9\cdot81 \times 0\cdot06}{15} = 0\cdot0785.$$

Therefore

$$v_1 = 0\cdot28 \text{ m/s}.$$

6.2. A point source of fluid has a discharge of 20 m^3/s into an infinite sea of the same fluid. Find the amplitude and direction of the velocity of the fluid at a point 0·5 m distant from the point source.

Answer. As only a single point source is under consideration, the flow will be symmetrical away from the point of the source. Therefore a Gaussian surface can be taken as the surface of a sphere centred on the source passing through the point in question. Then for a spherical surface of radius r, by application of Gauss's law,

$$Q = 4\pi r^2 v.$$

Therefore

$$v = \frac{Q}{4\pi r^2} = \frac{20}{4\pi \times 0\cdot25} = 6\cdot37 \text{ m/s},$$

and its direction is in a straight line away from the point of the source.

6.3. A line source of fluid discharges into an infinite sea of the same fluid. It could be realized by a thin perforated pipe. It has a flux of $0\cdot2$ (m³/s)/m. Find the velocity of fluid flow at a distance of $0\cdot4$ m from the line of the source.

Answer. Consider a cylindrical Gaussian surface surrounding the line source as shown in Fig. 2.4 (page 20). The radius of the cylinder is r and its length is l. Let the strength of the fluid source be q. Due to symmetry there will be no flow through the plane ends of the cylinder. Then, from Gauss's law,

$$ql = 2\pi r l v.$$

Therefore
$$v = \frac{q}{2\pi r} = \frac{0\cdot2}{2\pi \times 0\cdot4} = 0\cdot0796 \text{ m/s.}$$

6.4. Eight point sources of equal discharge of fluid are situated so that they form the corners of a cube. Obtain an expression for the velocity at the midpoint of any one of the edges of the cube.

Answer. By the principle of superposition, the velocity component due to each one of the sources may be calculated independently and all the velocities summed vectorially. It is left for the student to construct his own sketch of the cube and to calculate the distances of the various sources from the point. The velocities are obtained from the flux density relationship for a point source,

$$v = \frac{Q}{4\pi r^2}.$$

Taking a cube of unit side, the various velocities are given by

$$v = \frac{Q}{4\pi} \times \frac{4}{1} = \frac{Q}{\pi} \quad \text{occurs twice,}$$

$$v = \frac{Q}{4\pi} \times \frac{4}{5} = \frac{Q}{5\pi} \quad \text{occurs four times,}$$

$$v = \frac{Q}{4\pi} \times \frac{4}{9} = \frac{Q}{9\pi} \quad \text{occurs twice.}$$

By symmetry, the resultant velocity is perpendicular to the line of the edge on which the point lies and is at 135° to the plane of each adjacent face of the cube. The resultant velocity is given by

$$v = 4\left(\frac{Q}{5\pi} \times \frac{2}{\sqrt5} \times \frac{1}{\sqrt2}\right) + 2\left(\frac{Q}{9\pi} \times \frac{2\sqrt2}{3}\right) = 0\cdot23\,Q \text{ m/s.}$$

6.5. Find expressions for the stream functions due to a uniform flow of U m/s, a line source of strength q m²/s, and a doublet of strength c m³/s.

Answer. The uniform fluid field is equivalent to the uniform electrostatic field. The expressions for the flux function of a uniform electrostatic field are given in eqns. (3.6) and (3.7)(page 37). Alternatively, by reference to Fig. 6.2a, the stream function, which is the same as the flux function, is given by eqn. (6.2) which becomes

$$\psi = Uy = Ur \sin \theta.$$

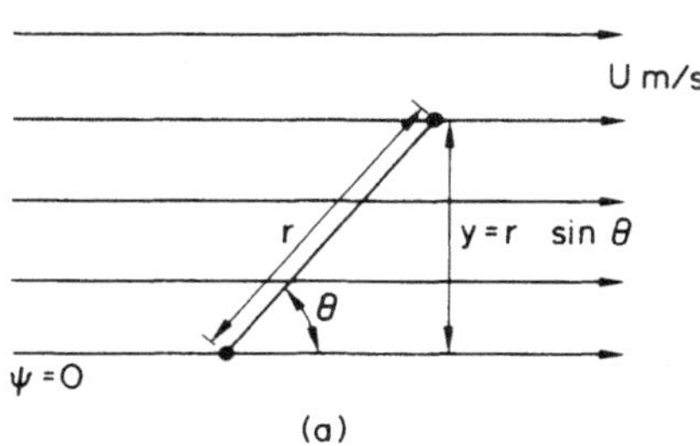

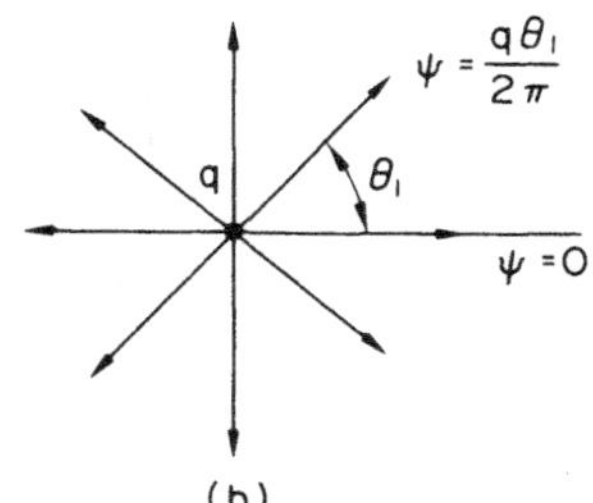

Fig. 6.2. Stream functions for (a) a uniform flow, (b) a line source.

For the line source, the flux function in the electrostatic field is given by eqn. (3.8)(page 37). In the fluid field, all the symbolism is the same. The flux function is illustrated in Fig. 6.2b. It is given by

$$\psi = \frac{q\theta}{2\pi}.$$

In the fluid field there is a possible realization for the doublet that was difficult to visualize in electrostatics. The line source is an approximation to the flow of fluid out of a perforated pipe. The doublet could consist of two perforated pipes next to one another so that one acts as a source and the other acts as a sink. In order to derive the flux function of the doublet, it must be considered as a line source close to but not coincident with an equal line sink. A line source and line sink of strength $\pm q'$ a distance d' apart

are shown in Fig. 6.3. The flux function of the line-source line-sink pair
is given by eqn. (3.10)(page 38) to be

$$\psi = -\frac{q\alpha}{2\pi}.$$

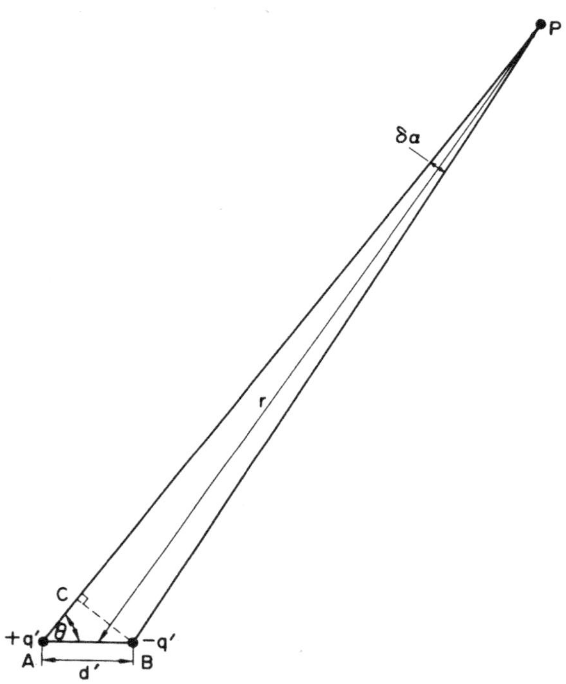

Fig. 6.3. Illustrating the geometry of a point P in the field
of a doublet.

For the doublet the angle α is small and made equal to $\delta\alpha$. Then, by geometry.

$$d' \sin\theta \approx r\,\delta\alpha.$$

Therefore
$$\psi = -\frac{q'}{2\pi}\frac{d' \sin\theta}{r} = -\frac{c \sin\theta}{2\pi r},$$

where $c = q'd'$ is the strength of the doublet. The result is the same as that
quoted as eqn. (3.12)(page 39).

6.6. A line source of strength q m²/s is combined with a uniform flow of
velocity U m/s. Obtain an expression for the position of the stagnation point,
i.e. the point having zero velocity in all directions.

Answer. A general flux line pattern resulting from a combination of a line source with a uniform flow is given in Fig. 6.4. The flux function

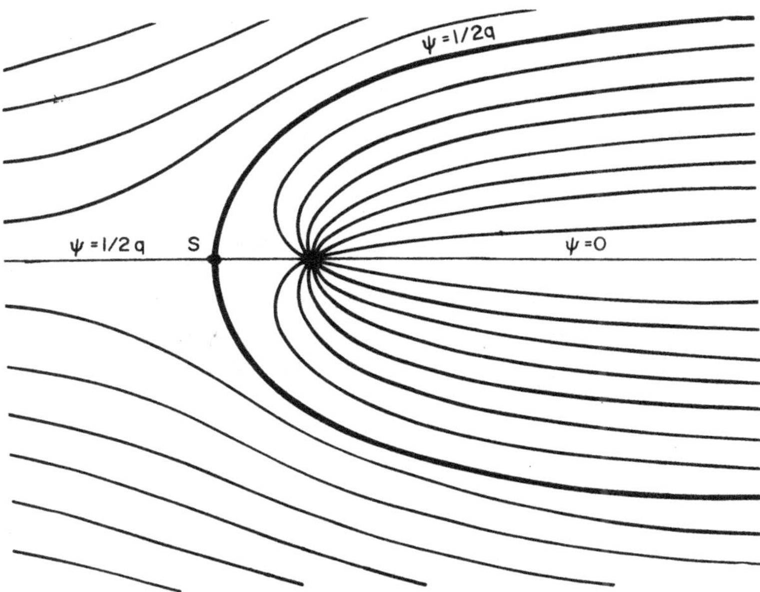

Fig. 6.4. The stream line pattern of the combination of a line
source with a uniform flow.

value of the line passing through the point S is given. S is the stagnation
point and it can be seen that there is no velocity at that point. To obtain
a mathematical expression for its position, we obtain an expression for the
flux function and then obtain the velocity by differentiating in accordance
with eqn. (6.4). If the zero of the coordinate system is taken to lie on the
line of the line source, the flux function of the combination is the sum of
the flux functions of the two component parts given in the solution to
the last example. Therefore

$$\psi = Ur \sin \theta + \frac{q\theta}{2\pi}.$$

Differentiating this expression with respect to r gives an expression for v_θ,
therefore

$$v_\theta = - \frac{\partial \psi}{\partial r} = - U \sin \theta.$$

Similarly, the radial component of the velocity is given by

$$v_r = \frac{1}{r} \frac{\partial \psi}{\partial \theta} = \frac{1}{r} \left(Ur \cos \theta + \frac{q}{2\pi} \right).$$

When $\theta = 0$ or $\theta = \pi$, $v_\theta = 0$. At the stagnation point, however, the radial component of the velocity is also zero. At the stagnation point, $\theta = \pi$, so that $\cos \theta = -1$. Therefore the position of the stagnation point is given by

$$r = \frac{q}{2\pi U}.$$

This is similar to the expression obtained in the solution to Example 3.6 except that it is now in fluid flow terminology.

6.7. A uniform stream of velocity 2·0 m/s is combined with a line source of strength 0·4 m²/s. Find the velocity on the stagnation stream line at a point defined by the line passing through the line of the line source perpendicular to the direction of undisturbed uniform flow.

Answer. The flux function of the combination is given by

$$\psi = Ur \sin \theta + \frac{q\theta}{2\pi}.$$

The flux line pattern is shown in Fig. 6.4. When $\theta = \pi$, $\psi = \frac{1}{2}q$. The point under consideration is on the line $\theta = \frac{1}{2}\pi$, therefore

$$\psi = Ur + \tfrac{1}{4}q.$$

Then on the stagnation stream line,

$$\tfrac{1}{2}q = Ur + \tfrac{1}{4}q.$$

Therefore
$$r = \frac{q}{4U}.$$

Differentiating the expression for the flux function in accordance with eqn. (6.4) gives the components of the velocity. They are

$$v_\theta = -\frac{\partial \psi}{\partial r} = -U \sin \theta = -U$$

when $\theta = \frac{1}{2}\pi$.

$$v_r = \frac{1}{r}\frac{\partial \psi}{\partial \theta} = U \cos \theta + \frac{q}{2\pi r} = \frac{2U}{\pi}$$

when $\theta = \frac{1}{2}\pi$ and $r = q/4U$.

Therefore $v = \sqrt{(v_r^2 + v_\theta^2)} = U \sqrt{(1 + \frac{4}{\pi^2})} = \sqrt{1\cdot4}\,U = 1\cdot2U = 2\cdot4$ m/s.

6.8. The boundary of a pier supporting a bridge in the middle of a river of approximately uniform depth is similar in shape to the stagnation stream line occurring from the combination of a line source with uniform flow. The pier is 3·06 m wide at its widest point and the river flow can be considered to be uniform at 0·54 m/s. Find the strength and position of the line source required to represent the shape of this boundary when combined with a uniform flow of 0·54 m/s.

Answer. The flux function of the combined flow has been derived in the
solution to Example 6.6 and is given by

$$\psi = Ur \sin \theta + \frac{q\theta}{2\pi}.$$

A typical flux line pattern is shown in Fig. 6.4. The dimension of the pier
at its widest point is considered to be a very long way downstream from the
stagnation point. At its widest point, the width is given by

$$\text{width} = 2r \sin \theta.$$

The flux function value of the boundary of the pier is $\frac{1}{2}q$. At the point a
long distance downstream, $r \to \infty$ and $\theta \to 0$, but the product $r \sin \theta$ is still a
finite quantity. Therefore, substituting into the expression for the flux
function,

$$\psi = \tfrac{1}{2}q = Ur \sin \theta + 0.$$

Therefore $q = 2Ur \sin \theta = (\text{width})\, U = 3 \cdot 06 \times 0 \cdot 54 = 1 \cdot 65\ (\text{m}^3/\text{s})/\text{m}.$
The expression for the distance between the line of the line source and the
position of the stagnation point is derived in the solution to Example 6.6.
Therefore

$$r = \frac{q}{2\pi U} = \frac{1 \cdot 65}{2\pi \times 0 \cdot 54} = 0 \cdot 487\ \text{m}.$$

6.9. The top half of the stagnation stream line shown in Fig. 6.4 is a
model of a vertical section through a cliff meeting the sea at the stagnation
point. The flux lines indicate the air flow over the cliff for a wind blowing
directly on to the shore. If the wind speed is $5 \cdot 0$ m/s and the distance
between the stagnation point and the line of the line source is 100 m, find
the strength of the line source to model this boundary. A glider having a
sinking speed in still air of $1 \cdot 5$ m/s flies above the cliff directly above the
line of the line source. Find the maximum height to which it will rise.

Answer. Using the formulae derived in the solution to the three previous
examples, the strength of the line source is given by

$$q = 2\pi Ur = 2\pi \times 5 \cdot 0 \times 100 = 1000\pi = 3140\ \text{m}^2/\text{s}.$$

The velocity in the vertical direction immediately above the origin of the
coordinate system which is the line of the line source, is given by

$$v_r = U \cos \theta + \frac{q}{2\pi r}.$$

But for the point under consideration, $\theta = \frac{1}{2}\pi$ and $\cos \theta = 0$, therefore

$$r = \frac{q}{2\pi v_r} = \frac{1000\pi}{2\pi \times 1\cdot 5} = 333 \text{ m.}$$

This is assuming that the lift provided by the wind exactly cancels the sinking speed of the glider. At a lower height, the glider will rise as the lift will be greater than the sinking speed, and at a greater height the glider will sink until the equilibrium height is reached.

6.10. A combination of a uniform field and a doublet gives the flow pattern around a cylindrical post in an otherwise uniform flow. Obtain an expression for the velocity on the surface of the cylinder and the force due to the pressure on the cylinder.

Answer. The combination of a uniform flow and a doublet gives a flow pattern which has a circular stream line which can represent a circular boundary. The flux function pattern has already been derived in the solution to Example 3.8. It is given in Fig. 3.8 (page 45). The expression for the flux function of the doublet has already been derived in the solution to Example 6.5. Therefore the combination of the uniform field and the doublet gives

$$\psi = Ur \sin\theta - \frac{c \sin\theta}{2\pi r}.$$

$\psi = 0$ on the circular stagnation stream line which is the surface of the cylinder. Therefore if a is the radius of the cylinder,

$$a^2 = \frac{c}{2\pi U}$$

and
$$\psi = Ur \sin\theta \left(1 - \frac{a^2}{r^2}\right).$$

On the wall of the cylinder, the radial component of the velocity will be zero. The circumferential component of the velocity is given by

$$v_\theta = -\frac{\partial\psi}{\partial r} = -U \sin\theta\left(1 + \frac{a^2}{r^2}\right) = -2U \sin\theta$$

when $r = a$ on the surface of the cylinder.

The pressure on the surface of the cylinder is obtained from the velocity according to Bernoulli's equation, eqn. (6.5). Therefore

$$P = P_o - \tfrac{1}{2}\rho v^2 = P_o - 2\rho U^2 \sin^2\theta.$$

This equation shows that the velocity pattern around the cylinder is symmetrical, which can also be observed from the flow field pattern in Fig.

3.8. Therefore the pressure distribution is symmetrical. At each point on
the cylinder there is a symmetrical point with an equal pressure so that the
net force on the cylinder is zero. For a circular cylinder in an ideal fluid
there is no force on the cylinder. This is called *D'Alembert's Paradox*. In
an ideal fluid there is symmetrical flow around the cylinder, and any force
on the front of the cylinder is balanced by an equal force on the back of the
cylinder. In a real fluid which has viscosity, a wake is formed behind the
cylinder which modifies the flow and gives rise to a force in the direction of
the uniform stream.

6.11. An infinitely long circular cylinder lies with its axis perpendicular
to the direction of flow of an infinite stream of uniform flow of an ideal
fluid. Find the doublet uniform stream combination which will produce the
same stream-line pattern when the undisturbed uniform flow velocity is 1·0 m/s
and the cylinder diameter is 0·5 m. Also find the maximum velocity on the
surface of the cylinder.

Answer. This solution uses the formulae derived in the solution to the
previous example. The doublet uniform stream combination to give the same
physical shape of flow is given by:

uniform flow velocity = U = undisturbed flow velocity = 1·0 m/s,
doublet strength = c = $2\pi U a^2$ = $2\pi \times 1\cdot0 \times (0\cdot25)^2$ = 0·39 m³/s.

An expression for the velocity on the wall of the cylinder has been derived
in the solution to the last example. It is given by

$$v_\theta = -\,2U \sin\theta,$$

which is a maximum when $\sin\theta = 1$, therefore

$$v_{\theta(\text{max})} = 2U = 2\cdot0 \text{ m/s}.$$

The negative sign occurs due to the relative directions of the circumferential
component of velocity when $\theta = \tfrac{1}{2}\pi$ and the positive x-direction which is the
datum direction for the uniform flow. It is immaterial to the answer to the
question, and has been omitted.

6.12. A cylinder of diameter 0·5 m is placed in the centre of a duct of
width 1·0 m through which water flows. Find the stream-line flow pattern
around the cylinder, and if the uniform water flow in the duct a great way
away from the cylinder is 2·0 m/s, find the maximum velocity on the surface
of the cylinder.

Answer. The flow pattern around the cylinder is obtained by sketching a pattern similar to Fig. 3.8 with straight stream lines at the position of the edge of the duct. The pattern is shown in Fig. 6.5. This pattern has eight flux tubes in the duct so that the flux function increment between each pair of lines is given by

$$\delta\psi = Uy = 2{\cdot}0 \times \frac{1{\cdot}0}{8} = 0{\cdot}25 \text{ m}^2/\text{s}.$$

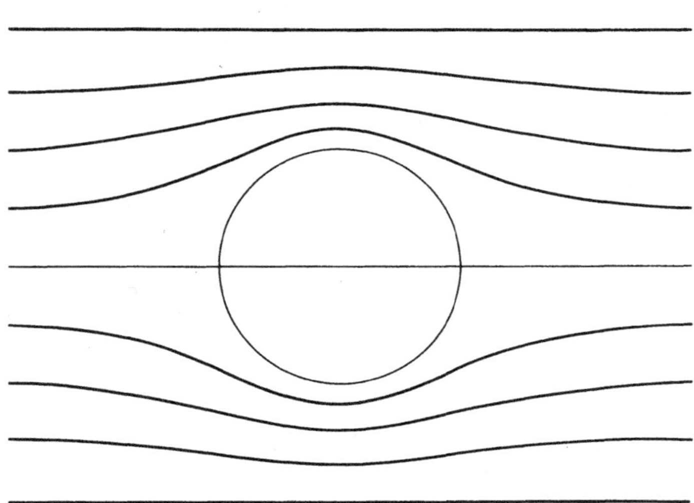

Fig. 6.5. The stream-line pattern for fluid flow past a cylindrical obstruction in a duct of uniform width.

From the flux pattern, the spacing between the flux lines at their nearest can be measured to be

$$\delta l = 0{\cdot}044 \times (\text{width of the duct}) = 0{\cdot}044 \text{ m}.$$

Therefore the velocity is given by

$$v = \frac{\delta\psi}{\delta l} = \frac{0{\cdot}25}{0{\cdot}044} = 5{\cdot}7 \text{ m/s}.$$

PROBLEMS

6.1. Find the velocity at the points $(2, 0)$, $(1, 1)$, and $(0, 1)$ m due to (a) a line source of strength $2 \cdot 0$ m²/s at $(1, 0)$ m, (b) a line source of strength $2 \cdot 0$ m²/s at $(-1, 0)$ m, and (c) the combination of both line sources (a) and (b).

$$\Big(\text{(a)} \; 0 \cdot 318 \mathbf{U}_x; \; 0 \cdot 318 \mathbf{U}_y; \; (-0 \cdot 159 \mathbf{U}_x + 0 \cdot 159 \mathbf{U}_y) \text{ m/s.}$$
$$\text{(b)} \; 0 \cdot 106 \mathbf{U}_x; \; (0 \cdot 127 \mathbf{U}_x + 0 \cdot 064 \mathbf{U}_y); \; (0 \cdot 159 \mathbf{U}_x + 0 \cdot 159 \mathbf{U}_y) \text{ m/s.}$$
$$\text{(c)} \; 0 \cdot 425 \mathbf{U}_x; \; (0 \cdot 127 \mathbf{U}_x + 0 \cdot 382 \mathbf{U}_y); \; 0 \cdot 382 \mathbf{U}_y \text{ m/s.} \Big)$$

6.2. Sketch the stream lines for the three conditions given in Problem 6.1.

6.3. A long circular cylinder, $0 \cdot 1$ m in diameter, lies with its axis normal to an infinitely wide stream of water flowing with a velocity of $3 \cdot 0$ m/s. Assuming ideal flow, find the doublet uniform stream combination which would produce the same stream-line pattern. If the pressure a long way upstream is 10^5 N/m², find the velocity and pressure at a point on the cylinder surface 90° from the downstream stagnation point. ($0 \cdot 047$ m³/s, $3 \cdot 0$ m/s; $6 \cdot 0$ m/s, $86 \cdot 5$ kN/m².)

6.4. For the situation of Problem 6.3, plot the stream line around the cylinder as far as the stream lines which lie $0 \cdot 1$ m on either side of the stagnation stream line a long way upstream.

6.5. The cylinder of Problem 6.3 is placed centrally in a $0 \cdot 2$ m wide duct through which water flows at $3 \cdot 0$ m/s. Assuming ideal fluid flow, find the stream-line pattern and the maximum velocity on the surface of the cylinder. ($8 \cdot 6$ m/s.)

SOLUTIONS

6.1. The student is invited to sketch out his own diagram which will aid the understanding of the solution of this problem. For the line sources, the velocity is the same as the flux density which is given by

$$v = \frac{q}{2\pi r}.$$

(a) $v_{(2,0)} = \frac{2 \cdot 0}{2\pi} = 0 \cdot 318$ in the direction of positive x. Therefore

$$\mathbf{v} = 0 \cdot 318\,\mathbf{U}_x.$$

$v_{(1,1)} = \frac{2 \cdot 0}{2\pi} = 0 \cdot 318$ in the direction of positive y. Therefore

$$\mathbf{v} = 0 \cdot 318\,\mathbf{U}_y.$$

$v_{(0,1)} = \frac{2 \cdot 0}{2\pi\sqrt{2}}$ at 135° to the positive x-axis.

Resolved this has equal horizontal and vertical components

$$v = v_{(0,1)} \times \frac{1}{\sqrt{2}} = \frac{2 \cdot 0}{4\pi} = 0 \cdot 159.$$

The two components are in the directions $-x$ and $+y$. Therefore

$$\mathbf{v} = -\,0 \cdot 159\,\mathbf{U}_x + 0 \cdot 159\,\mathbf{U}_y.$$

(b) By arguments similar to those used above,

$$\mathbf{v}_{(2,0)} = \frac{2 \cdot 0}{2\pi \times 3}\,\mathbf{U}_x = 0 \cdot 106\,\mathbf{U}_x,$$

$$\mathbf{v}_{(1,1)} = \frac{2 \cdot 0}{2\pi\sqrt{5}} \times \frac{2}{\sqrt{5}}\,\mathbf{U}_x + \frac{2 \cdot 0}{2\pi\sqrt{5}} \times \frac{1}{\sqrt{5}}\,\mathbf{U}_y = 0 \cdot 127\,\mathbf{U}_x + 0 \cdot 064\,\mathbf{U}_y,$$

$$\mathbf{v}_{(0,1)} = \frac{2 \cdot 0}{2\pi\sqrt{2}} \times \frac{1}{\sqrt{2}}\,\mathbf{U}_x + \frac{2 \cdot 0}{2\pi\sqrt{2}} \times \frac{1}{\sqrt{2}}\,\mathbf{U}_y = 0 \cdot 159\,\mathbf{U}_x + 0 \cdot 159\,\mathbf{U}_y.$$

(c) By the principle of superposition, the effect of the two sources can be calculated by the simple vector addition of each separately. The answers under this section are the sum of the answers for the velocities at each point of the two previous sections. Therefore

$$\mathbf{v}_{(2,0)} = 0 \cdot 424\,\mathbf{U}_x,$$

$$\mathbf{v}_{(1,1)} = 0 \cdot 127\,\mathbf{U}_x + 0 \cdot 382\,\mathbf{U}_y,$$

$$\mathbf{v}_{(0,1)} = 0 \cdot 318\,\mathbf{U}_y.$$

6.2. The sketches of the stream lines for the conditions (a) and (b) are

similar to the flux lines of a line source as given in Fig. 3.2 (page 37).
They are given in Fig. 6.6a and b. The stream lines for the combination of
(c) is obtained by summing the flux function patterns due to two line sources
in a manner similar to Fig. 3.14 (page 54). The solution is given in Fig.
6.6c.

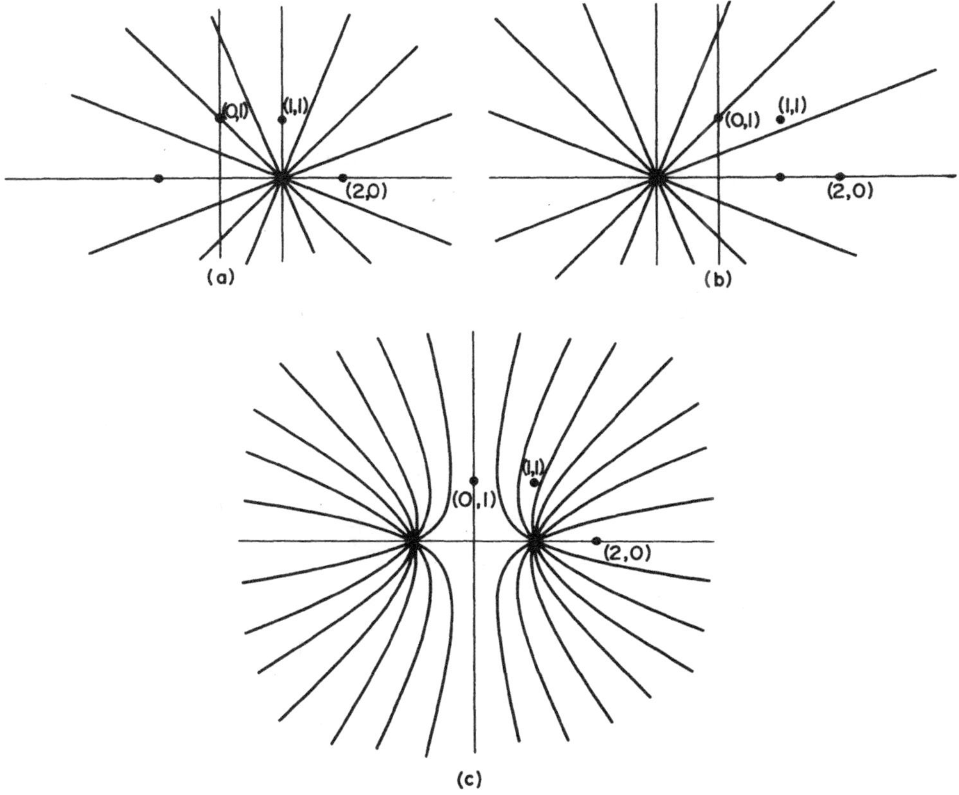

Fig. 6.6. The solution to Problem 6.2. Stream-line patterns for
(a) and (b) isolated line sources, (c) combination of two line
sources.

6.3. This problem is similar to Example 6.11. The uniform flow velocity
is the same as the undisturbed flow velocity which is given to be 3·0 m/s.
The doublet strength is given by

$$c = 2\pi U a^2 = 2\pi \times 3\cdot0 \times (0\cdot05)^2 = 0\cdot047 \text{ m}^3/\text{s}.$$

At the point 90° from the downstream stagnation point on the cylinder surface,

$$v_\theta = 2U = 6\cdot0 \text{ m/s}.$$

The pressure is given by Bernoulli's equation,

$$P + \tfrac{1}{2}\rho v^2 = \text{constant}.$$

Therefore

$$P = P_1 + \tfrac{1}{2}\rho U^2 - \tfrac{1}{2}\rho v^2,$$

$$P = 10^5 + \tfrac{1}{2} \times 10^3(3\cdot0^2 - 6\cdot0^2) = (100 - 13\cdot5) \times 10^3 \text{ N/m}^2,$$

$$P = 86\cdot5 \text{ kN/m}^2.$$

6.4. The required plot will be the same as the flux function plot given in Fig. 3.8 (page 45).

6.5. This is the problem of a cylinder in a duct of width twice the diameter of the cylinder. The flow pattern has already been derived in the solution to Example 6.12 and is shown in Fig. 6.5. For this problem, the flux increment between adjacent stream lines is given by

$$\delta\psi = \frac{3.0 \times 0.2}{8.} = 0\cdot075 \text{ m}^2/\text{s}.$$

The diameter of the cylinder is 100 mm, therefore the point of smallest spacing of stream lines gives

$$\delta l = 0\cdot0087 \text{ m}.$$

Therefore

$$\text{maximum velocity} = \frac{\delta\psi}{\delta l} = \frac{0\cdot075}{0\cdot0087} = 8.6 \text{ m/s}.$$

FURTHER PROBLEMS

6.6. A venturi meter in a circular pipe consists of a smooth taper to a circular section three-quarters of the original diameter followed by a further taper to a pipe of the original size. The pressure difference between the unrestricted flow and the narrow section of the venturi meter is measured as head of the working fluid. Find the relationship between the measured head and fluid velocity and the head for a flow of 3·0 m/s.   (990 mm.)

6.7. Two equal point sources of strength Q are situated at the points $(0, 1, 0)$ m and $(0, -1, 0)$ m. Find an expression for the velocity on the x-axis. $(Qx/2\pi(x^2 + 1)^{3/2}.)$

6.8. Four equal line sources of strength q are situated at the points (a, a), $(a, -a)$, $(-a, a)$, and $(-a, -a)$. Find an expression for the velocity on the plane of the x-axis. $(2qx^3/\pi(4a^4 + x^4).)$

6.9. A boundary in an otherwise uniform flow of ideal fluid is similar to the stagnation stream line shown in Fig. 6.4. If the undisturbed uniform

flow velocity is 3·0 m/s and the distance between the stagnation point and
the line of the line source is 1·0 m, find the variation of pressure on the
boundary. The fluid has a density of 1000 kg/m³.

6.10. Plot the flux function due to the combination of a line source of
strength 24 m²/s and a uniform stream of 0·133 m/s. If this pattern models
the flow over a cliff where it meets the sea, estimate some values of lift
experienced by an aeroplane due to the wind blowing from the sea on to the
cliff in terms of the wind speed far out to sea.

6.11. What is the shape and size of the stagnation stream line due to a
combination of a uniform stream of 10 m/s and a doublet of strength 240 m³/s?
(3·9 m.)

6.12. A long circular cylinder of diameter 1·0 m lies with its axis
perpendicular to the direction of flow of a uniform stream of velocity 5·0
m/s. Find the doublet strength needed to model the flow in an ideal fluid
and the maximum velocity occurring in the fluid.   (7·85 m³/s; 10·0 m/s.)

6.13. Construct a stream-line pattern for the combination of a line-source
line-sink pair with a uniform stream such that the line joining the line
source and line sink is parallel to the direction of flow of the uniform
stream. The strength of the line source is 4·0 m²/s and it is parallel to
and 0·4 m distant from an equal and opposite line sink. The undisturbed
uniform stream velocity is 5·0 m/s. The stagnation stream-line models a
boundary in the fluid; find the dimensions of the major and minor axes of the
boundary and the maximum velocity on the boundary.   (0·6 m × 0·4 m; 7·2 m/s.)

FURTHER SOLUTIONS

6.6. From Bernoulli's equation,

$$P_1 + \tfrac{1}{2}\rho v_1^2 = P_2 + \tfrac{1}{2}\rho v_2^2.$$

Therefore in terms of the head h,

$$P_1 - P_2 = \rho g h = \tfrac{1}{2}\rho(v_2^2 - v_1^2).$$

But $R_2 = \tfrac{3}{4}R_1$, therefore

$$v_2\pi(\tfrac{3}{4}R_1)^2 = v_1\pi R_1^2$$

or
$$v_2 = 1\cdot 78\, v_1.$$

Therefore
$$h = \frac{1}{2g}(1\cdot 78^2 - 1)v_1^2 = 0\cdot 11 v_1^2.$$

For a velocity of 3·0 m/s,

$$h = 0\cdot 99 \text{ m} = 990 \text{ mm.}$$

6.7. At any distance r from the source, the velocity is radial and given by

$$v = \frac{Q}{4\pi r^2}.$$

If the point is on the x-axis, $r^2 = x^2 + 1$. When both of the point sources are considered, the resultant velocity is in the direction of the x-axis and there is no component of the velocity in a perpendicular direction. Therefore the resultant velocity is given by

$$v_x = 2v \cos\theta = \frac{2Q}{4\pi r^2} \times \frac{x}{r} = \frac{Qx}{2\pi(x^2+1)^{3/2}}.$$

6.8. The method of solution of this problem is similar to the previous problem except that this problem has line sources so that the velocity due to an isolated source is given by

$$v = \frac{q}{2\pi r}.$$

Summing the effect of the four sources at any point x on the x-axis gives

$$v = \frac{2q}{2\pi r_1}\frac{(x-a)}{r_1} + \frac{2q}{2\pi r_2}\frac{(x+a)}{r_2}$$

$$= \frac{q}{\pi}\left(\frac{x-a}{a^2+(x-a)^2} + \frac{x+a}{a^2+(x+a)^2}\right) = \frac{2qx^3}{\pi(4a^4+x^4)}.$$

6.9. The flux pattern of the combination is discussed in the solution to

Example 6.6. The flux function of the combination is given by

$$\psi = Ur \sin \theta + \frac{q\theta}{2\pi}.$$

On the stagnation stream line, $\psi = \frac{1}{2}q$, therefore

$$\frac{1}{2}q = Ur \sin \theta + \frac{q\theta}{2\pi} \qquad \text{and} \qquad r = \frac{q(\pi - \theta)}{2\pi U \sin \theta}.$$

The components of the velocity are given by

$$v_\theta = -\frac{\partial \psi}{\partial r} = -U \sin \theta,$$

$$v_r = \frac{1}{r}\frac{\partial \psi}{\partial \theta} = \frac{1}{r}\left(Ur \cos \theta + \frac{q}{2\pi}\right) = U\left(\cos \theta + \frac{\sin \theta}{(\pi - \theta)}\right)$$

when substitution is made for the conditions on the boundary. Therefore the total velocity is given by

$$v^2 = U^2\left\{\sin^2 \theta + \left(\cos \theta + \frac{\sin \theta}{(\pi - \theta)}\right)^2\right\}.$$

The pressure variation on the boundary is given from Bernoulli's equation by

$$P_o - P = \frac{1}{2}\rho v^2 = \frac{1}{2}\rho U^2\left(1 + \frac{\sin 2\theta}{(\pi - \theta)} + \frac{\sin^2 \theta}{(\pi - \theta)^2}\right).$$

Inserting some numbers from the question gives;

when $\theta = 0$,

$$P_o - P = \frac{1}{2} \times 1000 \times (3\cdot0)^2 = 4500 \text{ N/m}^2;$$

when $\theta = \frac{1}{4}\pi$,

$$P_o - P = 4500\left(1 + \frac{1}{\frac{3}{4}\pi} + \frac{\frac{1}{2}}{(\frac{3}{4}\pi)^2}\right) = 6820 \text{ N/m}^2;$$

when $\theta = \frac{1}{2}\pi$,

$$P_o - P = 4500\left(1 + 0 + \frac{1}{(\frac{1}{2}\pi)^2}\right) = 6320 \text{ N/m}^2;$$

when $\theta = \frac{3}{4}\pi$,

$$P_o - P = 4500\left(1 - \frac{4}{\pi} + \frac{\frac{1}{2}}{(\frac{1}{4}\pi)^2}\right) = 2420 \text{ N/m}^2;$$

when $\theta = \pi$,

$$P_o - P = 4500 (1 - 2 + 1) = 0.$$

6.10. Taking a flux function increment between flux function lines of $1\cdot0$

m^2/s means that there will be twenty-four radial lines from the line source.
For the same flux function increment, the spacing of the parallel flux function
lines of the uniform flow is given by

$$\delta y = \frac{\delta \psi}{U} = \frac{1 \cdot 0}{0 \cdot 133} = 7 \cdot 5 \text{ m.}$$

A flux function plot of part of the field is given in Fig. 6.7.

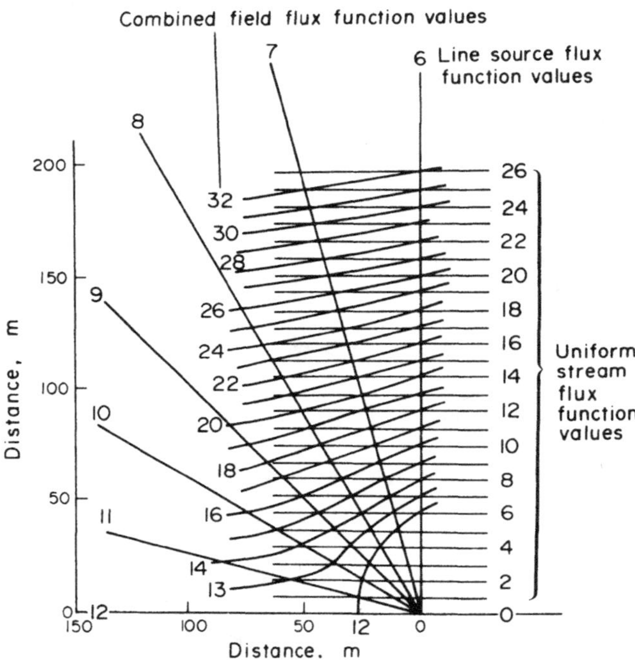

Fig. 6.7. The solution to Problem 6.10. A plot of part of the
flux function due to a combination of a line source with a
uniform stream.

For the model, the lift is going to be the vertical component of the
velocity. This can be obtained from the diagram by measuring the horizontal
distance between adjacent flux lines. It is given by the relationship

$$v_y = - \frac{\delta \psi}{\delta x},$$

where δx is the measured spacing between the lines and $\delta \psi = 7 \cdot 5\ U$, where
U is the uniform fluid flow velocity.

Some results measured from the diagram are given in Table 6.1.

TABLE 6.1

Height (m)	δx (m)	$\delta \psi$	v_y
50	13	$7 \cdot 5U$	$0.58U$
100	25	$7 \cdot 5U$	$0 \cdot 30U$
150	40	$7 \cdot 5U$	$0 \cdot 19U$
200	53	$7 \cdot 5U$	$0 \cdot 14U$

6.11. The combination of a uniform stream and a doublet gives a circular stagnation stream line. An expression for the radius of the stagnation stream line has been derived in the solution to Example 3.8 and is quoted in the solution to Example 6.10; quoting the formula derived there gives

$$a^2 = \frac{c}{2\pi U} = \frac{240}{2\pi \times 10} = 3 \cdot 82 \text{ m}^2.$$

Therefore $\qquad\qquad\qquad\qquad a = 1 \cdot 95 \text{ m}.$

The stagnation stream line is a circle of diameter $3 \cdot 9$ m.

6.12. Using the relationship for the radius of the circular stream line from the solution to the last problem,

$$c = 2\pi U a^2 = 2\pi \times 5 \cdot 0 \times (0 \cdot 5)^2 = 7 \cdot 85 \text{ m}^3/\text{s}.$$

The velocity on the surface of the cylinder has been derived in the solution to Example 6.10 to be

$$v_\theta = - 2U \sin \theta.$$

At its maximum, this expression becomes

$$v_{\theta \text{ max}} = 2U = 10 \text{ m/s}.$$

6.13. The stream-line pattern for the combination of a line source with a line sink is shown in Fig. 3.3 (page 38). If there are to be sixteen lines originating from each line source, then the increment of flux function between adjacent flux lines is $\delta\psi = 0 \cdot 25$ m^2/s. In order to have the same increment between flux function lines in the uniform field, the spacing of the parallel flux lines is given by

$$\delta y = \frac{\delta \psi}{U} = \frac{0 \cdot 25}{5 \cdot 0} = 0 \cdot 05 \text{ m}.$$

The stream-line pattern resulting from a combination of these two fields is the same as that shown in Fig. 3.9 (page 46). Measuring the dimensions of the stagnation stream line, which is elliptical, gives $0 \cdot 6$ m and $0 \cdot 4$ m for the

diameter along the major and minor axes respectively. The maximum velocity is
found by measuring the spacing of the flux function lines at the position of
the minor axes. The measured result is $\delta l = 0 \cdot 035$ m. Therefore

$$v = \frac{\delta \psi}{\delta l} = \frac{0 \cdot 25}{0 \cdot 035} = 7 \cdot 2 \text{ m/s}.$$

CHAPTER 7

Magnetic Flux

THEORY

The forces between magnets obey an inverse square law similar to that
governing forces between electric charges. For two isolated magnetic charges
(or poles) a distance d apart the force is given by Coulomb's law,

$$\mathbf{f} = \mathbf{U}_d \frac{q_1 q_2}{4\pi\mu_0 r^2}, \tag{7.1}$$

where q_1 and q_2 are point magnetic charges and μ_0 is a dimensional magnetic
constant similar to the permittivity. Although Coulomb's law was originally
formulated in terms of point magnetic charges, a magnetic charge has never
been isolated, and it is now generally accepted that the magnetic charge does
not exist as a separate entity. The ultimate indivisible unit of magnetism
is not a magnetic charge but a pair of equal and opposite magnetic charges
indissolubly fixed together. These magnetic dipoles act as both sources and
sinks of flux. In this book the symbol Ψ is used for magnetic flux to provide
continuity with the rest of the fields notation. However, the generally
accepted standard symbol for magnetic flux is Φ, whereas in this book the
symbol Φ is reserved for potential and is not used for magnetic flux. The
unit of magnetic flux is the Weber (Wb) and that of the flux density is Wb/m^2,
which has been given the name Tesla (T). The total flux is given by a
relationship similar to eqn. (2.7)(page 13),

$$\Psi = \iint_{\text{area}} \mathbf{B} \cdot d\mathbf{A}, \tag{7.2}$$

where $\mathbf{B}$ is the flux density. Because there are no isolated magnetic charges,
Gauss's law in magnetism becomes

$$\iint_{\text{area}} \mathbf{B} \cdot d\mathbf{A} = 0. \tag{7.3}$$

The magnetic field due to an electric current flowing in a coil of wire is
shown in Fig. 7.1. It is similar to that due to a permanent magnet shown in
Fig. 7.2. There is a force on a current carrying wire in a magnetic field,
and, therefore, a force between current-carrying wires. The force per unit
length between two parallel straight wires carrying currents I_1 and I_2, a

distance d apart, is given by

$$\mathbf{f} = \mu_o \frac{I_1 I_2}{2\pi d} \mathbf{U}_d . \tag{7.4}$$

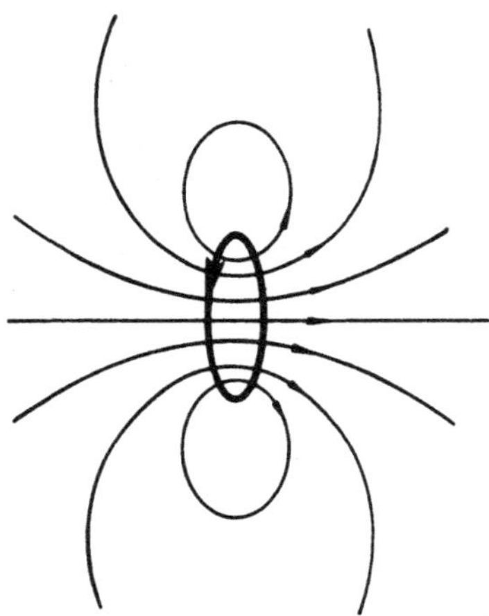

Fig. 7.1. Magnetic field due to an electric current flowing
in a coil of wire.

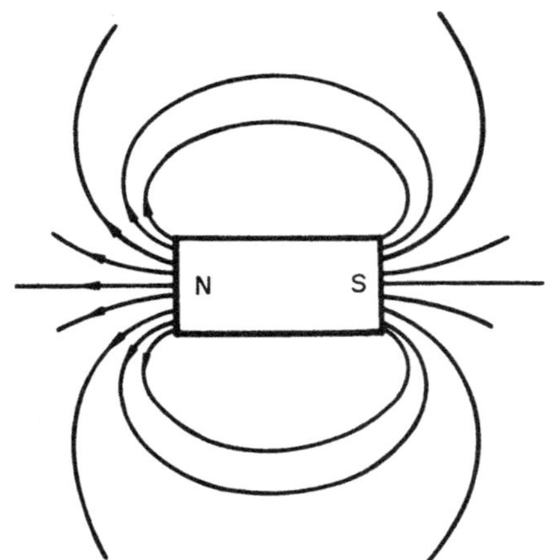

Fig. 7.2. Magnetic field due to a bar magnet.

μ_o is a fundamental magnetic constant which has the value $4\pi \times 10^{-7}$ H/m
and is called the permeability constant. It is similar to the permittivity

in electrostatics. The flux density a distance d from a long straight wire
carrying a current I is given by

$$B = \mu_o \, \frac{I}{2\pi d} \tag{7.5}$$

with the direction of the vector flux density making a tangent to a circle
around the line of the current. The general force relationship between an
electric current $\mathbf{I}$ and a flux density $\mathbf{B}$ is given by

$$\mathbf{f} = \mathbf{I} \times \mathbf{B}, \tag{7.6}$$

which is the force per unit length on the current-carrying wire. The direction
of magnetic forces can be resolved by consideration of the magnetic fluxes
due to the magnet or electric current. There is a force of attraction where
the two individual fluxes are opposed and a force of repulsion where the two
fluxes act in the same direction.

An electric current loop is equivalent to a magnetic dipole. The
equivalent dipole strength is given by

$$m = \mu_o IA, \tag{7.7}$$

where the current I is flowing in a single loop of wire having a cross-
sectional area A. The strength of a magnetic field can be measured by
measuring the torque acting on a dipole or a small loop of electric current.
The torque also depends on the relative angle that the axis of the dipole
makes with the direction of the magnetic field. If at each point of
measurement the maximum torque is taken, then the magnetic field intensity is
given by

$$\mathbf{H} = \lim_{m \to 0} \left(\frac{\left(\text{torque on dipole of moment } m\right)}{m} \right) \tag{7.8}$$

and
$$\mathbf{B} = \mu_o \mathbf{H}. \tag{7.9}$$

If a magnetic field is set up by an electric current loop in the presence
of a magnetic material such as iron, the resulting flux density due to the
magnetic field is larger than it would be if the same electric current flowed
in the same loop of wire in air. The increased effect of the magnetic
material is supplied mathematically by the relative permeability. Therefore
eqn. (7.9) becomes

$$\mathbf{B} = \mu \mathbf{H} = \mu_o \mu_r \mathbf{H}, \tag{7.10}$$

where the relative permeability and the permeability constant may be combined
in a total permeability. The relative permeability is defined by

$$\mu_r = \frac{\text{flux density in magnetic material}}{\text{flux density in air}} \qquad (7.11)$$

assuming that the same electric current and the same geometry is used each time. In terms of the internal magnetization of the material M the flux density becomes

$$\mathbf{B} = \mu_0 \mathbf{H} + \mathbf{M}. \qquad (7.12)$$

In this book, magnetization is measured in units of flux density (T).

EXAMPLES

7.1. Two equal bar magnets having magnetic moments m are situated a distance l apart. The line joining the magnets is parallel to the axis of each magnet. Obtain an expression for the force between the magnets.

Answer. Let each magnet consist of equal and opposite magnetic charges of q at a distance d apart. Then the magnetic moment is given by

$$m = qd.$$

If the magnets are in the position with equal poles nearest, then there will be repulsion between the two magnets. From Coulomb's law there will be two contributions to the force, one of repulsion due to two equal charges a distance l apart and one of attraction due to the further pole of the other magnet which is distant $(l + d)$ and another one of repulsion due to the two furthest poles distant $(l + 2d)$. Therefore the total force is given by

$$F = \frac{q^2}{4\pi\mu_0 l^2} - \frac{2q^2}{4\pi\mu_0 (l + d)^2} + \frac{q^2}{4\pi\mu_0 (l + 2d)^2}$$

$$= \frac{m^2 (3l^2 + 6ld + 2d^2)}{2\pi\mu_0 l^2 (l + d)^2 (l + 2d)^2}.$$

If the length of the magnet is small compared with the distance between the magnets, then the force simplifies to

$$f = \frac{3m^2}{2\pi\mu_0 l^4}.$$

7.2. A circular coil of diameter 20 mm is mounted with the plane of the coil perpendicular to the direction of a uniform magnetic flux density of 100 mT. Find the total flux threading the coil.

Answer. The flux is given by eqn. (7.2). As the flux density is uniform, the flux density is given by

$$\Psi = \mathbf{B} \cdot \mathbf{A}.$$

The vector area is normal to the plane of the area so that the vector flux density is parallel to the vector area and the required flux is the product of the flux density and the area. Therefore

$$\Psi = 10^{-1} \times \pi \times 10^{-4} = 3 \cdot 14 \times 10^{-5} \text{ Wb} = 31 \cdot 4 \text{ } \mu\text{Wb}.$$

7.3. A spherical surface is constructed so as to enclose one end of a bar magnet while leaving the other end outside the sphere. The magnetic moment of the magnet is $1 \cdot 0$ μWb m and it is 50 mm long. The diameter of the spherical surface is 30 mm. Find the total magnetic flux out of the spherical surface.

Answer. The spherical surface is a spherical Gaussian surface surrounding the charge at one end of a magnet. However, this is *not* similar to a spherical Gaussian surface surrounding an isolated flux source. If the sphere is surrounding the magnetic north pole which is a flux source, flux is coming out of the surface of the sphere which is external to the magnet. But the magnetic form of Gauss's law states that there is no net flux out of any closed surface. In the case of this permanent magnet there is an equal flux flow into the sphere for that portion of the spherical surface which is inside the body of the magnet. Therefore the answer to this question is that, according to Gauss's law for magnetism, there is no flux out of the spherical surface.

7.4. Two parallel electric wires 10 mm apart are carrying electric currents of $1 \cdot 0$ A and $5 \cdot 0$ A in the same direction. Find the direction and value of the force between them.

Answer. The force between two parallel electric currents is given by eqn. (7.4). Inserting numbers from the problem gives

$$f = 4\pi \times 10^{-7} \times \frac{1 \cdot 0 \times 5 \cdot 0}{2\pi \times 10^{-2}} = 10^{-4} \text{ N/m} = 0 \cdot 1 \text{ mN/m}.$$

The direction of the force may be obtained from Fig. 7.3. If the uniform flux density B is the flux density due to a current in a wire to the left of the picture flowing in the same direction as the current I in the diagram and the dashed line is the self flux due to the current shown, following the rule that parallel fluxes repel and opposing fluxes attract gives the direction of the force shown in the diagram. Hence the two wires in the question will attract one another with a force of 0·1 mN/m.

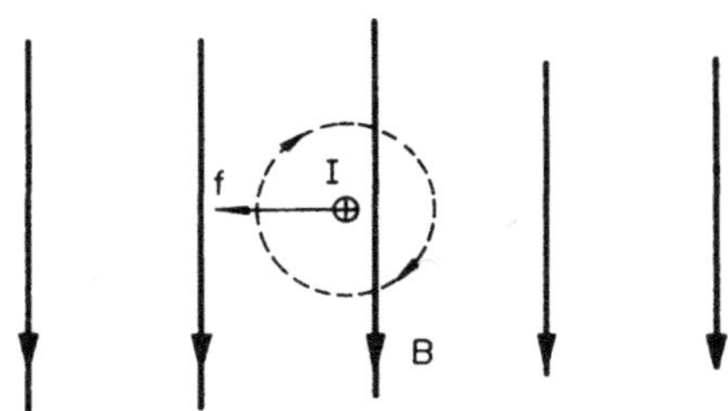

Fig. 7.3.   Current filament in a magnetic flux, showing the
relationship between the self-magnetic flux and the force.

7.5. A long straight wire carrying an electric current of 100 A lies
perpendicularly to the earth's magnetic field of 50 μT. Find the force
acting on the wire.

Answer.   The earth's magnetic field can be assumed to be a uniform field.
The force is given by eqn. (7.6) and **I** and **B** are perpendicular. Inserting
numbers from the problem gives

$$f = IB = 100 \times 50 \times 10^{-6} = 5 \cdot 0 \times 10^{-3} \text{ N/m} = 5 \cdot 0 \text{ mN/m.}$$

7.6. Find the overall force acting on a circular coil of diameter $2R$
carrying a current I in a uniform flux density B and the tension in the wire
due to the magnetic flux. The plane of the coil is perpendicular to the
direction of the magnetic field.

Answer.   The net overall force on the coil is zero. The force on each
element of the coil is directed either radially towards the centre of the
coil or radially outwards from the centre of the coil. Consider half of the
coil as shown in Fig. 7.4. If the relative directions of the current and
magnetic field are such as to make the force radially outwards, the forces
acting on a short length $R\delta\theta$ of the coil are shown in Fig. 7.4. The radial
force is $IBR\delta\theta$. Considering half of the coil as shown in the diagram,

$$2T = \int_{0}^{\pi} IBR \sin\theta \, d\theta = 2IBR.$$

Therefore $T = IBR$ N.

7.7. By sketching flux lines prove that two bar magnets parallel to one
another with their like poles nearest will repel one another.

Answer.   The result to this problem is sketched in Fig. 7.5. The lines

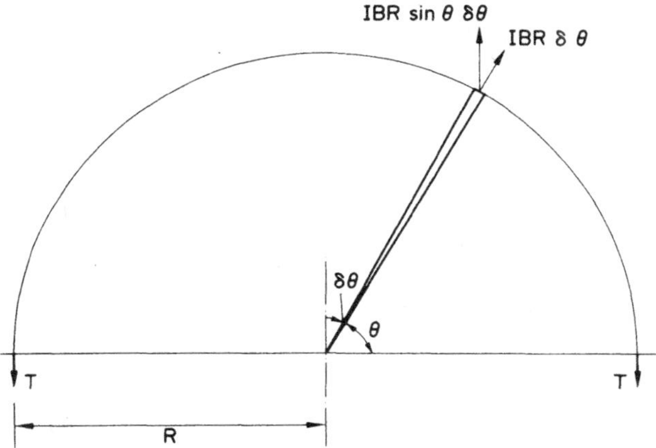

Fig. 7.4. The forces acting on half of a circular coil carrying
a current I situated in a uniform magnetic flux density B.

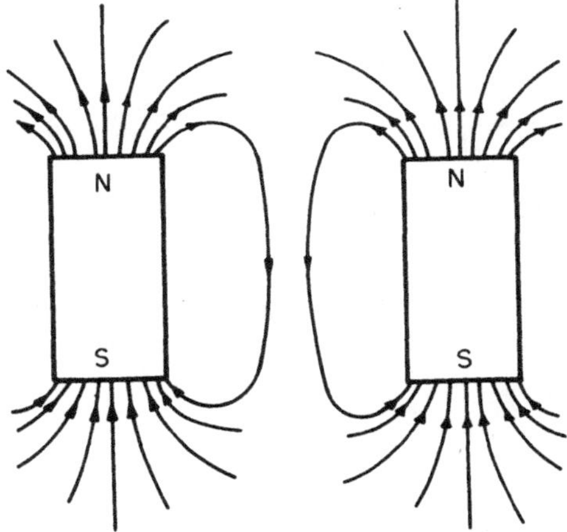

Fig. 7.5. Two permanent magnets in a position of mutual repulsion
showing the magnetic flux.

of flux will originate from the north pole of the magnet which acts similarly
to a source and return to the south pole of the magnet which acts like a sink.
It is seen that in between the magnets the flux lines due to each are parallel
and in the same general direction so that the magnets repel one another.

7.8. A wire carrying an electric current is laid across the top of a bar

magnet as shown in Fig. 7.6. Find the direction of the action of the force
on the wire when the direction of the current and the polarity of the magnet
are shown in the figure.

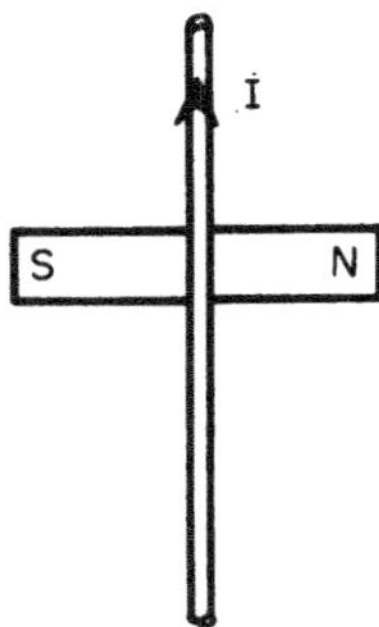

Fig. 7.6. An electric wire across the top of a permanent magnet.

Answer. There are two ways of finding the direction of the force on the
wire. Application of eqn. (7.6) by consideration of the direction of the
vectors in the problem shows that the force is one of repulsion tending to
lift the current-carrying wire off the top of the magnet. The flux density B
is that due to the magnet whose vector direction is parallel to the axis of
the magnet going from north to south. An alternative approach to the solution
and one that is possibly less prone to error due to misinterpretation of the
relative direction of the vectors in a vector product, is to sketch the
magnetic flux lines due to the magnet and the electric current. Such a sketch
is shown in Fig. 7.7 which is a section through the magnet and the wire taken

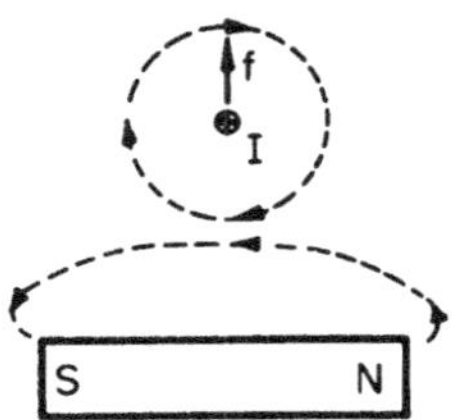

Fig. 7.7.   Showing the magnetic flux lines due to an electric
wire and a permanent magnet.

perpendicular to the diagram of Fig. 7.6. The wire has been lifted away from
the magnet so as to leave a space in which to draw the flux lines. It is seen
that the two fluxes are parallel, so confirming that the force is one of

repulsion.

7.9. A small bar magnet having a magnetic moment of 9×10^{-9} Wb m is suspended at its centre of gravity by a light torsionless string a distance of 10 mm vertically above a long straight horizontal wire carrying a current of $1 \cdot 0$ A. Find the frequency of oscillation of the magnet about its equilibrium position assuming that the motion is undamped. The moment of inertia of the magnet is 6×10^{-9} kg m^2.

Answer. Assume that the bar magnet consists of a positive and negative magnetic charge of q a distance d apart such that $m = qd$. The force exerted on each of these magnetic charges is given by

$$F = qH.$$

If the magnet is disturbed from its equilibrium position through an angle θ, then the restoring torque is given by

$$\text{torque} = Fd \sin \theta = qdH \sin \theta = mH \sin \theta.$$

From the mechanics of an oscillating body, the restoring torque is equivalent to the product of the moment of inertia and the angular acceleration. Therefore

$$\text{torque} = I \ddot{\theta}.$$

Therefore

$$I\ddot{\theta} = mH \sin \theta \approx mH\theta$$

for small angles θ. In terms of the electric current i, the magnetic field intensity due to a long straight wire is given by

$$H = \frac{i}{2\pi r}.$$

For simple harmonic motion of an oscillating body, the frequency is given by

$$f = \frac{1}{2\pi}\sqrt{\left(\frac{\ddot{\theta}}{\theta}\right)},$$

$$\frac{\ddot{\theta}}{\theta} = \frac{mi}{2\pi r I} = \frac{9 \times 10^{-9} \times 1 \cdot 0}{2\pi \times 10^{-2} \times 6 \times 10^{-9}} = 23 \cdot 9.$$

Therefore

$$f = \frac{1}{2\pi}\sqrt{23 \cdot 9} = 0 \cdot 78 \text{ Hz}.$$

7.10. By calculating the torque exerted on each in a uniform magnetic field of flux density B, prove the equivalence between a small current-carrying coil of area A and a magnetic dipole of moment m to be $m = \mu_{o}IA$, where I is the current in the coil of one turn.

Answer. In the solution to the last example it was shown that in a
magnetic field the torque acting on a small bar magnet (which is the same as
a magnetic dipole) is given by

$$\text{torque} = mH \sin \theta = \frac{mB}{\mu_o} \sin \theta.$$

In order to find the torque on a current-carrying coil, we will calculate the
torque acting on a rectangular loop of wire carrying a current I. A section
through a rectangular loop is shown in Fig. 7.8. Only the current flowing

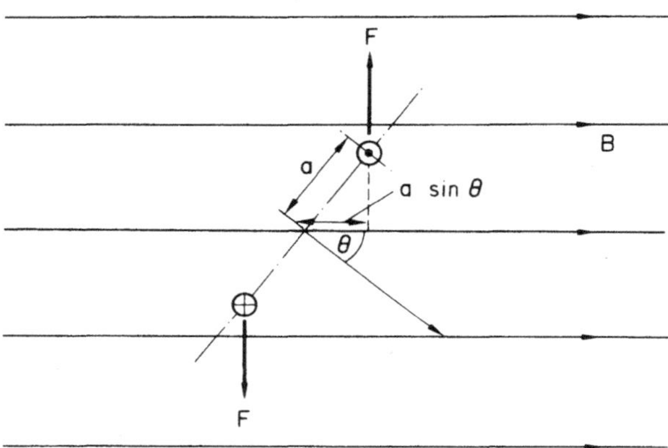

Fig. 7.8. A section through a rectangular current loop showing
the forces exerted by a uniform magnetic field.

through the horizontal parts of the loop will contribute to the torque on the
loop. The current flowing in the other parts of the loop gives rise to forces
acting in the plane of the loop which cancel overall. If the length of the
loop in the direction perpendicular to the paper in the diagram is l, the
force per unit length of each horizontal wire is given by eqn. (7.6), whence
the total force on each wire is given by

$$F = IBl.$$

The area of the loop is given by

$$A = 2al.$$

Therefore the torque is given by

$$\text{torque} = 2Fa \sin \theta = 2IBla \sin \theta = IAB \sin \theta.$$

Equating the two expressions for the torque gives

$$m = \mu_{o} IA.$$

7.11. Discuss the analogies between field quantities in the magnetic field and the electrostatic field.

Answer. There are direct analogies between Coulomb's law in the two field systems. The magnetic pole strength is the charge in the magnetic field. Unfortunately, the physicist who tries to separate the poles of a magnet obtains, not a pair of single poles, but a pair of magnets. The ultimate indivisible unit of magnetism is a magnetic dipole. However, although there are no isolated magnetic charges which can provide a source of magnetic flux, the dipoles, being pairs of equal and opposite magnetic charges, act as both sources and sinks of flux. There is a useful analogy between the electrostatic charge and the magnetic pole strength. The solutions to Examples 7.9 and 7.10 show how the concept of magnetic charge may usefully be applied to the solution of certain problems involving permanent magnets.

There is an equivalence between Gauss's law in the two systems except that in the magnetic field there are no isolated charges so that the net enclosed charge is always zero. The equivalents are eqns. (2.12) (page 14) and (7.3). There is also an equivalence between the flux density and field intensity in the two systems. There is a direct analogy between eqns. (2.10) (page 14) and (7.9). The magnetic flux density is directly analogous to the electric flux density; **B** is analogous to **D**. The magnetic field intensity is directly analogous to the electric field intensity; **H** is analogous to **E**. However, the analogy ends at the equivalence between the two field descriptions. This analogy is most useful in transferring results from one field analogy to another, but it fails when there is some attempt to apply the analogy to the physical system which field theory is describing. The field quantities can only be observed by measuring some other quantity. In electrostatics, the force on an electric charge can be measured so that the electric field intensity becomes the observable quantity. The existence of the flux density is only inferred from the field intensity and the charge distribution. In magnetism it is the flux density which can be measured from the force exerted on a current-carrying wire. The existence of the field intensity has to be inferred. Some of the field quantities are very helpful in an understanding of the mathematical analysis applicable to field theory problems and to the solution of numerical problems, but they are very loosely related to measurable quantities so that their physical existence is in doubt.

7.12. An insulated wire carrying a current of 6·28 mA is threaded through

a small hole in a block of steel. Calculate the magnetic flux density in the
steel at a distance of 0·1 m from the wire assuming that the diameter of the
wire is negligibly small. The permeability of the steel is given by Fig. 7.9.

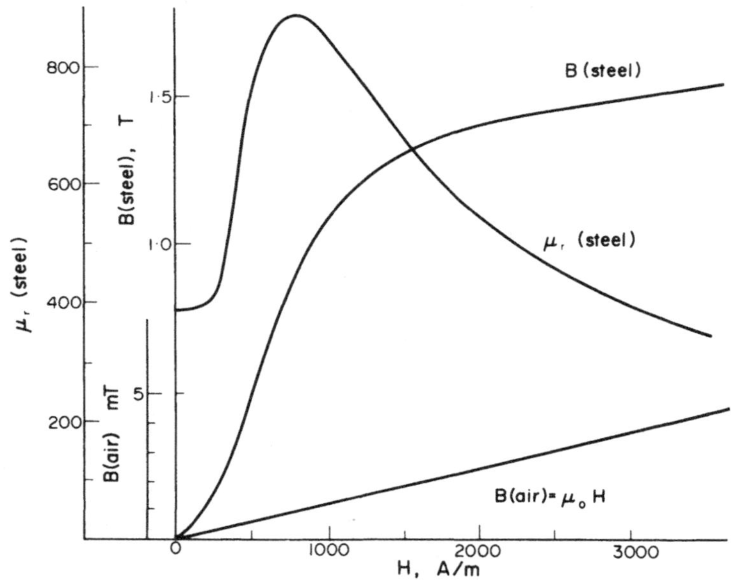

Fig. 7.9. A graphical plot of the relationship $B = \mu H$ for a
typical steel showing the variation of relative permeability
with variation of H.

Answer. Assuming a long straight wire, the field intensity is given by
a relationship similar to eqn. (7.5), therefore

$$H = \frac{I}{2\pi r} = \frac{6\cdot28 \times 10^{-3}}{2\pi \times 0\cdot1} = 10^{-2} \text{ A/m.}$$

The flux density is obtained from eqn. (7.10) and the value of the relative
permeability from Fig. 7.9. As the value of the field intensity is small, the
value of the relative permeability is that at $H = 0$. From the graph it is 390.
Therefore

$$B = \mu_0\mu_r H = 4\pi \times 10^{-7} \times 390 \times 10^{-2} = 4\cdot9 \times 10^{-6} \text{ T} = 4\cdot9 \ \mu\text{T.}$$

PROBLEMS

7.1. A d.c. power transmission line consists of two parallel cables 10 m
apart. Calculate the force between the cables if the current is 10,000 A in
both the forward and return cables.   (2·0 N/m.)

7.2. A coaxial cable consists of an inner conductor surrounded by
insulation which is in turn surrounded by a concentric conducting sheath. If
the direct current in the cable is I and it causes an equal and opposite return
current in the sheath, prove that the bursting pressure on the sheath is
given by

$$p = \frac{\mu_o I^2}{8\pi^2 r^2} \text{ N/m}^2 .$$

7.3. By the method of dimensions, check the validity of eqn. (7.6).

7.4. If the earth's magnetic field is 50 μT, find the maximum possible
force on a long straight wire carrying a current of 10,000 A due to the earth's
magnetic field.   (0·5 N/m.)

7.5. A small bar magnet is suspended at its centre of gravity by a light
torsionless string above a long straight horizontal wire carrying a direct
electric current. Determine the position of the magnet relative to the wire.

7.6. A coil of area 10^{-4} m^2 has 1000 turns of fine wire carrying a current
of 10 mA. It is suspended from one end by a torsion wire having a constant
torque of 20 μN m. Determine the angle that the coil will adopt relative to a
uniform magnetic flux density of 0·1 T.   (11·5°.)

7.7. A long closely wound solenoid is mounted on a non-magnetic former
and the flux density inside is found to be 2·5 mT. A bar of steel whose
magnetic properties are shown in Fig. 7.9 is inserted into the solenoid.
Estimate the flux density inside the steel.   (1·4 T.)

7.8. A galvanometer has a rectangular coil suspended in a radial magnetic
field so that the magnetic field always acts across the plane of the coil. If
the coil is 10 mm by 10 mm side and has 1000 turns, and if the magnet provides
a constant flux density of 0·3 T, find the torque exerted on the coil for a
current of 10 mA.   (300 μN m.)

7.9. Prove, by the method of dimensions, that the equivalent dimensions
of the Weber are V s.

7.10. An insulated copper wire is threaded through a small hole in a block
of steel and carries a current of 1·0 A. Calculate the magnetic flux density
in the steel a distance of 10 mm from the centre of the wire. Assume that the
diameter of the wire is negligibly small. The permeability of the steel is
given by Fig. 7.9.   (7·8 mT.)

SOLUTIONS

7.1. The force between two parallel cables carrying electric currents is given by eqn. (7.4). As the wires may be assumed to be forming forward and return paths for the current, the direction of the currents in the two wires will be opposing and the force will be one of repulsion. The value of the force is given by

$$f = \frac{\mu_0 I_1 I_2}{2\pi d} = \frac{4\pi \times 10^{-7} \times 10^4 \times 10^4}{2\pi \times 10} = 2{\cdot}0 \text{ N/m.}$$

7.2. The magnetic flux density due to the long straight centre conductor is given by eqn. (7.5). At the outer sheath, there is an equal and opposite current, and the magnetic flux outside is zero. Therefore the average flux density at the sheath is half that just inside the sheath. It is given by

$$B = \tfrac{1}{2}\mu_0 \frac{I}{2\pi r} = \frac{\mu_0 I}{4\pi r}.$$

The force is given by eqn. (7.6). Assuming a uniform distribution of current in the sheath, the current in a small element $r\delta\theta$ of the sheath is given by

$$\delta I = \frac{I}{2\pi r} (r\delta\theta) = \frac{I\delta\theta}{2\pi}.$$

The force on the current element δI is given by

$$\delta f = \delta IB = \frac{I\delta\theta}{2\pi} \frac{\mu_0 I}{4\pi r} \text{ N/m.}$$

The pressure is given by

$$p = \frac{\delta f}{r\delta\theta} = \frac{\mu_0 I^2}{8\pi^2 r^2} \text{ N/m}^2.$$

7.3. Equation (7.6) is

f = I × B.

The dimensions of the right-hand side of this equation are

$$\text{(A) } (VTL^{-2}) = (\text{energy}) L^{-2} = (M L^2 T^{-2}) (L^{-2}) = M T^{-2}.$$

The dimensions of the left-hand side are force per unit length, which gives

$$(M L T^{-2}) L^{-1} = M T^{-2}.$$

Therefore the dimensions of each side of eqn. (7.6) are the same. This result has assumed that the dimensions of magnetic flux, the weber, is voltage × time.

7.4. The maximum possible force occurs when the current is perpendicular to the magnetic flux density as shown by eqn. (7.6). Therefore the force is given by

$$f = IB = 10^4 \times 50 \times 10^{-6} = 0 \cdot 5 \text{ N/m}.$$

7.5. This problem is similar to Example 7.9. The magnet will align itself to the position of maximum attraction between the magnet and the magnetic flux of the current in the wire. This is with the magnet perpendicular to the wire and with the opposite orientation to that shown for the similar problem in Fig. 7.6. The freely suspended magnet will align itself along the flux lines due to the current in the wire so that the magnet appears to act as a source and sink for the flux lines surrounding the wire.

7.6. The torque due to a current-carrying coil in a magnetic field has been derived in the solution to Example 7.10 and is given by

$$\text{torque} = IAB \sin \theta.$$

The coil will orientate itself so that this torque is equal to that given by the suspension. Therefore

$$\sin \theta = \frac{\text{torque}}{IAB} = \frac{20 \times 10^{-6}}{10^3 \times 10^{-2} \times 10^{-4} \times 0 \cdot 1} = 0 \cdot 20.$$

Therefore

$$\theta = 11 \cdot 5^{\circ}.$$

7.7. If the solenoid is uniform and long, it may be assumed that the flux density inside it is approximately uniform. Then the field intensity inside is also uniform and is given by

$$H = \frac{B}{\mu_o} = \frac{2 \cdot 5 \times 10^{-3}}{4\pi \times 10^{-7}} = 2 \cdot 0 \times 10^3 \text{ A/m.}$$

When the bar of steel is inserted, it must be assumed that the field intensity is unchanged. Then the flux density is obtained from Fig. 7.9. It is seen to be $1 \cdot 4$ T.

7.8. The torque acting on the coil will be the same as the maximum torque acting on a similar coil suspended in a uniform field. The force exerted on one of the straight arms of the coil of 1000 turns is given by

$$f = lIB = 10^{-2} \times 10 \times 10^{-3} \times 10^3 \times 0 \cdot 3 = 3 \cdot 0 \times 10^{-2} \text{ N.}$$

Therefore the total torque, due to both arms, is given by

$$\text{torque} = 3 \cdot 0 \times 10^{-2} \times 10^{-2} = 3 \cdot 0 \times 10^{-4} \text{ N m} = 300 \text{ } \mu\text{N m.}$$

7.9. The necessary equivalence is obtained from eqn. (7.7),

$$m = \mu_0 I A \text{ Wb m.}$$

The dimensions of the permeability constant are obtained from the force between two parallel current-carrying wires, eqn. (7.4) whence the dimensions of μ_0 are

$$\frac{\text{force}}{(\text{current})^2} = M\, L\, T^{-2}\, A^{-2}.$$

Therefore the dimensions of magnetic flux or charge are

$$(M\, L\, T^{-2}\, A^{-2})\ (A)\ (L^2)/(L) = M\, L^2\, T^{-2}\, A^{-1} = (\text{energy})\, A^{-1}$$

$$= (V\, A\, T)\, A^{-1} = V\, T.$$

Therefore the dimensions of Wb are V s.

7.10. This problem is similar to Example 7.12. Therefore

$$H = \frac{I}{2\pi r} = \frac{1 \cdot 0}{2\pi \times 10^{-2}} = 15 \cdot 9 \text{ A/m.}$$

We can use the value of relative permeability appropriate to small fields. From the graph it is 390, therefore

$$B = \mu_0 \mu_r H = 4\pi \times 10^{-7} \times 390 \times 15 \cdot 9 = 7 \cdot 8 \times 10^{-3} \text{ T} = 7 \cdot 8 \text{ mT.}$$

FURTHER PROBLEMS

7.11. Two short bar magnets of dimensions 10 mm × 10 mm × 30 mm are placed one on each side of a non-magnetic sheet 10 mm thick. Estimate the maximum force of attraction between the magnets if they both have a magnetic moment of 0·7 µWb m. What is the accuracy of the calculation?   (0·18 N ± 40%.)

7.12. A d.c. transmission cable consists of two parallel wires 0·3 m apart each carrying a current of 10^3 A, flowing in opposite directions. Find the direction and strength of the force between the cables.   (0·67 N/m.)

7.13. A wire carrying an electric current is mounted between the poles of a permanent magnet, as shown in Fig. 7.10. If the diameter of the cross-section of the poles is 20 mm and the magnetic flux density in the gap between the poles is uniform at 0·36 T and zero outside, find a relationship between the force on the wire and the electric current flowing through it.
$(F = 0\cdot0072I.)$

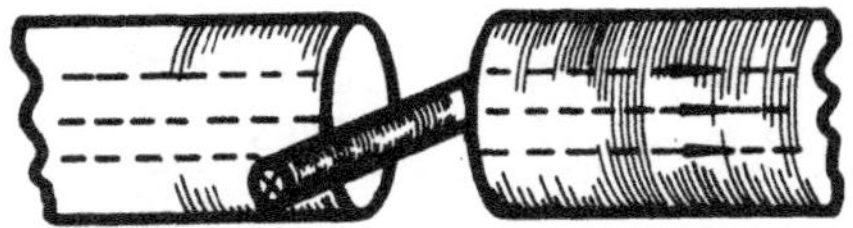

Fig. 7.10. A wire mounted between the poles of a permanent magnet.

7.14. A d.c. cable consists of a coaxial cable with the current return through the outer sheath. If the maximum current rating of the cable is 80 A and the diameter of the sheath is 15 mm, find the maximum bursting pressure on the cable due to magnetic effects.   (1·8 N/m^2.)

7.15. A small bar magnet of magnetic moment m is rotating at a steady speed in a uniform magnetic field. Obtain an expression of the variation of the torque on the magnet as it rotates due to the magnetic field and the mean power required to overcome this torque.   (0.)

7.16. A small bar magnet having a magnetic moment of 2·0 × 10^{-7} Wb m is pivoted at its centre of gravity 1·0 m vertically below a long straight horizontal wire carrying a direct current of 100 A. Find the period of oscillation of the magnet about its equilibrium position assuming that the motion is undamped. The moment of inertia of the magnet is 4·0 × 10^{-7} kg m^2.
(2·2 s.)

7.17. The strength of a magnetic field can be measured by suspending a small current-carrying coil in the field and measuring the torque exerted on the coil. Such a coil has 500 turns and an area of 80×10^{-6} m^2. In a certain uniform field it was found that a current of 5·5 mA gave a maximum torque of 40 µN m. What is the strength of the field?   (0·18 T.)

7.18. A long closely wound coil provides an approximately uniform magnetic field inside it. Such a coil wound on a non-magnetic former is found to have a flux density inside it of 1·2 mT for a current of 1·0 A. If a bar of steel whose properties are given in Fig. 7.9 is inserted into the coil, estimate the flux density inside the steel for the same current in the coil.   (1·05 T.)

7.19. Four long straight wires are parallel and spaced so that they are situated at the corners of a square of side 1·0 m. They all carry equal electric currents of I A. Find the force on one wire due to the current in the other wires. $(4·24 \times 10^{-7} \ I^2$ N/m.)

7.20. A galvanometer has a rectangular coil of size 5 mm × 20 mm pivoted about the centre of the shorter side. It is mounted in a radial magnetic field so that the magnetic field always acts across the plane of the coil. If the coil has 2000 turns and the magnet provides a constant flux density of 0·25 T, find the torque exerted on the coil for a coil current of 25 µA. (1·25 µN m.)

FURTHER SOLUTIONS

7.11. Let each magnet consist of equal and opposite magnetic charges q a distance d apart. Then the magnetic moment is given by

$$m = qd.$$

For the maximum force of attraction between the magnets, the opposite poles will be nearest one another. From Coulomb's law there will be two contributions to the force, one of attraction due to equal and opposite charges a distance l apart, where l is the distance between the centre line of the magnets, and one of repulsion due to the further pole of the other magnet which is distant $\sqrt{(l^2 + d^2)}$ and acting at an angle to the force of attraction. As each of these forces acts once on each pole, the total force is given by

$$F = \frac{2q^2}{4\pi\mu_0 l^2} \times \frac{2q^2}{4\pi\mu_0 (l^2 + d^2)} \times \frac{l}{\sqrt{(l^2 + d^2)}}.$$

$$= \frac{m^2}{2\pi\mu_0} \left(\frac{1}{l^2 d^2} - \frac{l}{d^2 (l^2 + d^2)^{3/2}} \right).$$

If the pole is at a distance of half the thickness of the magnet from its end, then $d = 20$ mm, but if the pole is right on the end of the magnet, $d = 30$ mm. It seems reasonable that the pole of the magnet will lie somewhere between these positions which will give the uncertainty in the calculated value:
$l = 20$ mm $= 0\cdot02$ m.

If $d = 0\cdot02$ m,

$$F = \frac{0\cdot49 \times 10^{-12}}{2\pi \times 4\pi \times 10^{-7}} \left(\frac{1}{4\cdot0 \times 10^{-4} \times 4\cdot0 \times 10^{-4}} - \frac{2\cdot0 \times 10^{-2}}{4\cdot0 \times 10^{-4} \times 10^{-6} \times 8\sqrt{8}} \right)$$

$$= 0\cdot250 \text{ N}.$$

If $d = 0\cdot03$ m,

$$F = \frac{0\cdot49 \times 10^{-12}}{2\pi \times 4\pi \times 10^{-7}} \left(\frac{1}{4\cdot0 \times 10^{-4} \times 9\cdot0 \times 10^{-4}} - \frac{2\cdot0 \times 10^{-2}}{4\cdot0 \times 10^{-4} \times 10^{-6} \times 13\sqrt{13}} \right)$$

$$= 0\cdot106 \text{ N}.$$

The average value is $F = 0\cdot18$ N, and the uncertainty is $\pm\, 0\cdot072$ N, which is the same as $\pm 40\%$.

7.12. From eqn. (7.4)

$$f = \mu_o \frac{I_1 I_2}{2\pi d} = \frac{4\pi \times 10^{-7} \times 10^6}{2\pi \times 0\cdot3} = 0\cdot67 \text{ N/m.}$$

As the currents are flowing in opposite directions, the force is one of repulsion.

7.13. From eqn. (7.6) all the vectors are perpendicular, therefore

$$F = IBl = I \times 0\cdot36 \times 0\cdot02 = 0\cdot0072I \text{ N.}$$

7.14. An analytical solution to a similar problem is given in the solution to Problem 7.2. From the formula derived there,

$$p = \frac{\mu_o I^2}{8\pi^2 r^2} = \frac{4\pi \times 10^{-7} \times 6400}{8\pi^2 \times 56\cdot25 \times 10^{-6}} = 1\cdot81 \text{ N/m}^2.$$

7.15. In the solution to Example 7.10 the torque on a magnet in a uniform field is found to be

$$\text{torque} = \frac{mB}{\mu_o} \sin \theta.$$

As the magnet rotates the torque will vary between positive and negative values sinusoidally, so that the average torque will be zero. Therefore the mean power expended due to this torque will also be zero.

7.16. In the solution to Example 7.9 the frequency of oscillation of a small magnet is derived. It is

$$f = \frac{1}{2\pi} \sqrt{\left(\frac{mi}{2\pi rI} \right)} = \frac{1}{2\pi} \sqrt{\left(\frac{2\cdot0 \times 10^{-7} \times 100}{2\pi \times 1\cdot0 \times 4\cdot0 \times 10^{-7}} \right)} = 0\cdot45 \text{ Hz.}$$

Therefore period $= \dfrac{1}{f} = 2\cdot2$ s.

7.17. In the solution to Example 7.10 the torque on a coil in a uniform field is found to be

$$\text{torque} = IAB \sin \theta.$$

For the maximum value of torque, $\sin \theta = 1$, therefore

$$B = \frac{\text{torque}}{IA} = \frac{40 \times 10^{-6}}{5\cdot5 \times 10^{-3} \times 500 \times 80 \times 10^{-6}} = 0\cdot18 \text{ T.}$$

7.18. The field intensity inside the coil is given by

$$H = \frac{B}{\mu_o} = \frac{1\cdot2 \times 10^{-3}}{4\pi \times 10^{-7}} = 955 \text{ A/m.}$$

For this value of field intensity, the flux density inside the steel is found from the magnetization curve given in Fig. 7.9, therefore

$$B_{steel} = 1 \cdot 05 \text{ T.}$$

7.19. The force between parallel current-carrying wires is given by eqn. (7.4). In this problem the force can be calculated due to each wire individually, and by the principle of superposition the total force will be the vector sum of the individual forces. The geometry is similar to that shown in Fig. 2.10 (page 30). Due to symmetry, the resultant force on one wire will act in the direction of the diagonal of the square. Then resolving the forces in the direction of the diagonal gives

$$f = \frac{\mu_0 I^2}{2\pi} \left(\frac{1}{1} \times \frac{1}{\sqrt{2}} + \frac{1}{1} \times \frac{1}{\sqrt{2}} + \frac{1}{\sqrt{2}} \right) = \frac{4\pi \times 10^{-7} I^2}{2\pi} \times \frac{3}{\sqrt{2}} = 4 \cdot 24 \times 10^{-7} I^2 \text{ N/m.}$$

7.20. From eqn. (7.6),

$$F = lIB = 20 \times 10^{-3} \times 25 \times 10^{-6} \times 2000 \times 0 \cdot 25 = 250 \times 10^{-6} \text{ N.}$$

$$\text{Torque} = 250 \times 10^{-6} \times 5 \cdot 0 \times 10^{-3} = 1 \cdot 25 \times 10^{-6} \text{ N m.}$$

PART II

Potential

CHAPTER 8

Electric Potential

THEORY

The electric potential difference between two points in an electric field is defined as being numerically equal to the work done in moving a unit charge between the two points. The formal definition is:

$$\text{potential rise, } A \text{ to } B = \lim_{\delta q \to 0} \left(\frac{\text{work done by external agency in moving } \delta q \text{ from } A \text{ to } B}{\delta q} \right). \qquad (8.1)$$

If the field is as shown in Fig. 8.1, the potential difference is given by

$$\text{potential difference} = \int_{A}^{B} - \mathbf{E} \cdot d\mathbf{l}. \qquad (8.2)$$

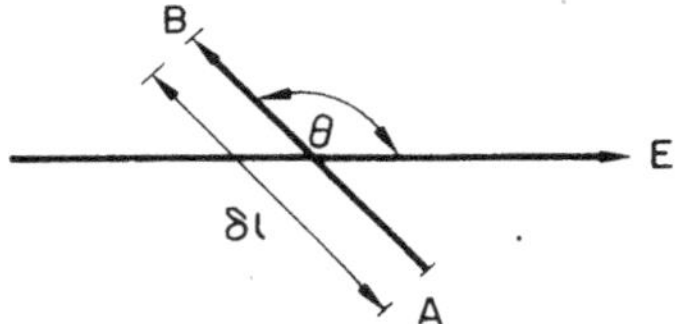

Fig. 8.1. Illustrating potential rise in an electric field.

The potential difference between any two points in an electrostatic field is independent of the path taken in the measurement of the potential difference. Therefore

$$\oint \mathbf{E} \cdot d\mathbf{l} = 0. \qquad (8.3)$$

Such a field is called a *conservative field*, and possesses a *scalar potential function*. In the electrostatic field, the scalar potential function is *voltage*. Therefore in terms of the potential function eqn. (8.2) becomes

$$\Phi = \phi_B - \phi_A = \int_{A}^{B} - \mathbf{E} \cdot d\mathbf{l}. \qquad (8.4)$$

The *absolute potential* is defined to be the potential function value relative to a zero of potential at an infinite distance away from the source or sources of flux.

The differential form of the relationship given in eqn. (8.4) is

$$E_x = -\frac{\partial \phi}{\partial x}, \qquad E_y = -\frac{\partial \phi}{\partial y}. \tag{8.5}$$

The total field is the vector sum of these two components,

$$\mathbf{E} = -\mathbf{U}_x \frac{\partial \phi}{\partial x} - \mathbf{U}_y \frac{\partial \phi}{\partial y} = -\nabla \phi = -\operatorname{grad} \phi. \tag{8.6}$$

In cylindrical polar coordinates,

$$\mathbf{E} = -\mathbf{U}_r \frac{\partial \phi}{\partial r} - \mathbf{U}_\theta \frac{1}{r} \frac{\partial \phi}{\partial \theta}. \tag{8.7}$$

Equipotential lines are lines of constant potential function. They are orthogonal to the lines of constant flux function.

The capacitance of a system of charged bodies is defined by

$$\text{capacitance} = C = \frac{Q}{V} = \frac{\Psi}{\phi} = \frac{\text{flux}}{\text{potential difference}}. \tag{8.8}$$

The general effect of a boundary between two materials of different permittivities on a field is that the normal component of the flux density and the tangential component of the field intensity is the same on each side of the boundary. The energy stored in an electrostatic field is given by

$$W = \iiint\limits_{\text{volume}} \tfrac{1}{2}DE\, dv \text{ J}. \tag{8.9}$$

EXAMPLES

8.1. Prove that the potential difference between two points in a uniform electrostatic field is independent of the path traversed in calculating the potential difference.

Answer. Consider movement along the path adb in the parallel field shown in Fig. 8.2. To find the potential difference, move a search charge δq from a to b along the path adb. Then the potential difference is given by eqn. (8.2). This is equal to the work done in moving the small charge from a to b divided by the size of the charge. Each element of the path δl can be split into components perpendicular and parallel to the field. Movement along the component perpendicular to the field contributes nothing to the rise in potential so that the rise in potential is the same as that given by the work done in moving from a to c. Were the path of integration to be changed to acb, no work would be done in moving from c to b, and the potential difference

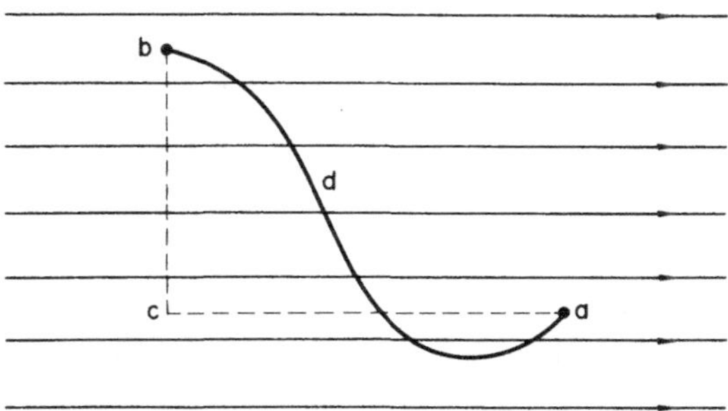

Fig. 8.2. A path in a parallel electric field.

between b and a is the same as the potential difference between c and a.
Therefore the potential difference between b and a is always the same as that
between c and a irrespective of the path used to go from a to b.

8.2. Two parallel charged plates are a distance d apart. They have equal
and opposite charges of uniform charge density $\pm\sigma$ C/m². Find the potential
difference between the plates.

Answer. The situation of the two parallel charged plates is shown in
Fig. 8.3 with the flux lines and equipotential lines also shown. The potential

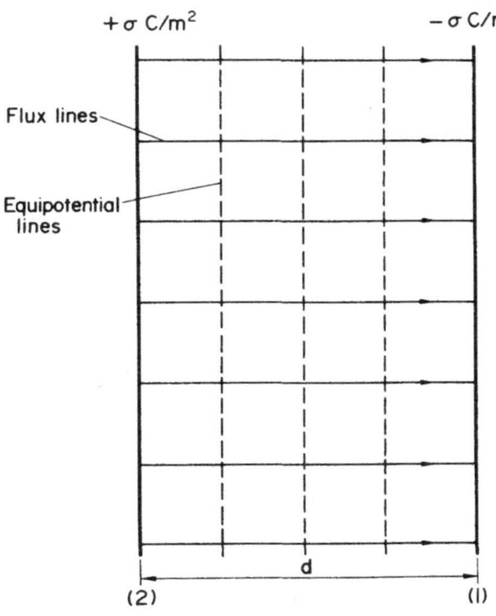

Fig. 8.3. Flux lines and equipotential lines between two parallel
charged plates.

rise from plate (1) to plate (2) is given by eqn. (8.4).

$$\Phi = \phi_2 - \phi_1 = V_2 - V_1 = - \int_1^2 \mathbf{E} \cdot d\mathbf{l}.$$

The field intensity between two equal and opposite plane charge densities is given by

$$E = \frac{D}{\varepsilon_0} = \frac{\sigma}{\varepsilon_0}.$$

The potential rise occurs when moving in the opposite direction to the field intensity. Mathematically this means that the directions of the $\mathbf{E}$ and $d\mathbf{l}$ vectors are parallel but in the opposite directions so that the scalar product of the two vectors in the integral is negative. Then the potential rise from plate (1) to plate (2) is given by

$$\phi_2 - \phi_1 = \frac{\sigma}{\varepsilon_0} \int_0^d dl = \frac{\sigma d}{\varepsilon_0}.$$

Because the field intensity is constant and the path of integration is taken parallel to the field intensity vector, the potential rise may be expressed in another way.

$$\phi_2 - \phi_1 = Ed = \frac{\sigma d}{\varepsilon_0}.$$

The potential difference between the plates is numerically identical. Therefore

$$\Phi = \frac{\sigma d}{\varepsilon_0}.$$

If plate (1) is defined to be at zero potential, $\phi_1 = 0$, and plate (2) is then at a potential of $\phi_2 = \sigma d/\varepsilon_0$.

8.3. A parallel-plate capacitor consists of two circular metal plates 0·30 m diameter, 10 mm apart. If it is given a charge of 10 nC on one plate and an equal and opposite charge of -10 nC on the other plate, find the potential difference between the plates.

Answer. It must be assumed that the electric charge will distribute itself uniformly over the inside faces of each of the plates of the capacitor and that any non-uniform field effects at the edges of the plates may be ignored. Then the charge density on the plates is given by

$$\sigma = \frac{\text{charge}}{\text{area}}$$

and the potential difference between the plates may be obtained from the formula derived in the answer to the last example. Therefore

$$\Phi = \phi_2 - \phi_1 = \frac{\sigma d}{\varepsilon_o} = \frac{10 \times 10^{-9} \times 10^{-2} \times 36\pi \times 10^9}{\pi \times 0.15 \times 0.15} = 160 \text{ V}.$$

8.4. A uniform electrostatic field has the electric intensity vector acting in the x-direction of strength E_x. Find expressions in both rectangular and polar coordinates for the potential at any point in the field where the origin has zero potential function value.

Answer. The uniform field and a point $P(x, y)$ or (r, θ) is illustrated in Fig. 8.4. The rise in potential in going from O to P is the same as the

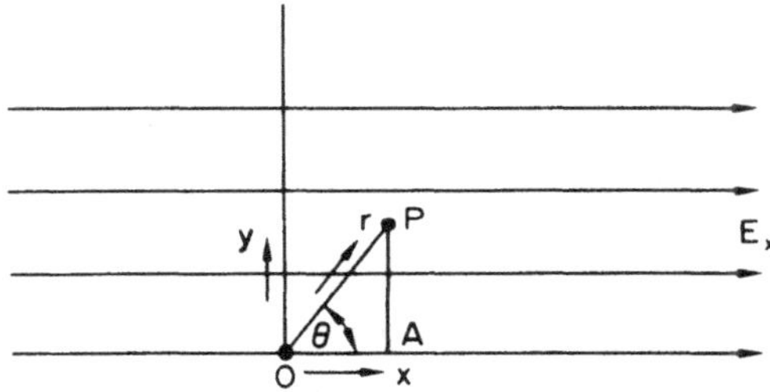

Fig. 8.4. Illustrating the calculation of potential function in
a uniform field.

rise in potential in going from O to A because the rise in potential is independent of the path taken and there is no rise in potential when moving perpendicular to the field from A to P. The value of the potential function is given by eqn. (8.4),

$$\Phi = \phi - 0 = \int_O^P - \mathbf{E} \cdot d\mathbf{l} = - E_x x = - E_x r \cos \theta.$$

8.5. The potential function of a field is given by

$$\phi = \frac{Q}{4\pi \varepsilon_o r}.$$

Find expressions for the components of the electric field intensity vector.

Answer. As the potential function is specified in terms of a radial distance, the electric field intensity will be found from eqn. (8.7).
Therefore

$$E_r = - \frac{\partial \phi}{\partial r} = \frac{Q}{4\pi \varepsilon_o r^2} \qquad \text{and} \qquad E_\theta = - \frac{1}{r} \frac{\partial \phi}{\partial \theta} = 0.$$

A radial field intensity of the above value together with no circumferential component is the field due to a point charge of strength Q.

8.6. The potential function of a field is given by

$$\phi = 8r \sin \theta \text{ V}.$$

Find an expression for the field intensity in the field.

Answer. As the potential function is specified in polar coordinates, the field intensity will be obtained using eqn. (8.7).

$$\mathbf{E} = - \mathbf{U}_r \frac{\partial \phi}{\partial r} - \mathbf{U}_\theta \frac{1}{r} \frac{\partial \phi}{\partial \theta} = - \mathbf{U}_r 8 \sin \theta - \mathbf{U}_\theta 8 \cos \theta.$$

If this field intensity is resolved into components parallel to the rectangular coordinates,

$$E_x = 0, \qquad E_y = -8 \text{ V/m}.$$

This is a uniform field with the field intensity vector directed parallel to the negative y-axis.

8.7. Find the capacitance of a parallel-plate capacitor having a plate area of A and a separation between the plates of d.

Answer. The field between the plates of a parallel-plate capacitor is the same as the field due to a plane charge source parallel to a plane charge sink of the same strength. The system is illustrated in Fig. 8.3. It is necessary to assume that the field is uniform in the region between the plates and that there is no distortion of the field at the edge of the plates. The assumption introduces little error provided that $d^2 < A$. In problems involving capacitance it is usually necessary to apply suitable charges to the system of conductors forming the capacitor to calculate the potential difference set up by these charges and then to derive the capacitance from the ratio of the flux to the potential difference. It is found in these calculations that the originally specified charge cancels out in the expression for the capacitance. Let us specify a total charge of $\pm Q$ on each of the plates. Then, in the region between the plates,

$$D = \sigma = \frac{Q}{A}.$$

Since the electric field is uniform,

$$\Phi = Ed = \frac{Dd}{\varepsilon_0} = \frac{Qd}{A\varepsilon_0}.$$

Therefore the capacitance is given by

$$C = \frac{\psi}{\phi} = \frac{Q}{\phi} = \frac{QA\varepsilon_0}{Qd} = \frac{\varepsilon_0 A}{d}.$$

8.8. A parallel-plate capacitor consists of two sheets of copper foil, each of area $0 \cdot 1$ m^2, separated by a $2 \cdot 0$ mm thick sheet of plastic having a relative permittivity of $2 \cdot 1$. What is its capacitance?

Answer. An expression for the capacitance of a parallel-plate capacitor having air between the plates is given in the solution to the last example. In this case there is a dielectric material between the plates so that ε_0 in the answer to the last example has to be replaced by $\varepsilon_0 \varepsilon_r$ to give the answer to this problem. Otherwise the problem is the same. Therefore

$$C = -\frac{\varepsilon_0 \varepsilon_r A}{d} = \frac{2 \cdot 1 \times 0 \cdot 1}{36\pi \times 10^9 \times 2 \cdot 0 \times 10^{-3}} = 9 \cdot 3 \times 10^{-10} \text{ F} = 930 \text{ pF}.$$

8.9. From a consideration of the electric field between two parallel uniformly charged surfaces with equal and opposite charge densities, derive expressions for the flux density and field intensity on each side of a boundary between air and a dielectric material of relative permittivity ε_r when the boundary is parallel to an equipotential line.

Answer. Consider the dielectric material filling half the space between two parallel charged surfaces as shown in Fig. 8.5. It is assumed that the two parallel charged sheets are of infinite extent so that there is a unif rm flux density between them. The flux density between the charged sheets will be uniform and equal to the charge density on the sheets as derived in Chapter 2 (page 15), therefore

$$D = \sigma.$$

It has been shown in Chapter 4 (page 64) that the effect of a dielectric material can be described mathematically by the introduction of a relative permittivity. Since the boundary of the dielectric is parallel to the charge sheets, the dielectric can have no influence on the charge distribution on the sheets so that the flux density is unaltered by the presence of the dielectric material. However, the field intensity inside the dielectric is reduced in proportion to the relative permittivity according to eqn. (4.7). Therefore the flux densities are the same on each side of the boundary,

$$D_a = D_b = D = \sigma,$$

and the field intensities are proportional to the permittivity

$$E_a = \frac{D}{\varepsilon_0}, \qquad\qquad E_b = \frac{D}{\varepsilon_0 \varepsilon_r},$$

because the relative permittivity of air is 1.

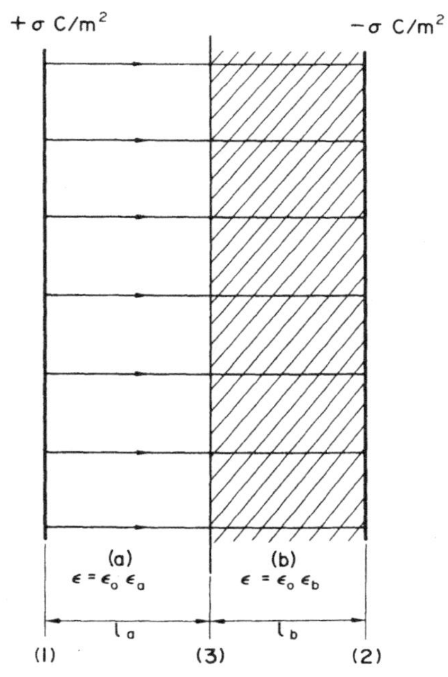

Fig. 8.5. A boundary parallel to an equipotential line between air and a dielectric material in the region between two parallel charge sheets.

8.10. From a consideration of the electric field between two parallel charged plates, derive expressions for the flux density and field intensity on each side of a boundary between air and a dielectric material of relative permittivity ε_r when the boundary is parallel to a flux line.

Answer. The parallel conductor system is shown in Fig. 8.6 with the dielectric material completely filling the gap between the plates so that the boundary is along a line of flux. As the source and sink for the field are conductors, they will be at the same potential all over and any variation from a uniform field pattern will have to be generated by a redistribution of the charges on the plates. It is assumed that there is a uniform charge density in each area of the plates but that these charge densities are different. Because the plates have a constant potential difference between them, the electric field intensity is constant in the region between the plates. It is given by

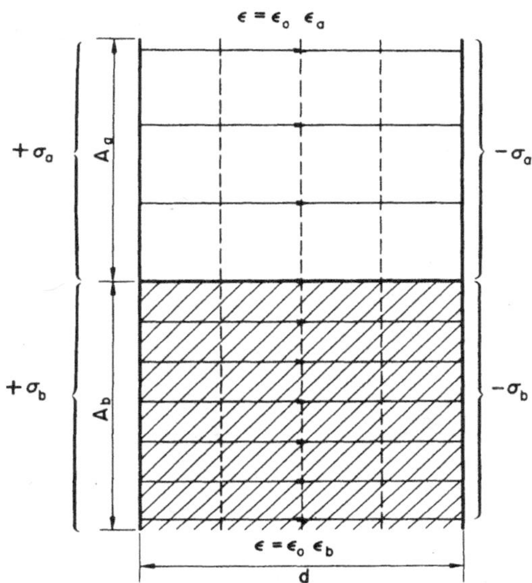

Fig. 8.6. A boundary parallel to a flux line between air and a
dielectric material in the region between two parallel
charge sheets.

$$E_a = E_b = E = \frac{\Phi}{d}.$$

The flux density is now determined from eqn. (4.7)(page 64), therefore

$$D_a = \epsilon_0 E = \epsilon_0 \frac{\Phi}{d}, \qquad\qquad D_b = \epsilon_0 \epsilon_r E = \epsilon_0 \epsilon_r \frac{\Phi}{d}.$$

8.11. If Fig. 8.5 represents a parallel-plate capacitor with two different
dielectric materials in the space between the plates, find the capacitance of
the system if the area of the plates is A.

Answer. The capacitance is determined from eqn. (8.8). The total flux
is given by

$$\Psi = DA = \sigma A.$$

Because the field intensity is uniform in each material, the potential
difference is given by the product of the field intensity and the distance.
Therefore the potential difference between each of the boundaries is given by

$$\phi_1 - \phi_3 = \Phi_a = E_a l_a = \frac{D}{\epsilon_0 \epsilon_a} l_a,$$

$$\phi_3 - \phi_2 = \Phi_b = E_b l_b = \frac{D}{\varepsilon_o \varepsilon_b}\, l_b.$$

The total potential difference between the plates (1) and (2) is the sum of
the potential difference across the two dielectric materials. Therefore

$$\Phi = \Phi_a + \Phi_b = D \left(\frac{l_a}{\varepsilon_o \varepsilon_a} + \frac{l_b}{\varepsilon_o \varepsilon_b} \right).$$

Therefore the capacitance is given by

$$C = \frac{\Psi}{\Phi} = \frac{A}{\left(\dfrac{l_a}{\varepsilon_o \varepsilon_a} + \dfrac{l_b}{\varepsilon_o \varepsilon_b} \right)} = \frac{\varepsilon_o \varepsilon_a \varepsilon_b A}{\varepsilon_b l_a + \varepsilon_a l_b}.$$

8.12. It is desired to increase the capacitance of a parallel-plate
capacitor by inserting a material of relative permittivity 5·0 into the space
between the plates. There is only sufficient dielectric material to fill
half the space between the plates. What is the most efficient shape of
dielectric to use?

Answer. Since the parallel-plate capacitor has a uniform field
between the plates, the most efficient dielectric shape will be with the
boundary either parallel to or perpendicular to the plates. The dielectric
will either be in the shape shown in Fig. 8.5 or that shown in Fig. 8.6.
For the shape shown in Fig. 8.5, the capacitance has been derived in the
answer to the last example, therefore for this shape

$$C = \frac{\varepsilon_o \varepsilon_a \varepsilon_b A}{\varepsilon_b l_a + \varepsilon_a l_b}.$$

But $\varepsilon_a = 1$, $\varepsilon_b = 5 \cdot 0$, $l_a = l_b = \tfrac{1}{2}d$, therefore

$$C = \frac{5 \varepsilon_o A}{3d} = 1 \cdot 67\, \frac{\varepsilon_o A}{d}.$$

For the shape shown in Fig. 8.6, which is the other alternative, the field
quantities have been derived in the solution to Example 8.10. Then the total
flux is given by

$$\Psi = A_a D_a + A_b D_b = \varepsilon_o \frac{\Phi}{d} (\varepsilon_a A_a + \varepsilon_b A_b).$$

But $\varepsilon_a = 1$, $\varepsilon_b = 5 \cdot 0$, $A_a = A_b = \tfrac{1}{2}A$, therefore

$$\Psi = \varepsilon_o\, \tfrac{1}{2}A\, \frac{\Phi}{d}\, (1 + 5) = \frac{3\varepsilon_o A \Phi}{d}$$

and
$$C = \frac{\Psi}{\Phi} = \frac{3\varepsilon_o A}{d}.$$

Therefore the system shown in Fig. 8.6 is the best way of using the dielectric material which increases the capacitance of the capacitor by three compared with there being no dielectric material.

8.13 Obtain from first principles an expression for the energy stored in the field of a parallel-plate capacitor.

Answer. The energy will be calculated using two methods. Firstly, the energy will be found from the mechanical energy needed to separate the plates of a parallel-plate capacitor, and, secondly, the energy will be found from the electrical energy needed to charge the capacitor.

Firstly - using the concept of mechanical energy - take a parallel-plate capacitor as shown in Fig. 8.3 with a charge density of $\sigma = Q/A$ on each plate. Then the force on the plate is given by eqn. (4.5) so that the total force on the plate is given by

$$F = \tfrac{1}{2}DEA = \tfrac{1}{2} \frac{\sigma^2}{\varepsilon_o} A.$$

If, while remaining electrically isolated, the spacing between the plates is increased from d to $d + \delta d$, the increase in stored energy is equal to the work done on the force, therefore

$$\delta W = \tfrac{1}{2} \frac{\sigma^2}{\varepsilon_o} A\delta d.$$

As the field between the plates is uniform, the energy will be uniformly distributed in the field so that the total energy is given by

$$W = \int_0^d \delta W = \tfrac{1}{2} \frac{\sigma^2}{\varepsilon_o} Ad = \frac{Q^2 d}{2\varepsilon_o A} = \frac{Q^2}{2C}.$$

Secondly - using the electrical energy used to charge the capacitor - assume that an electrical circuit is connected to charge up the capacitor. The voltage supplied by the battery will increase as the charge on the plates increases. The potential difference is related to the charge by the capacitance

$$V = \frac{Q}{C}.$$

Let the e.m.f. of the battery move a small element of charge δQ from the negative plate of the capacitor to the positive. The work done is given by

$$\delta W = V \delta Q.$$

The total work done in setting up the field is the integral of this expression. Therefore

$$W = \int_0^Q V dQ = \frac{1}{C} \int_0^Q Q dQ = \frac{Q^2}{2C}.$$

8.14. The base of a thunder-cloud may be assumed to be a charged plane area which forms a parallel-plate capacitor with the surface of the earth as the other conductor. If the potential difference between the cloud and earth is 200 MV, the base area of the cloud is 10^6 m^2 and the height of the base of the cloud is 3·0 km, find the energy stored in the electric field between the base of the cloud and earth.

Answer. The energy may be determined by using eqn. (8.9), therefore

$$W = \tfrac{1}{2}DE(\text{volume}) = \tfrac{1}{2}\varepsilon_0 E^2 Ad = \frac{(2 \times 10^8)^2 \times 10^6 \times 3 \times 10^3}{2 \times 36\pi \times 10^9 \times (3 \times 10^3)^2} = 5 \cdot 9 \times 10^7 \text{ J} = 59 \text{ MJ.}$$

PROBLEMS

8.1. A parallel-plate capacitor consists of two circular metal plates each of diameter 0·4 m, which are mounted with a 5·0 mm gap between them. The dielectric is air. If the potential difference between the plates is 9·0 V, find the charge on the plates. Work from first principles and neglect . fringing. Hence find the capacitance of the capacitor. (2·0 nC; 222 pF.)

8.2. A dielectric material of relative permittivity 9·0 is inserted to completely fill the gap between the plates of the capacitor of Problem 8.1. Calculate the new value of the capacitance and the new charge or potential difference if (a) the potential difference is maintained constant, (b) the charge is maintained constant. (2000 pF; 18 nC; 1·0 V.)

8.3. The base of a thunder-cloud may be regarded as a charged plane area of 10^8 m^2. The height of the cloud base above the earth is 1·0 km and the potential difference between them is 10^8 V. Calculate the total charge in the base of the thunder-cloud, the capacitance of the system, and the energy stored in the space between the cloud and the earth. (88 C; 0·88 µF; 4·4 GJ.)

8.4. A parallel-plate capacitor has a separation of 2·0 mm between the plates. A sheet of mica 1·0 mm thick, having a relative permittivity of 6·0, is placed between them. Find the mechanical force per unit area on each plate when a potential difference of 1·0 kV is maintained between them. (3·25 N/m².)

8.5. If the breakdown strength of air is 3·0 MV/m and that of mica is 50 MV/m, what voltage will the plates of the parallel-plate capacitor of Problem 8.4 withstand (a) without the mica in the gap, (b) with the 1·0 mm thick mica in the gap, (c) with 2·0 mm thick mica in the gap? (6 kV; 3·5 kV; 100 kV.)

8.6. A capacitor is formed by two parallel conducting plates, each of area 0·05 m², a distance 5·0 mm apart. Neglecting fringing, determine the charge on the plates, the force between them, and the energy stored in the electric field, when the plates are connected to a 200 V supply.
(17·7 nC; 354 μN; 1·77 μJ.)

8.7. Taking the capacitor of Problem 8.6, while the plates are still connected to a 200 V supply, the plate separation is increased to 10 mm. Find the energy now stored in the field and specify the energy interchanges that have taken place during the movement of the plates. (0·88 μJ.)

8.8 Taking the capacitor of Problem 8.6, the plate separation is increased to 10 mm while the plates are connected to a 200 V supply, the supply is disconnected, and the plate separation is increased to 15 mm. Find the charge on the plates, the potential difference between them, and the energy stored in the electric field. Where does the increase in energy come from?
(8·8 nC; 300 V; 1·32 μJ.)

8.9. For the parallel-plate capacitor with mixed dielectric shown in Fig. 8.5, sketch the equipotential lines for equal increments of potential if $\varepsilon_b = 2\varepsilon_a$.

8.10. Prove that the force between two parallel plates, in terms of the potential difference between them, is given by

$$F = \frac{\varepsilon_o \varepsilon_r A V^2}{2d^2},$$

where V is the potential difference between them, d is the spacing between the plates, A is the area of the plates, and ε_r is the relative permittivity of the dielectric.

SOLUTIONS

8.1. It is first necessary to assume that there is a surface charge density on the plates of $\pm\sigma$ C/m^2. Then the flux density between the plates is σ C/m^2 because it is a uniform field. The field intensity is given by

$$E = \frac{\sigma}{\varepsilon_0}$$

and the potential difference by the application of eqn. (8.4). Therefore

$$\Phi = \int_1^2 -\mathbf{E} \cdot d\mathbf{l} = \int_0^d \frac{\sigma}{\varepsilon_0}\, dl = \frac{\sigma d}{\varepsilon_0} = V.$$

The charge on the plates is given by

$$Q = \sigma \text{ (area)} = \frac{\varepsilon_0 V}{d} \text{ (area)}.$$

Inserting numbers from the problem gives

$$Q = \frac{9\cdot0 \times \pi \times 0\cdot04}{36\pi \times 10^9 \times 5\cdot0 \times 10^{-3}} = 2\cdot0 \times 10^{-9} \text{ C} = 2\cdot0 \text{ nC}.$$

The capacitance is given by eqn. (8.8),

$$C = \frac{\Psi}{\Phi} = \frac{Q}{V} = \frac{2\cdot0 \times 10^{-9}}{9\cdot0} = 2\cdot22 \times 10^{-10} \text{ F} = 222 \text{ pF}.$$

8.2. The effect of the material is to replace ε_0 by $\varepsilon_0\varepsilon_r$ in the expressions in the solution to Problem 8.1. Therefore

$$C = 222 \times 9\cdot0 \text{ pF} = 2000 \text{ pF}.$$

(a) The potential difference remains constant, therefore

$$Q = VC = 9\cdot0 \times 2\cdot0 \times 10^{-9} \text{ C} = 18 \text{ nC}.$$

(b) The charge remains constant, therefore

$$V = \frac{Q}{C} = \frac{2\cdot0 \times 10^{-9}}{2\cdot0 \times 10^{-9}} = 1\cdot0 \text{ V}.$$

8.3. The space between the base of the thunder-cloud and the surface of the earth will be a region of uniform field. Therefore, if the surface charge density is σ C/m^2, the potential difference is the same as that derived in the solution to Example 8.2,

$$\Phi = \frac{\sigma d}{\varepsilon_0}.$$

The total charge is given by

$$Q = \sigma(\text{area}) = \frac{\Phi\varepsilon_0}{d}(\text{area}) = \frac{10^8 \times 10^8}{10^3 \times 36\pi \times 10^9} = 88\cdot4 \text{ C.}$$

The capacitance is given by

$$C = \frac{\Psi}{\Phi} = \frac{88\cdot4}{10^8} \text{ F} = 0\cdot884 \text{ }\mu\text{F.}$$

The energy stored in a capacitor has been derived in the solution to Example 8.13, therefore the energy is given by

$$W = \tfrac{1}{2}CV^2 = \tfrac{1}{2} \times 88\cdot4 \times 10^{-8} \times 10^{16} = 44\cdot2 \times 10^8 \text{ J} = 4\cdot42 \text{ GJ.}$$

8.4. It is necessary to assume a surface charge density of σ C/m^2 on the plates of the capacitor. Then there will be a uniform flux density between the plates so that $D = \sigma$ C/m^2 and $E = \sigma/\varepsilon_0\varepsilon_r$ V/m. The potential difference is found from eqn. (8.4), therefore

$$\Phi = \int_0^{d_1} \frac{\sigma}{\varepsilon_0}dl + \int_{d_1}^{(d_2 + d_1)} \frac{\sigma}{\varepsilon_0\varepsilon_r}dl = \frac{\sigma d_1}{\varepsilon_0} + \frac{\sigma d_2}{\varepsilon_0\varepsilon_r} = \frac{\sigma}{\varepsilon_0}\left(d + \frac{d_2}{\varepsilon_r}\right).$$

Inserting numbers from the problem gives

$$\sigma = \frac{10^3}{36\pi \times 10^9(10^{-3} + 10^{-3}/6)} = \frac{1}{42\pi \times 10^3}.$$

The mechanical force per unit area on the charged surface is given by eqn. (4.4)(page 64), therefore

$$f = \frac{\sigma^2}{2\varepsilon_0} = \frac{36\pi \times 10^9}{2 \times (42\pi \times 10^3)^2} = 3\cdot25 \text{ N/m}^2.$$

8.5. The breakdown strength is the maximum permissible value of the E field. When there is air present somewhere between the plates of the capacitor, then the breakdown strength of air will be the governing factor. When the gap is completely filled with mica, then the breakdown strength of mica is the governing factor. The analytical formulae for a similar problem have already been obtained in the solution to Problem 8.4. Therefore when V is the potential difference between the plates:

(a) All air,

$$E = \frac{\sigma}{\varepsilon_0} = \frac{V}{d}.$$

Therefore $\quad V = Ed = 3\cdot0 \times 10^6 \times 2\cdot0 \times 10^{-3} = 6\cdot0 \times 10^3 \text{ V} = 6\cdot0 \text{ kV.}$

(b) Half-mica, half-air,

$$E = \frac{\sigma}{\varepsilon_0} = \frac{V}{d_1 + d_2/\varepsilon_r}.$$

Therefore

$$V = E(d_1 + d_2/\varepsilon_r) = 3 \cdot 0 \times 10^6 \times (1 + \tfrac{1}{6}) \times 10^{-3} = 3 \cdot 5 \times 10^3 \text{ V} = 3 \cdot 5 \text{ kV}.$$

(c) All mica,

$$E = \frac{\sigma}{\varepsilon_0 \varepsilon_r} = \frac{V}{d}.$$

Therefore $V = Ed = 50 \times 10^6 \times 2 \cdot 0 \times 10^{-3} = 1 \cdot 0 \times 10^5 \text{ V} = 100 \text{ kV}.$

8.6. Let there be a surface charge density on the plates of σ C/m^2. Because there is a uniform field, the potential difference between the plates is given by

$$\Phi = Ed = \frac{\sigma d}{\varepsilon_0}.$$

Therefore $\sigma = \dfrac{200}{36\pi \times 10^9 \times 5 \cdot 0 \times 10^{-3}} = 3 \cdot 54 \times 10^{-7} \text{ C/m}^2.$

The total charge is given by

$$Q = \sigma(\text{area}) = 3 \cdot 54 \times 10^{-7} \times 0 \cdot 05 = 1 \cdot 77 \times 10^{-8} \text{ C} = 17 \cdot 7 \text{ nC}.$$

The force per unit area is given by eqn. (4.4) (page 64), therefore the total force is given by

$$F = \tfrac{1}{2}\sigma^2(\text{area})/\varepsilon_0 = \tfrac{1}{2}(3 \cdot 54 \times 10^{-7})^2 \times 0 \cdot 05 \times 36\pi \times 10^9 = 3 \cdot 54 \times 10^{-4} \text{ N} = 253 \text{ }\mu\text{N}.$$

The energy stored in the field is given by eqn. (8.9). As there is a uniform field, integration over the volume is easy and substitution for D and E in terms of the charge density gives,

$$W = \tfrac{1}{2}\sigma^2(\text{volume})/\varepsilon_0 = \tfrac{1}{2}(3 \cdot 54 \times 10^{-7})^2 \times 0 \cdot 05 \times 5 \cdot 0 \times 10^{-3} \times 36\pi \times 10^9$$

$$= 1 \cdot 77 \times 10^{-6} \text{ J} = 1 \cdot 77 \text{ }\mu\text{J}.$$

8.7. If the potential difference between the plates is maintained while they are moved apart, the charge will change. Therefore

$$\sigma = \frac{\Phi \varepsilon_0}{d} = \frac{200}{10^{-2} \times 36\pi \times 10^9} = 1 \cdot 77 \times 10^{-7} \text{ C/m}^2.$$

The energy is given by

$$W = \tfrac{1}{2}\sigma^2(\text{volume})/\varepsilon_0 = \tfrac{1}{2}(1 \cdot 77 \times 10^{-7})^2 \times 0 \cdot 05 \times 10^{-2} \times 36\pi \times 10^9$$

$$= 8 \cdot 8 \times 10^{-7} \text{ J} = 0 \cdot 88 \text{ }\mu\text{J}.$$

The stored energy is reduced even though work has been done in moving the plates further apart against their mutual force of attraction. The charge on the plates is halved and energy is given up in moving the excess charge into the electrical supply.

8.8. The first part of this problem is the same as Problem 8.7, therefore the charge on the plates is the same as that given there. Therefore

$$Q = \sigma(\text{area}) = 1 \cdot 77 \times 10^{-7} \times 0 \cdot 05 = 8 \cdot 8 \times 10^{-9} \text{ C} = 8 \cdot 8 \text{ nC}.$$

If the supply is disconnected, the charge will remain constant. Therefore the potential difference is given by

$$\Phi = Ed.$$

But the field intensity will remain constant as the plates are moved because the charge density remains constant. Therefore

$$\Phi = \frac{200}{10} \times 15 = 300 \text{ V}.$$

The energy is given by

$$W = \tfrac{1}{2}QV = \tfrac{1}{2} \times 8 \cdot 8 \times 10^{-9} \times 300 = 1 \cdot 32 \times 10^{-6} \text{ J} = 1 \cdot 32 \ \mu\text{J}.$$

This shows another form in which the energy may be calculated. In this case the stored energy increases because of the work done in moving the plates further apart against the force of attraction between them. There is no electrical connection to the system so that the only energy interchange is through the mechanical energy used in moving the plates apart.

8.9. For this problem, let $\varepsilon_a = 1$ and let $\varepsilon_b = \varepsilon_r = 2 \cdot 0$. If the equal increments of potential difference are chosen so that there is not an integral number of equipotential drops in the length l_a, the answer to this problem will be as shown in Fig. 8.7. If, as an alternative, it is chosen so that there are exactly two equipotential drops in the length l_a, then there will only be one in the length l_b and the diagram will be slightly ambiguous. For this second possibility, if $\delta\Phi$ is the potential difference between equipotential lines, then

$$E_a l_a = 2\delta\Phi, \qquad\qquad E_b l_b = \delta\Phi,$$

and there will only be one equipotential line in the diagram other than the boundaries.

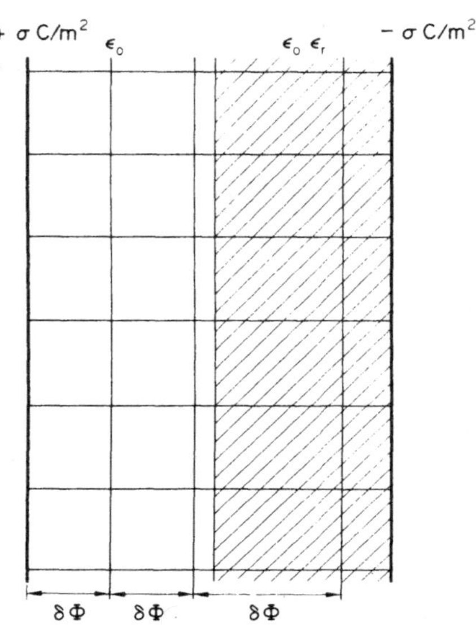

Fig. 8.7. The field pattern for equal increments of potential difference with different media in a uniform field. The solution to Problem 8.9.

8.10. From the solution to Problem 8.1 the surface charge density is given by

$$\sigma = \frac{V\epsilon_0\epsilon_r}{d}.$$

The mechanical force per unit area on the charged surface is given by eqn. (4.4)(page 64), therefore the total force is given by

$$F = \frac{\sigma^2 A}{2\epsilon_0\epsilon_r} = \frac{\epsilon_0\epsilon_r A V^2}{2d^2}.$$

FURTHER PROBLEMS

8.11. Two parallel charged plates are a distance 0·1 m apart. They have equal and opposite uniform charge densities of 10 nC/m². Find the potential difference between the plates.   (113 V.)

8.12. A parallel-plate capacitor consists of two metal plates 5·0 mm apart. If the potential difference between the plates is 120 V, find the charge density on each plate.   (0·21 µC/m².)

8.13. A parallel-plate capacitor consists of two metal plates each of area 0·1 m², separated by a 1·0 mm thick sheet of plastic of relative permittivity 2·0. Find the capacitance of the capacitor.   (1770 pF.)

8.14. Prove that the potential function of a point charge of strength q is

$$\phi = \frac{q}{4\pi\varepsilon_0 r}.$$

8.15. A particular electric field has a potential function

$$\phi = \frac{c \cos \theta}{2\pi\varepsilon_0 r}.$$

Find an expression for the amplitude and direction of the flux density vector. $(c/(2\pi r^2)$, angle 2θ with respect to the x-axis.)

8.16. A parallel-plate capacitor consists of two metal plates each having a surface coating of 1·0 mm of dielectric material of relative permittivity 9·0 and another coating of 0·5 mm of material of relative permittivity 2·0 on top of that. If each plate is a square of side 0·2 m and they are mounted 10 mm apart with the surface coating on the inside faces, find the capacitance of the capacitor.   (46 pF.)

8.17. It is desired to increase the capacitance of a parallel-plate capacitor by inserting a material of relative permittivity 5·0 into the space between the plates. There is only sufficient dielectric material to fill one-third of the space between the plates. Find the most efficient configuration of dielectric to use and find the percentage increase in capacitance using this configuration.   (133%.)

8.18. Prove that the energy stored in a unit length of coaxial cable with a charge of q C/m on its inner conductor is given by

$$W = \frac{q^2}{4\pi\varepsilon_0} \log_e \frac{b}{a},$$

where a is the radius of the inner conductor and b is the inside radius of the outer conductor.

8.19. A capacitor is formed by two parallel conducting plates, each of area $0 \cdot 2$ m^2, 10 mm apart. (a) Neglecting fringing, determine the charge on the plates, the force between them, and the energy stored in the electric field when they are connected to a 120 V supply. (b) A sheet of glass $5 \cdot 0$ mm thick and relative permittivity $4 \cdot 0$ is now inserted into the space between the plates while they are still connected to the 120 V supply. What are the new values of charge, force, and energy? (c) The plates are now disconnected from the supply and allowed to close up until they touch each side of the glass sheet. Find the new value of the potential difference and the stored energy.
$\big($(a) $21 \cdot 2$ nC; 127 μN, $1 \cdot 27$ μJ; (b) 34 nC; 326 μN; $2 \cdot 04$ μJ; (c) 24 V, $0 \cdot 41$ μJ.$\big)$

8.20. A particular electrostatic voltmeter operates by measuring the separation between the plates of a parallel-plate capacitor when one plate is suspended by a system exerting a constant force on the plate. Find the relationship between the separation and the potential difference for this system when the plates are of area A and the support exerts a constant force F in addition to supporting any gravitational forces on the plate.
$(d = V\sqrt{(\varepsilon_0 A/2F)}.)$

FURTHER SOLUTIONS

8.11. Since it may be assumed that there is a uniform field between the charged plates and using the result given in the solution to Example 8.2,

$$\Phi = Ed = \frac{\sigma d}{\varepsilon_o} = 10 \times 10^{-9} \times 0 \cdot 1 \times 36\pi \times 10^9 = 113 \text{ V}.$$

8.12. Similarly to the solution to Problem 8.11,

$$\sigma = \frac{\Phi\varepsilon_o}{d} = \frac{120}{5 \cdot 0 \times 10^{-3} \times 36\pi \times 10^9} = 2 \cdot 1 \times 10^{-7} \text{ C/m}^2 = 0 \cdot 21 \text{ } \mu\text{C/m}^2.$$

8.13. Using the result in the solution to Example 8.7,

$$C = \frac{\varepsilon_o \varepsilon_r A}{d} = \frac{2 \cdot 0 \times 0 \cdot 1}{10^{-3} \times 36\pi \times 10^9} = 1 \cdot 77 \times 10^{-9} \text{ F} = 1770 \text{ pF}.$$

8.14. In order to find the potential function for any field it is necessary to use eqn. (8.4). A diagram showing circles of equipotential lines around a point charge is shown in Fig. 8.8. The equipotential surfaces

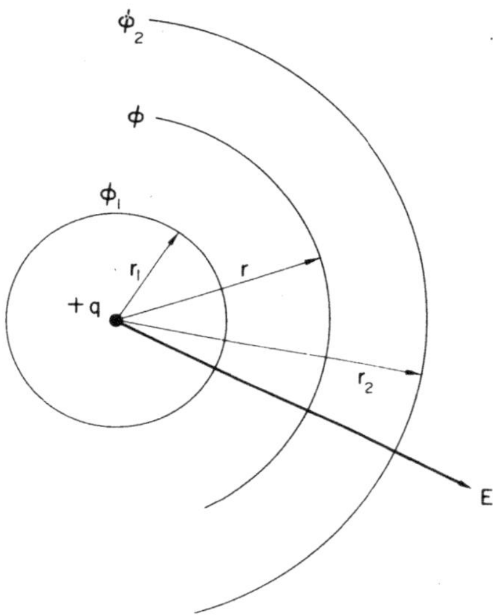

Fig. 8.8. Some equipotential lines due to a point charge.

around the point charge will be spherical because they are orthogonal to the direction of the field intensity vector which is radial. We will obtain the potential difference between two points on a radial flux line out of the point source at radius r_1 and r_2 respectively. The field intensity due to a point source is given by

$$\mathbf{E} = \mathbf{U}_r \frac{q}{4\pi\epsilon_0 r^2}.$$

The rise in potential in moving from radius r_2 to radius r_1 is

$$\phi_1 - \phi_2 = \int_2^1 -\mathbf{E}\cdot d\mathbf{l} = \int_{r_2}^{r_1} -\frac{q}{4\pi\epsilon_0 r^2}\, dr = \frac{q}{4\pi\epsilon_0}\left(\frac{1}{r_1} - \frac{1}{r_2}\right).$$

If the zero of potential function is taken to be at an infinite distance away, then $\phi_2 = 0$ when $r_2 = \infty$. Then at any other radius r, the potential function is given by

$$\phi = \frac{q}{4\pi\epsilon_0 r}.$$

8.15.   As the potential function is given in polar coordinates, the field intensity will be obtained from eqn. (8.7). Therefore

$$E_r = -\frac{\partial\phi}{\partial r} = \frac{c\cos\theta}{2\pi\epsilon_0 r^2}, \qquad\qquad E_\theta = -\frac{1}{r}\frac{\partial\theta}{\partial\theta} = \frac{c\sin\theta}{2\pi\epsilon_0 r^2}.$$

Therefore the resultant amplitude of the field intensity is given by

$$E = \frac{c}{2\pi\epsilon_0 r^2} \qquad \text{and} \qquad D = \frac{c}{2\pi r^2}.$$

The angle with the x-axis is given by

$$\alpha = \theta + \tan^{-1}\frac{E_\theta}{E_r} = 2\theta.$$

This is the field due to a doublet of strength c.

8.16.   Let the plates of the capacitor have a charge density of σ C/m^2. Then

$$\Phi = \frac{\sigma}{\epsilon_0}\left(\frac{2\cdot0}{9\cdot0} + \frac{1\cdot0}{2\cdot0} + 7\cdot0\right) \times 10^{-3} = \frac{\sigma \times 10^{-3} \times 7\cdot72}{\epsilon_0}.$$

Therefore

$$C = \frac{\sigma A}{\Phi} = \frac{\epsilon_0 \times 0\cdot04}{10^{-3} \times 7\cdot72} = 4\cdot6 \times 10^{-11}\ \text{F} = 46\ \text{pF}.$$

8.17.   This problem is similar to Example 8.12 and the formulae derived in the solution to that example will be quoted here. When the boundary is parallel to the equipotential,

$$C = \frac{\epsilon_0 \epsilon_a \epsilon_b}{\epsilon_b l_a + \epsilon_a l_b}$$

For this problem, $\varepsilon_a = 1$, $\varepsilon_b = 5{\cdot}0$, $l_a = \tfrac{2}{3}d$, $l_b = \tfrac{1}{3}d$, therefore

$$C = \frac{15\varepsilon_o A}{11d}.$$

When the boundary is parallel to the flux line,

$$C = \frac{\varepsilon_o}{d}(\varepsilon_a A_a + \varepsilon_b A_b).$$

For this problem, $\varepsilon_a = 1$, $\varepsilon_b = 5{\cdot}0$, $A_a = \tfrac{2}{3}A$, $A_b = \tfrac{1}{3}A$, therefore

$$C = \frac{7\varepsilon_o A}{3d}.$$

Therefore the system similar to that shown in Fig. 8.6 gives the largest capacitance. Without any dielectric present, the capacitance is given by

$$C_o = \frac{\varepsilon_o A}{d}.$$

The percentage increase in capacitance is given by

$$\frac{C - C_o}{C_o} \times 100 = 133 \text{ per cent.}$$

The system having the dielectric material completely filling the gap between the plates for one-third of the area of the plates is the most efficient giving a 133 per cent increase in capacitance. The alternative system having a reduced thickness of dielectric over the whole area of the plates only gives 36 per cent increase in capacitance.

8.18. The geometry of the problem is shown in Fig. 8.9. The value of the

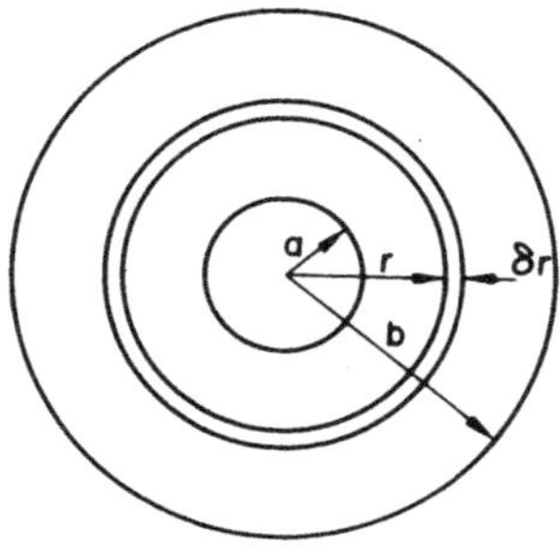

Fig. 8.9. Ilustrating the solution to Problem 8.18, the geometry of a coaxial cable.

field in the space between the two conductors is given by eqn. (2.15) (page 14). The stored energy is given by eqn. (8.9). For the small annular element of volume of radial thickness δr,

$$\delta W = \frac{1}{2\varepsilon_0} \left(\frac{q}{2\pi r}\right)^2 2\pi r \delta r.$$

Integrating this expression between the limits of the field gives the total energy stored,

$$W = \frac{q^2}{4\pi\varepsilon_0} \int_a^b \frac{1}{r}\, dr = \frac{q^2}{4\pi\varepsilon_0} \log_e \frac{b}{a}.$$

8.19. (a) $Q = \sigma Q = \dfrac{\Phi\varepsilon_0 A}{d} = \dfrac{120 \times 0.2}{10^{-2} \times 36\pi \times 10^9} = 2.12 \times 10^{-8}$ C $= 21.2$ nC,

$$F = \frac{\sigma^2 A}{2\varepsilon_0} = \frac{\Phi^2\varepsilon_0 A}{2d^2} = \frac{120 \times 120 \times 0.2}{2 \times 10^{-4} \times 36\pi \times 10^9} = 1.27 \times 10^{-4}\ \text{N} = 127\ \mu\text{N}.$$

$$W = \tfrac{1}{2}Q\Phi = \tfrac{1}{2} \times 2.12 \times 10^{-8} \times 120 = 1.27 \times 10^{-6}\ \text{J} = 1.27\ \mu\text{J}.$$

(b) Making use of Fig. 8.5,

$$\Phi = \frac{\sigma}{\varepsilon_0}\left(l_a + \frac{l_b}{\varepsilon_r}\right) = \frac{\sigma}{\varepsilon_0}\left(5 + \frac{5}{4}\right) \times 10^{-3} = \frac{6.25 \times 10^{-3}\sigma}{\varepsilon_0}.$$

Therefore

$$Q = \sigma A = \frac{\Phi\varepsilon_0 A}{6.25 \times 10^{-3}} = \frac{120 \times 0.2}{6.25 \times 10^{-3} \times 36\pi \times 10^9} = 3.4 \times 10^{-8}\ \text{C} = 34\ \text{nC},$$

$$F = \frac{\sigma^2 A}{2\varepsilon_0} = \frac{\Phi^2\varepsilon_0 A}{2 \times (6.25 \times 10^{-3})^2} = 3.26 \times 10^{-4}\ \text{N} = 326\ \mu\text{N}.$$

$$W = \tfrac{1}{2}Q\Phi = \tfrac{1}{2} \times 3.4 \times 10^{-8} \times 120 = 2.04 \times 10^{-6}\ \text{J} = 2.04\ \mu\text{J}.$$

(c) The total charge is unchanged, therefore $Q = \sigma A$ and

$$\Phi = \frac{\sigma d}{\varepsilon_0\varepsilon_r} = \frac{Qd}{A\varepsilon_0\varepsilon_r} = \frac{3.4 \times 10^{-8} \times 5.0 \times 10^{-3} \times 36\pi \times 10^9}{0.2 \times 4.0} = 24\ \text{V},$$

$$W = \tfrac{1}{2}Q\Phi = \tfrac{1}{2} \times 3.4 \times 10^{-8} \times 24 = 4.1 \times 10^{-7}\ \text{J} = 0.41\ \mu\text{J}.$$

8.20.   There is a constant force F. For the same situation, an expression has been derived for the force F as the solution to Problem 8.10. From that solution

$$F = \frac{\varepsilon_0\varepsilon_r A V^2}{2d^2}.$$

Rearranging this equation gives

$$\frac{d^2}{V^2} = \frac{\varepsilon_o A}{2F}.$$

Therefore

$$d = V \sqrt{\left(\frac{\varepsilon_o A}{2F}\right)}.$$

The displacement is directly proportional to the voltage.

CHAPTER 9

Potential Function

THEORY

In most of the practical problems which can be solved with the aid of field theory, it is the potential difference which is known and it is the charge or flux which is to be found. However, mathematically it is easier to calculate an expression for the potential difference in terms of some charge or flux distribution than to calculate the charge distribution from the potential difference. The potential function is calculated from the flux distribution through the field intensity from eqn. (8.4). The usefulness of potential function is similar to that of flux function; potential function values due to a number of different sources may be summed at each point in the field so as to obtain the potential function pattern due to the combined field. A simple arithmetical sum of potential function values replaces the much more complicated vector addition of the various field intensities at each point. Expressions for the potential function of various simple charge distributions are given below. Their derivation is given in the answer to Example 9.4, where the student is invited to derive these expressions for himself.

Point charge of strength Q,

$$\phi = \frac{Q}{4\pi\varepsilon_0 r}. \tag{9.1}$$

Dipole of strength m,

$$\phi = \frac{m \cos\theta}{4\pi\varepsilon_0 r^2}. \tag{9.2}$$

Line charge of strength q,

$$\phi_1 - \phi_2 = \frac{q}{2\pi\varepsilon_0} \log_e \frac{r_2}{r_1}. \tag{9.3}$$

Doublet of strength c,

$$\phi = -\frac{c \cos\theta}{2\pi\varepsilon_0 r}. \tag{9.4}$$

The *method of images* may be used to determine the field whenever a charge is situated near to a plane conducting surface. The field due to any combination of charges near to a plane conducting surface may be determined by replacing the conducting surface with a mirror system of charges on the opposite side of the surface the same perpendicular distance from the surface and of opposite charge to each of the original charges.

Because the flux function lines and the equipotential lines are everywhere perpendicular to one another in the field, their roles may be interchanged. Such systems are the *duals* of each other. The consideration

of duals is useful when a shape that is a flux boundary for a simple charge
distribution is presented as a potential boundary in a problem or vice versa.

EXAMPLES

9.1.   A spherically symmetric charge system consists of a solid
conducting sphere of radius 0·2 mm at the centre of a sphere of dielectric
material of relative permittivity 6·0 and outer radius 2·0 mm which is
surrounded by a layer of dielectric material of relative permittivity 2·0 to
an outer radius 4·0 mm. If a charge of 1·0 nC is placed on the solid
conducting sphere, find the absolute potential of the surface of the
conducting sphere and of the outer surface of the outer dielectric material.

Answer. The charge uniformly distributed on the surface of the solid
conducting sphere may be replaced by a point charge of the same value. A
general field due to a point charge is shown in Fig. 8.8. In order to find
the absolute potential, it is necessary to consider that the zero of potential
is at an infinite distance away from the point charge so that the potential
difference is obtained by integrating from an outer radius of infinity to the
point under consideration. In this problem there are three regions of field
having different relative permittivities. The boundaries between the different
materials lie along equipotential lines so that the flux density is the same
on each side of the boundary. The flux density due to a point charge is
given by eqn. (2.2)(page 12). Therefore, working in symbols,

$$D = \frac{Q}{4\pi r^2} \quad \text{and} \quad E = \frac{Q}{4\pi \varepsilon_0 \varepsilon_r r^2}.$$

The potential difference between any two radii r_1 and r_2 is given by
substitution into eqn. (8.4)(page 141), therefore

$$\Phi = \phi_1 - \phi_2 = \int_2^1 - \mathbf{E} \cdot d\mathbf{l} = \int_{r_2}^{r_1} - \frac{Q}{4\pi \varepsilon_0 \varepsilon_r r^2} \, dr = \frac{Q}{4\pi \varepsilon_0 \varepsilon_r} \left(\frac{1}{r_1} - \frac{1}{r_2} \right).$$

Inserting numbers from the problem, the absolute potential of the outer
surface of the outer sphere is given by

$$\Phi_1 = \frac{10^{-9}}{4\pi \varepsilon_0} \left(\frac{1}{4\cdot 0 \times 10^{-3}} - \frac{1}{\infty} \right) = \frac{10^{-9} \times 36\pi \times 10^9}{4\pi \times 4\cdot 0 \times 10^{-3}} = 2\cdot 25 \times 10^3 \text{ V} = 2\cdot 25 \text{ kV}.$$

The absolute potential of the surface of the inner sphere is given by

$$\Phi_2 = \text{(p.d. across the inner dielectric)}$$
$$+\text{(p.d. across the outer dielectric)} + \Phi_1.$$

Therefore

$$\Phi_2 = \frac{Q}{4\pi\varepsilon_0 \times 6.0}\left(\frac{1}{0.2} - \frac{1}{2.0}\right) \times 10^3 + \frac{Q}{4\pi\varepsilon_0 \times 2.0}\left(\frac{1}{2.0} - \frac{1}{4.0}\right) \times 10^3 + \Phi_1$$

$$= \frac{10^{-9} \times 36\pi \times 10^9 \times 10^3}{4\pi}(0.75 + 0.125) + 2.25 \times 10^3$$

$$= (7.88 + 2.25) \times 10^3 \text{ V} = 10.13 \text{ kV.}$$

9.2.   Find the capacitance of an isolated metal sphere having a radius of 1.0 m.

Answer. Normally the capacitance is only measurable for a system of at least two isolated conductors. However, the definition of capacitance gives

$$\text{capacitance} = \frac{\text{flux}}{\text{potential difference}}.$$

In the case of the isolated metal sphere, if it is given a charge Q, it is possible to specify an absolute potential for the surface of the sphere so that we can imagine the capacitor consisting of the sphere and an earth an infinite distance away as the other conductor. It is possible to find the capacitance of this sphere-earth system. The absolute potential of a spherical system has been derived in the solution to the last example, therefore

$$\Phi = \frac{Q}{4\pi\varepsilon_0}\left(\frac{1}{r} - \frac{1}{\infty}\right) = \frac{Q}{4\pi\varepsilon_0 r}.$$

The capacitance is given by eqn. (8.8)(page 142), therefore

$$C = \frac{Q}{\Phi} = 4\pi\varepsilon_0 r = \frac{4\pi \times 1.0}{36\pi \times 10^9} = 1.11 \times 10^{-10} \text{ F} = 111 \text{ pF.}$$

9.3.   If the electric breakdown strength of air is 3.0 MV/m, determine the highest voltage that can be applied to a spherical conductor of radius 20 mm before breakdown of the air around the conductor begins to occur.

Answer. Let the spherical conductor have a charge Q. It can be assumed that the system will behave as if there were a point charge Q at the centre of the spherical conductor. Then from the solution to the last example, the absolute potential of the surface of the sphere is given by

$$\Phi = \frac{Q}{4\pi\varepsilon_0 r}.$$

Breakdown will occur if the maximum field intensity in the field exceeds the breakdown strength of the air. For the spherical conductor, the maximum field intensity occurs on the surface of the sphere, therefore

$$E_{max} = \frac{Q}{4\pi\varepsilon_0 r^2}.$$

Therefore

$$\Phi = E_{max}\, r = 3 \cdot 0 \times 10^6 \times 20 \times 10^{-3} = 60 \times 10^3 \text{ V} = 60 \text{ kV}.$$

9.4.   Derive, from first principles, the potential function expressions given in eqns. (9.1) - (9.4).

Answer. (1) A point charge of strength Q. If the zero of potential function is taken to be at an infinite distance away from the point charge, the value of the potential function is the same as the absolute potential in the same field. An expression for the absolute potential in the field due to a point charge has been derived in the solutions to Examples 9.1 and 9.2. Therefore eqn. (9.1) is the same as the expression for the absolute potential given in the solution to Example 9.2.

(2) A dipole of strength m. A dipole is two equal and opposite point charges a small distance apart such that the product of their charge strength and their separation is constant and equal to the strength of the dipole. Therefore if the magnitude of the charge is q' and the distance between the charges is d', the product $q'd' = m$ remains finite even though the charge q' becomes infinitely large and the separation d' becomes infinitely small. Consider Fig. 9.1, which shows a point a distance r from a dipole of moment $m = q'd'$. The potential function at P due to the dipole will be the sum of the potential function values due to each point charge which are given by eqn. (9.1). Therefore the potential function of the combination is given by

$$\phi = \frac{q'}{4\pi\varepsilon_0}\left(\frac{1}{r_1} - \frac{1}{r_2}\right) = \frac{q'}{4\pi\varepsilon_0}\left(\frac{r_2 - r_1}{r_1 r_2}\right).$$

If the distance d' is small compared with r, then

$$\frac{r_2 - r_1}{r_1 r_2} \approx -\frac{d'\cos\theta}{r^2}$$

and the potential function is given by

$$\phi = -\frac{m\cos\theta}{4\pi\varepsilon_0 r^2}$$

which is eqn. (9.2).

(3) A line charge of strength q. Consider a line filament of uniform charge density. This is the two-dimensional problem with the geometry

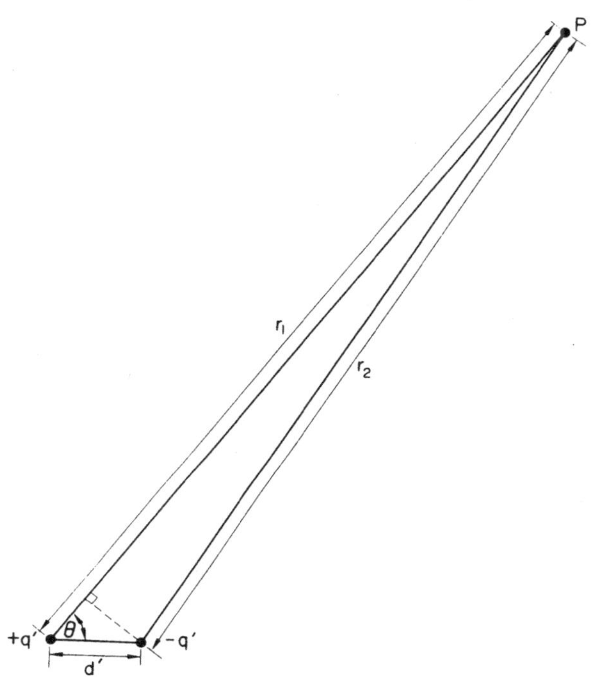

Fig. 9.1. A point in the field of a dipole.

shown in Fig. 8.8 (page 162). The flux density from such a line source is
given in eqn. (2.15)(page 14). Therefore

$$E = \frac{q}{2\pi\varepsilon_0 r}.$$

First the potential difference will be found using eqn. (8.4)(page 141),

$$\Phi = \phi_1 - \phi_2 = \int_2^1 - \mathbf{E} \cdot d\mathbf{l} = \int_{r_2}^{r_1} \frac{q}{2\pi\varepsilon_0 r}\, dr = \frac{q}{2\pi\varepsilon_0} \log_e \frac{r_2}{r_1}.$$

From this result it is seen that no meaningful result can be obtained by
taking $r_2 = \infty$, so that there is no absolute value of potential for the line
charge. The potential function has to be quoted for the zero of potential at
a particular radius. Therefore, if $\phi = 0$ at $r_2 = a$, then

$$\phi = \frac{q}{2\pi\varepsilon_0} \log_e \frac{a}{r}.$$

(4) Doublet of strength c. A doublet consists of a line

source parallel to a line sink of equal strength where the distance between
them becomes infinitely small and the strength of the sink becomes infinitely
large; the product of the strength of the source and the separation between
the source and sink remains constant and equal to the strength of the doublet.
Consider the line source and line sink a distance $2d$ apart as shown in Fig.
9.2. Let the zero of potential function be at a distance s from the doublet.

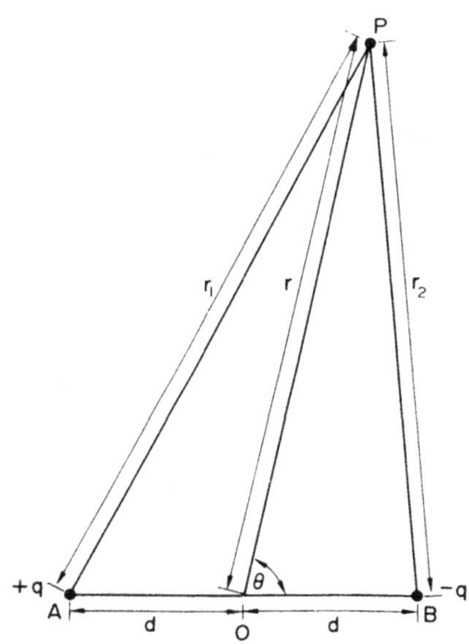

Fig. 9.2.   Illustrating the geometry for the calculation of the
potential function of a doublet.

If s is sufficiently large, there will be a negligible difference between s_1,
the large distance to the line source, and s_2, the large distance to the line
sink. The potential function at the point is given by the sum of the
potential functions due to the source and sink individually, therefore

$$\phi = \frac{q}{2\pi\varepsilon_0} \log_e \frac{s_1}{r_1} + \frac{-q}{2\pi\varepsilon_0} \log_e \frac{s_2}{r_2} = \frac{q}{2\pi\varepsilon_0} \log_e \frac{s_1 r_2}{r_1 s_2}.$$

Because s is very large, $s_1 \approx s_2$ and the potential function becomes

$$\phi = \frac{q}{2\pi\varepsilon_0} \log_e \frac{r_2}{r_1}.$$

For the doublet, $r_1 \approx r_2$, and the expression of $\log_e(r_2/r_1)$ becomes difficult
to visualize. In order to put this expression into a different form as in

eqn. (9.4), reference must be made to Fig. 9.2. Applying the cosine law to triangles APO and BPO gives

$$r_1^2 = r^2 + d^2 + 2rd \cos \theta,$$

$$r_2^2 = r^2 + d^2 - 2rd \cos \theta.$$

Subtracting these equations and dividing through by r_1^2 gives

$$\frac{r_2^2}{r_1^2} = 1 - \frac{4rd \cos \theta}{r_1^2}.$$

As the second term on the right-hand side of this equation is small, it will introduce negligible error to substitute $r_1 \approx r$ in this term, giving

$$\frac{r_2^2}{r_1^2} = 1 - \frac{4d \cos \theta}{r}.$$

The expression for the potential function now has to be put into the form

$$\phi = \frac{q}{4\pi\varepsilon_0} \log_e \frac{r_2^2}{r_1^2}.$$

Substitution gives an expression of the form, $\log_e(1-\delta)$. However, $\delta \ll 1$ and the expression may be simplified to give

$$\log_e(1-\delta) = -\delta.$$

Therefore the potential function becomes

$$\phi = -\frac{qd \cos \theta}{\pi\varepsilon_0 r} = -\frac{c \cos \theta}{2\pi\varepsilon_0 r},$$

which is eqn. (9.4).

9.5.   Find the capacitance of 100 m length of coaxial cable which consists of a circular inner conductor of 1·6 mm diameter surrounded by insulation of relative permittivity 2·0 surrounded by a conducting sheath of inside diameter 17·6 mm. The cable is circular in section and uniform along its length.

Answer. As the system is circularly symmetric in a two-dimensional system, it is necessary to postulate a uniform charge density on the centre conductor of q C/m. The potential difference between two radii in the field due to a line charge has been derived in the solution to the last example to be

$$\Phi = \frac{q}{2\pi\varepsilon_0\varepsilon_r} \log_e \frac{r_2}{r_1}.$$

The total flux in a length d of the cable is

$$\Psi = qd.$$

Therefore the capacitance is given by substitution into eqn. (8.8)(page 142),

$$C = \frac{\Psi}{\Phi} = \frac{2\pi\varepsilon_0\varepsilon_r d}{\log_e(r_2/r_1)}.$$

Inserting numbers into this equation, we have $\varepsilon_r = 2\cdot 0$, $(r_2/r_1) = 11$, and $d = 100$ m, therefore

$$C = \frac{2\pi \times 2\cdot 0 \times 100}{36\pi \times 10^9 \times 2\cdot 4} = 4\cdot 63 \times 10^{-9} \text{ F} = 4630 \text{ pF}.$$

9.6.   A long straight conductor of radius a having a charge density q runs parallel to a plane conducting surface (the ground) with its centre a distance d above the surface. Assume that the radius of the conductor is negligibly small compared with its height above the ground. Find (a) the flux density at the surface of the ground immediately below the conductor, (b) the potential difference between the surface of the conductor and the surface of the ground, and (c) the capacitance per unit length of the conductor-ground system.

Answer. The line charge parallel to a plane conducting surface can be analysed using the method of images. Take an image charge of equal and opposite strength on the opposite side of the plane conducting surface and an equal length away and remove the plane conducting surface. The line charge with its image is shown in Fig. 9.3.

(a) The flux density immediately below the conductor at the surface of the conducting plane will now be given by the vector sum of the flux densities due to the line charge and its image. These flux densities are parallel, so that

$$D = \frac{q}{2\pi d} - \frac{-q}{2\pi d} = \frac{q}{\pi d} \text{ C/m}^2.$$

(b) Referring to Fig. 9.2, the potential function value at any point in the field due to a line source parallel to an equal and opposite line sink has been derived in the solution to Example 9.4(4) to be

$$\phi = \frac{q}{2\pi\varepsilon_0} \log_e \frac{r_2}{r_1}.$$

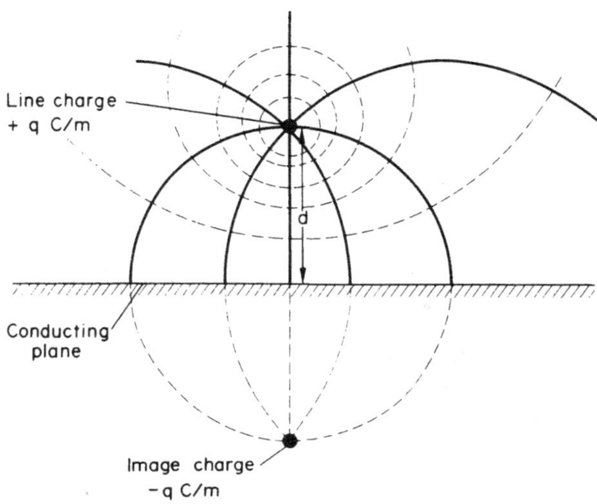

Fig. 9.3. Flux lines and equipotential lines for a line source
parallel to and near to a perfectly conducting plane surface.

At the surface of the conducting wire, $r_2 = 2d - a$ and $r_1 = a$. On the
conducting ground, $r_1 = r_2$ and the potential value is zero. Therefore the
potential difference between the surface of the wire and the surface of the
ground is given by

$$\Phi = \frac{q}{2\pi\epsilon_0} \log_e \frac{2d - a}{a} \approx \frac{q}{2\pi\epsilon_0} \log_e \frac{2d}{a}.$$

(c) The capacitance is given by substitution into eqn. (8.8). For a
unit length of conductor, the flux is given by

$$\Psi = q.$$

Therefore
$$C = \frac{\Psi}{\Phi} = \frac{2\pi\epsilon_0}{\log_e (2d/a)}.$$

9.7.   Find an exact expression for the capacitance of a right circular
cylinder of radius p whose axis is parallel to and distant s from a plane
conducting surface.

Answer. The equipotential lines due to a line source parallel to an
equal line sink are circular. Therefore the circular cylinder parallel to a
plane surface may be represented by a line source and its image line sink as
shown in Fig. 9.4. The surface of the cylinder coincides with one of the
circular equipotential lines shown in the figure and the plane surface

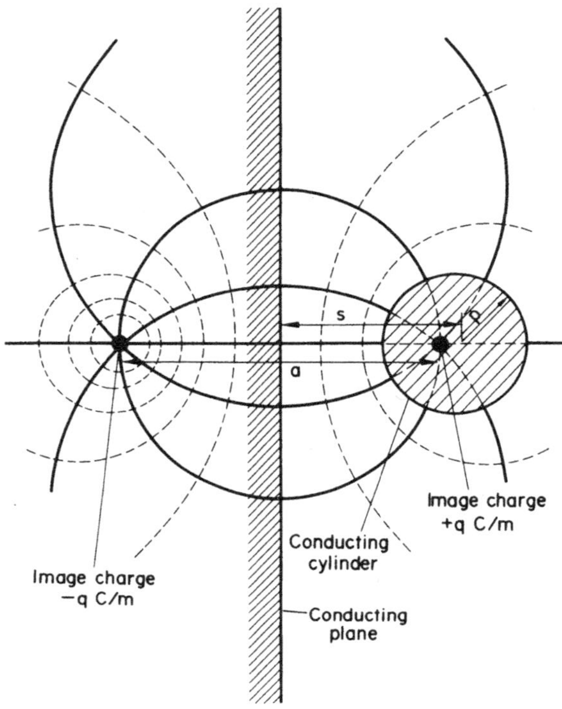

Fig. 9.4. Flux lines and equipotential lines for a line source
parallel to the axis of a perfectly conducting cylinder, which
is the same field pattern as for a perfectly conducting
cylinder parallel to a plane conducting surface.

coincides with the equipotential line half-way between the line source and
its image; both source and sink are images. The difficulty is that the
centre of the equipotential circles does not coincide with the line of the
line source. It is necessary to find an expression for the centre of the
equipotential circle. Consider a line source distant a from an equal line
sink as shown in Fig. 9.5. The equation for the equipotential line is

$$\phi = \frac{q}{2\pi\varepsilon_0} \log_e \frac{r_2}{r_1},$$

where ϕ is a constant. Rearranged, this becomes

$$\frac{r_2^2}{r_1^2} = \exp \frac{4\pi\varepsilon_0 \phi}{q} = c \text{ (say).}$$

From the geometry of Fig. 9.5,

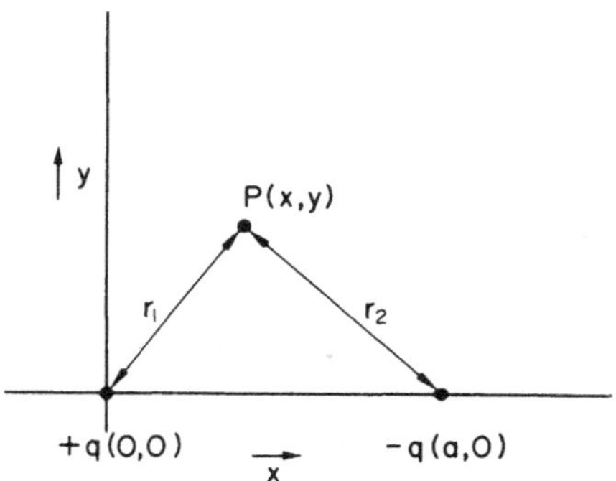

Fig. 9.5. Dimensions of a point relative to a line source and
a line sink.

$$r_1^2 = x^2 + y^2, \qquad\qquad r_2^2 = (a - x)^2 + y^2.$$

Therefore
$$(a - x)^2 + y^2 = c(x^2 + y^2),$$

which is the equation of a circle. Rewriting this equation gives

$$\left(x + \frac{a}{c - 1} \right)^2 + y^2 = \frac{ca^2}{(c - 1)^2},$$

which shows that the circle has a radius of $ac^{\frac{1}{2}}/(c - 1)$ and centre
coordinates $\{- a/(c - 1), 0\}$.

From the geometry of the problem, the radius of the cylinder is p and
the position of its centre relative to the coordinate system in Fig. 9.5 is
$\tfrac{1}{2}a - s$. Therefore

$$p = \frac{ac^{\frac{1}{2}}}{c - 1} \qquad \text{and} \qquad \tfrac{1}{2}a - s = - \frac{a}{c - 1}$$

From the second of these equations,

$$c - 1 = \frac{2a}{2s - a}.$$

Therefore
$$c = \frac{2s + a}{2s - a}.$$

Substituting into the other of these equations gives

$$p^2 \left(\frac{2a}{2s - a} \right)^2 = a^2 \left(\frac{2s + a}{2s - a} \right).$$

Therefore

$$4a^2p^2 = a^2(4s^2 - a^2)$$

and

$$a^2 = 4(s^2 - p^2).$$

On the surface of the cylinder,

$$r_1 = p - \frac{a}{c-1} = \tfrac{1}{2}a - s + p,$$

$$r_2 = a - r_1 = \tfrac{1}{2}a + s - p.$$

Substituting for the value of a gives

$$\frac{r_2}{r_1} = \frac{\sqrt{(s+p)} + \sqrt{(s-p)}}{\sqrt{(s+p)} - \sqrt{(s-p)}}.$$

On the surface of the plane conducting surface, $r_1 = r_2$, so that this has a potential function value of zero and the potential difference between the cylinder and the plane is given by

$$\phi = \frac{q}{2\pi\varepsilon_0} \log_e \frac{\sqrt{(s+p)} + \sqrt{(s-p)}}{\sqrt{(s+p)} - \sqrt{(s-p)}}.$$

Therefore the capacitance is given by

$$C = 2\pi\varepsilon_0 \bigg/ \log_e \frac{\sqrt{(s+p)} + \sqrt{(s-p)}}{\sqrt{(s+p)} - \sqrt{(s-p)}}.$$

9.8.   A right circular cylinder of radius a is mounted with its axis perpendicular to the direction of a uniform electrostatic field. The surface of the cylinder forms a flux boundary to the electrostatic field which will be distorted by the presence of the cylinder. The electrostatic field in the absence of the cylindrical boundary has a uniform flux density D. Making use of potential function expressions, obtain an expression for the flux density on the surface of the cylinder.

Answer. A circular flux boundary is obtained analytically by considering the combination of a uniform field with a doublet, as shown in the solution to Example 3.8 and Fig. 3.8 (page 45). The potential function of a uniform field is given in the solution to Example 8.4 and that of a doublet is given by eqn. (9.4). The combination gives

$$\phi = -\frac{D}{\varepsilon_0} r \cos\theta - \frac{c \cos\theta}{2\pi\varepsilon_0 r} = -\frac{D \cos\theta}{\varepsilon_0 r}\left(r^2 + \frac{c}{2\pi D}\right).$$

The electric field intensity is obtained from eqn. (8.7)(page 142). The radial component of the field intensity is zero on the surface of the circular flux boundary. At any radius r,

$$E_r = -\frac{\partial \phi}{\partial r} = \frac{D \cos \theta}{\epsilon_0}\left(1 - \frac{c}{2\pi D r^2}\right).$$

When $r = a$, $E_r = 0$, therefore

$$a^2 = \frac{c}{2\pi D}$$

and the potential function becomes

$$\phi = -\frac{D \cos \theta}{\epsilon_0 r}\left(r^2 + a^2\right).$$

The circumferential component of the field intensity is given by

$$E_\theta = -\frac{1}{r}\frac{\partial \phi}{\partial \theta} = -\frac{D \sin \theta}{\epsilon_0}\left(1 + \frac{a^2}{r^2}\right).$$

When $r = a$ on the surface of the cylinder, the flux density becomes

$$D_\theta = \epsilon_0 E_\theta = -2D \sin \theta.$$

9.9.   A perfectly conducting boundary in an otherwise uniform electrostatic field has the shape shown by the stagnation stream line in Fig. 6.4 (page 101). In general terms, obtain expressions for both the potential function and flux function of the field.

Answer. The boundary concerned is the flux boundary due to the combination of a uniform field with a line source. As in this problem the same shape boundary is a potential boundary, it will be necessary to take the dual of the field due to a combination of the line source and the uniform field. The flux function of the combination is given in the solution to Example 3.6 (page 42), therefore the potential function of the dual is the same except for division by the permittivity constant.

$$\phi = \frac{D}{\epsilon_0}\, r \sin \theta + \frac{q\theta}{2\pi\epsilon_0}.$$

From this expression, the distance a from the stagnation point to the origin of the line source has been shown in Example 3.6 to be

$$a = \frac{q}{2\pi D}.$$

The potential function of the combination of a uniform field and a line source is given by the sum of the expressions given in the solution to Example 8.4 (page 145) and eqn. (9.3),

$$\phi = \frac{D}{\varepsilon_o} \, r \, \cos \theta + \frac{q}{2\pi\varepsilon_o} \, \log_e \frac{a}{r},$$

where the zero of the potential function of the line source is taken at a
radius a from the line of the source. When $\cos \theta = -1$, substitution of the
value of a into the expression shows that $\phi = 0$ at the stagnation point.
Therefore from the dual

$$\psi = Dr \, \cos \theta + \frac{q}{2\pi} \, \log_e \frac{a}{r}.$$

Substituting the value for a into the two expressions gives a convenient
form for both the potential function and the flux function of the dual as
required by the question.

$$\phi = \frac{D}{\varepsilon_o} \, (r \, \sin \theta + a\theta),$$
$$\psi = D \left(r \, \cos \theta + a \, \log_e \frac{a}{r} \right).$$

9.10.   A uniform conducting cylinder is placed in a uniform electric
field so that the axis of the cylinder lies on one of the equipotential
planes of the field and has the same potential as that plane. The field at a
large distance from the cylinder is of strength E V/m and the cylinder
diameter is $2a$. Assume that the field is of infinite extent and that the
cylinder is infinitely long. Find the maximum field strength on the surface
of the cylinder and also at a distance of a from the surface of the cylinder.

   Answer. The flux lines for a uniform field in the presence of a
circular flux boundary are shown in Fig. 3.8 (page 45). This is the field
due to the combination of a doublet with a uniform field. The field shown
in Fig. 3.8 will also be the potential line pattern for a uniform field in
the presence of a circular potential boundary. The two fields will be the
dual of one another. The flux function for the combination of a uniform
field with a doublet is given in the solution to Example 3.8,

$$\psi = Dr \, \sin \theta - \frac{c \, \sin \theta}{2\pi r},$$

and the radius of the circular flux line is given by

$$a^2 = \frac{c}{2\pi D}.$$

Therefore the flux function of the combination comes out to be

$$\psi = Dr \, \sin \theta \left(1 - \frac{a^2}{r^2} \right).$$

Therefore for the dual which is needed to describe the field of the problem specified in this example,

$$\phi = Er \, \sin \theta \left(1 - \frac{a^2}{r^2}\right).$$

The maximum field intensity occurs where the equipotential lines are closest together. Reference to Fig. 3.8 shows that this occurs at $\theta = \pm\frac{1}{2}\pi$, i.e. $\sin \theta = \pm 1$. The field intensity is obtained from the potential function by differentiating according to eqn. (8.7). Therefore

$$E_r = -\frac{\partial \phi}{\partial r} = -E \, \sin \theta \left(1 + \frac{a^2}{r^2}\right),$$

$$E_\theta = -\frac{1}{r}\frac{\partial \phi}{\partial \theta} = -E \, \cos \theta \left(1 - \frac{a^2}{r^2}\right).$$

When $\theta = \frac{1}{2}\pi$, $E_\theta = 0$ and $\sin \theta = 1$.

When $r = a$, $E_r = -2E$,

When $r = 2a$, $E_r = -1\frac{1}{4}E$.

These are the two maximum field intensities at the specified radii.

PROBLEMS

9.1.   If a charge of 100 pC is placed on the 2·0 mm diameter spherical
conductor at the centre of the spherical system shown in Fig. 9.6, determine

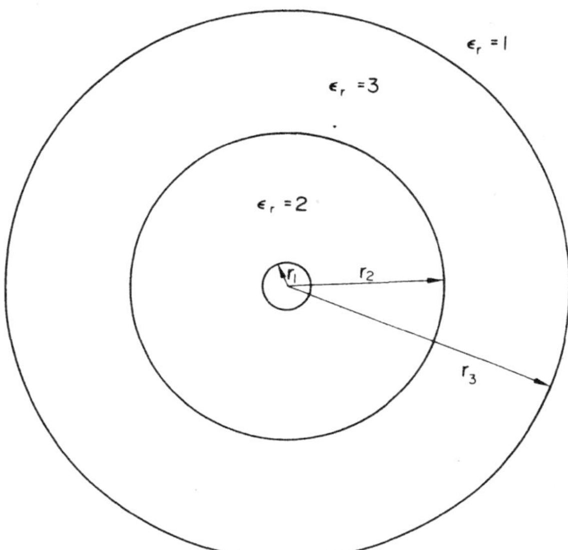

Fig. 9.6.   Spherical conductor at the centre of two concentric
thick spherical shells.

its potential if r_1 = 1·0 mm, r_2 = 10 mm, and r_3 = 20 mm (a) with respect to
a point an infinite distance away, and (b) with respect to a point 1·0 mm from
its centre.   (465·0 V; 464·1 V.)

9.2.   A spherical volume of radius 1·0 m has a uniform charge density
of 1·0 μC/m³. What is the absolute potential on the surface and at a radius
of 0·5 m, assuming an air dielectric?   (37·7 kV; 51·8 kV.)

9.3.   If the electric breakdown strength of air is 3·0 MV/m, determine
the highest voltage that can be applied to a spherical conductor, of diameter
2·0 mm, before breakdown of the air around the conductor (corona) begins to
occur.   (3·0 kV.)

9.4.   A coaxial cable has a core of diameter 4·0 mm and a conducting
sheath of internal diameter of 10 mm separated by an insulator of relative
permittivity 3·5. Calculate the capacitance per unit length of the cable,
and the maximum potential gradient when the core is held at 200 V, and the
casing is earthed.   (212 pF/m; 109 kV/m.)

9.5.   For the coaxial cable of Problem 9.4, find the new values of capacitance and maximum voltage gradient if the outer $1 \cdot 0$ mm of the insulator is replaced by an air gap.   (132 pF/m; 118 kV/m.)

9.6.   A wire of diameter $2 \cdot 0$ mm runs for $1 \cdot 0$ km at a height of $3 \cdot 0$ m above ground. Assuming that the ground is a perfect conductor, find the capacitance of the system.   (6400 pF.)

9.7.   Find the capacitance of a concentric cable having a core of radius $5 \cdot 0$ mm and an earthed sheath of internal radius 50 mm and air dielectric in between.   (24 pF/m.)

9.8.   A concentric cable has a core of radius a and an earthed sheath of internal radius b. Find the maximum value of the electric field intensity in terms of the potential difference V across the cable. Corona discharge from a high potential surface may be assumed to create a uniform conducting layer around the surface. Show that, as the potential V of the core is raised, the corona will be stable if $b/a > e$, but the corona will be unstable causing sparkover if $b/a < e$.

9.9.   Derive an expression for the capacitance of a capacitor formed by two concentric spheres. The radii of the spheres are a and b, where $b > a$, and the space between them is filled with an insulator of relative permittivity ε_r. $(4\pi\varepsilon_0\varepsilon_r ab/(b - a).)$

9.10.   In Problem 9.9, if a = 50 mm and b = 200 mm, find the greatest electric field strength in the dielectric when the potential difference between the spheres is 20 kV.   (533 kV/m.)

SOLUTIONS

9.1.   (a) Let the charge on the central conductor be Q C. As it is a spherically symmetrical system, the flux density at any radius r is given by

$$D = \frac{Q}{4\pi r^2} \; C/m^2.$$

The potential difference between the inner radius r_1 and any radius r_4 outside the spherical shells, is given by

$$\Phi = \int_{r_4}^{r_1} - \mathbf{E} \cdot d\mathbf{l}$$

$$= - \int_{r_4}^{r_3} \frac{Q}{4\pi\varepsilon_0\varepsilon_{r3}r^2} \, dr - \int_{r_3}^{r_2} \frac{Q}{4\pi\varepsilon_0\varepsilon_{r2}r^2} \, dr - \int_{r_2}^{r_1} \frac{Q}{4\pi\varepsilon_0\varepsilon_{r1}r^2} \, dr$$

$$= \frac{Q}{4\pi\varepsilon_0}\left\{ \frac{1}{\varepsilon_{r3}}\left(\frac{1}{r_3} - \frac{1}{r_4}\right) + \frac{1}{\varepsilon_{r2}}\left(\frac{1}{r_2} - \frac{1}{r_3}\right) + \frac{1}{\varepsilon_{r1}}\left(\frac{1}{r_1} - \frac{1}{r_2}\right)\right\}.$$

Inserting numbers from the problem when $r_4 = \infty$ gives

$$\Phi = \frac{100 \times 10^{-12} \times 36\pi \times 10^9}{4\pi}\left\{\left(\frac{1}{2 \times 10^{-2}} - \frac{1}{\infty}\right) + \frac{1}{3}\left(\frac{1}{10^{-2}} - \frac{1}{2 \times 10^{-2}}\right)\right.$$

$$\left. + \frac{1}{2}\left(\frac{1}{10^{-3}} - \frac{1}{10^{-2}}\right)\right\}$$

$$= 0 \cdot 9(50 + 17 + 450) = 465 \; V.$$

Φ is the potential of the surface of the conducting sphere.

(b) The potential of the point $1 \cdot 0$ m from its centre is

$$\Phi_1 = 0 \cdot 9\left(\frac{1}{1} - \frac{1}{\infty}\right) = 0 \cdot 9 \; V.$$

The required potential difference is given by $\Phi - \Phi_1 = 464 \cdot 1$ V.

9.2.   External to the sphere, the flux density is given by eqn. (2.13). Therefore the absolute potential on the surface of the sphere is given by

$$\Phi_1 = - \int_{\infty}^{R} \frac{\rho R^3}{3\varepsilon_0 r^2} \, dr = \frac{\rho R^3}{3\varepsilon_0 R} = \frac{10^{-6} \times 36\pi \times 10^9 \times 1 \cdot 0}{3}$$

$$= 37 \cdot 7 \times 10^3 \; V = 37 \cdot 7 \; kV.$$

Inside the sphere, the flux density is given by eqn. (2.14). Therefore the potential difference between the surface of the sphere and some radius r_2

inside is given by

$$\Phi_2 = -\int_R^{r_2} \frac{\rho r}{3\varepsilon_0}\,dr = \frac{\rho}{6\varepsilon_0}(R^2 - r_2^2) = \frac{10^{-6} \times 36\pi \times 10^9}{6}(1\cdot 0 - 0\cdot 25)$$

$$= 14\cdot 1 \times 10^3 \text{ V} = 14\cdot 1 \text{ kV}.$$

Therefore the total absolute potential of the point inside the surface of the sphere is given by

$$\Phi_1 + \Phi_2 = 37\cdot 7 + 14\cdot 1 = 51\cdot 8 \text{ kV}.$$

9.3.   This problem is similar to Example 9.3. Using the formula derived in the solution to that example gives

$$\Phi = E_{max}r = 3\cdot 0 \times 10^6 \times 10^{-3} = 3\cdot 0 \times 10^3 \text{ V} = 3\cdot 0 \text{ kV}.$$

9.4.   The potential difference between two radii in the field due to a line charge q has been derived in the solution to Example 9.4,

$$\Phi = \frac{q}{2\pi\varepsilon_0\varepsilon_r}\,\log_e\frac{r_2}{r_1}.$$

Therefore the capacitance is given by

$$C = \frac{\Psi}{\Phi} = \frac{2\pi\varepsilon_0\varepsilon_r}{\log_e(r_2/r_1)} = \frac{2\pi \times 3\cdot 5}{36\pi \times 10^9 \times \log_e 2\cdot 5} = 2\cdot 12 \times 10^{-10} \text{ F/m} = 212 \text{ pF/m}.$$

The maximum field intensity occurs at the boundary of the centre conductor, therefore

$$E_{max} = \frac{q}{2\pi r_1\varepsilon_0\varepsilon_r} = \frac{\Phi}{r_1\,\log_e(r_2/r_1)} = \frac{200}{2.0 \times 10^{-3} \times \log_e 2\cdot 5}$$

$$= 109 \times 10^3 \text{ V/m} = 109 \text{ kV/m}.$$

9.5.   If the outer $1\cdot 0$ mm of insulator is replaced by an air gap, the radius of the change in dielectric is $4\cdot 0$ mm. The potential difference is given by

$$\Phi = \frac{q}{2\pi\varepsilon_0}\left(\frac{1}{\varepsilon_r}\log_e\frac{r_3}{r_1} + \log_e\frac{r_2}{r_3}\right) = \frac{36\pi \times 10^9 q}{2\pi}\left(\frac{1}{3\cdot 5}\log_e\frac{4}{2} + \log_e\frac{5}{4}\right)$$

$$= 18 \times 10^9 \times 0\cdot 421q.$$

Therefore the capacitance is given by

$$C = \frac{\Psi}{\Phi} = 1\cdot 32 \times 10^{-10} \text{ F/m} = 132 \text{ pF/m}.$$

The maximum field intensity in the insulator occurs at the inner radius and is given by

$$E_{max} = \frac{q}{2\pi\varepsilon_0\varepsilon_r r_1} = \frac{\Phi}{\varepsilon_r \times 0\cdot421 \times r_1} = \frac{200}{3\cdot5 \times 0\cdot421 \times 2\cdot0 \times 10^{-3}}$$

$$= 6\cdot8 \times 10^4 \text{ V/m} = 68 \text{ kV/m}.$$

The maximum field intensity in the air gap will be at the outer surface of the insulator, therefore

$$E_{max} = \frac{\Phi}{0\cdot421 \times r_3} = \frac{200}{0\cdot421 \times 4\cdot0 \times 10^{-3}} = 1\cdot18 \times 10^5 \text{ V/m} = 118 \text{ kV/m}.$$

Surprisingly enough, the maximum electric field intensity occurs in the air gap at the larger radius. Also the maximum field intensity with the partially filled gap is higher than the maximum field intensity obtained in answer to the last problem where the gap between the conductors was completely filled with insulation.

9.6.   This problem is similar to Example 9.6. Inserting numbers into the expression for the capacitance derived there gives

$$C = \frac{2\pi\varepsilon_0 l}{\log_e(2d/a)} = \frac{10^3}{18 \times 10^9 \times \log_e(6\cdot0/0\cdot001)} = 6\cdot4 \times 10^{-9} \text{ F} = 6400 \text{ pF}.$$

9.7.   This is similar to Problem 9.4. Inserting numbers into the expressions derived in the solution to that problem gives

$$C = \frac{2\pi\varepsilon_0}{\log_e(r_2/r_1)} = \frac{1}{18 \times 10^9 \times \log_e(50/5)} = 2\cdot41 \times 10^{-11} \text{ F/m} = 24\cdot1 \text{ pF/m}.$$

9.8.   Again this problem is similar to Problem 9.4. In the solution to that problem, the maximum field intensity is derived to be

$$E_{max} = \frac{V}{a \log_e(b/a)}.$$

The creation of corona effectively increases the inner radius a. If E_{max} increases with increase in a, the corona is unstable and, if it decreases in a, it is stable. Differentiating gives

$$\frac{\partial E_{max}}{\partial a} = \left(\frac{V}{a}\right)\left(\frac{1}{\log_e(b/a)}\right)^2\left(\frac{1}{a}\right) - \left(\frac{V}{\log_e(b/a)}\right)\left(\frac{1}{a^2}\right)..$$

The limit is when

$$\frac{\partial E_{max}}{\partial a} = 0.$$

Therefore $\log_e(b/a) = 1$, or $b/a = e$.

If $\dfrac{b}{a} > e$, $\dfrac{\partial E_{max}}{\partial a}$ is negative and the corona is stable.

If $\dfrac{b}{a} < e$, $\dfrac{\partial E_{max}}{\partial a}$ is positive and the corona is unstable.

9.9. For a spherically symmetrical system having a total charge on the inner sphere of Q C, the field intensity is given by

$$E = \frac{Q}{4\pi\epsilon_0\epsilon_r r^2}.$$

The potential difference for such a system has been derived in the solution to Example 9.1, therefore

$$\Phi = \frac{Q}{4\pi\epsilon_0\epsilon_r}\left(\frac{1}{a} - \frac{1}{b}\right) = \frac{Q(b - a)}{4\pi\epsilon_0\epsilon_r ab}.$$

Therefore the capacitance is given by

$$C = \frac{Q}{\Phi} = 4\pi\epsilon_0\epsilon_r\left(\frac{ab}{b - a}\right).$$

9.10. Continuing from the solution to the previous problem, the maximum value of field intensity occurs at the inner radius, therefore

$$E_{max} = \frac{Q}{4\pi\epsilon_0\epsilon_r a^2} = \frac{\Phi}{(b - a)}\left(\frac{b}{a}\right) = \frac{20 \times 10^3}{(200 - 50) \times 10^{-3}}\left(\frac{200}{50}\right)$$

$$= 5\cdot33 \times 10^5 \text{ V/m} = 533 \text{ kV/m}.$$

FURTHER PROBLEMS

9.11.   A capacitor is formed by two concentric spherical surfaces. The inner sphere is of radius 10 mm and surrounded by a dielectric material to a radius of 20 mm whose relative permittivity is 6·0 and breakdown strength is 50 MV/m. The inner radius of the outer spherical surface is 50 mm and the rest of the gap between the surfaces is filled with air whose breakdown strength is 3·0 MV/m. Find the maximum permitted potential difference between the electrodes without electrical breakdown occurring. Express the potential difference as a percentage of the maximum potential difference for the same spherical surface without dielectric material present. (46 kV; 192 per cent.)

9.12.   Find the capacitance of an isolated metal sphere having a diameter of 0·45 m.   (25 pF.)

9.13.   Find the capacitance of 25 m length of cable which consists of a circular inner conductor of 4·0 mm diameter surrounded by a coaxial circular metal sheath of inside diameter 10 mm having an insulating material of relative permittivity 5·0.   (7600 pF.)

9.14.   Obtain an expression for the potential difference between the conductors of a coaxial cable having an inner conductor radius a and an outer conductor of inner radius b in terms of the maximum electric field intensity in the insulation of the cable. $(E_{max}a \, \log_e(b/a).)$

9.15.   An electric cable consists of a concentric circular inner conductor of diameter 30 mm surrounded by insulation with a circular earthed conducting sheath having an inside radius of 150 mm. The relative permittivity of the insulation is 5·0 and its breakdown strength is 20 MV/m. There is also a 1·0 mm thick air gap just inside the sheath with a breakdown strength of 3·0 MV/m; find the maximum permissible potential difference on the cable. (208 kV.)

9.16.   A long straight conductor of radius 1·0 mm is supported at a height of 10 m above a perfectly conducting ground. Find the capacitance of the system if the conductor is 7·0 km long.   (0·039 µF.)

9.17.   Obtain an expression for the potential function due to a long straight conductor, having a charge density of q C/m, which lies parallel to a plane conducting surface. The conductor lies at the point $(0, d)$ in a two-dimensional rectangular coordinate system and the conducting surface lies in the plane $y = 0$. $\left(\phi = 9 \times 10^9 q \, \log_e [\{x^2 + (d + y)^2\} / \{x^2 + (d - y)^2\}].\right)$

9.18.   A conducting cylinder of radius 1·0 m having a charge density of 10 nC/m lies parallel to a plane conducting surface. The axis of the cylinder is a distance of 3·0 m from the surface. Find the potential difference between the cylinder and the surface.   (317 V.)

9.19.   A telegraph line consists of two parallel wires, of diameter 2.0 mm, 50 mm apart. Find the capacitance per unit length of the line   (7·1 pF/m.)

9.20.   A thunder-cloud can be considered to generate a uniform field between the cloud base and earth. The potential difference between such a cloud base at a height of 1·0 km and earth is 10^8 V. Find the maximum field intensity set up in the field by a long straight circular conductor which is parallel to the ground. The radius of the conductor may be assumed to be small compared with its height above ground.   (200 kV/m.)

FURTHER SOLUTIONS

9.11.   This problem is similar to Example 9.1. The electric field
intensity due to a total charge Q on the inner sphere is given by

$$E = \frac{Q}{4\pi\varepsilon_0\varepsilon_r r^2}$$

and the potential difference by

$$\Phi = \frac{Q}{4\pi\varepsilon_0}\left\{\frac{1}{\varepsilon_r}\left(\frac{1}{r_1} - \frac{1}{r_2}\right) + \left(\frac{1}{r_2} - \frac{1}{r_3}\right)\right\}$$

$$= \frac{Q}{4\pi\varepsilon_0}\left\{\frac{1}{6\cdot0}\left(\frac{1}{10^{-2}} - \frac{1}{2\cdot0\times10^{-2}}\right) + \left(\frac{1}{2\cdot0\times10^{-2}} - \frac{1}{5\cdot0\times10^{-2}}\right)\right\}$$

$$= \frac{Q}{4\pi\varepsilon_0}\times 38\cdot3.$$

Therefore
$$Q = \frac{4\pi\varepsilon_0\Phi}{38\cdot3}.$$

In the dielectric,
$$E_{max} = \frac{\Phi}{38\cdot3 \times 6\cdot0 \times 10^{-4}}.$$

Therefore

$$\Phi_1 = 50 \times 10^6 \times 38\cdot3 \times 6\cdot0 \times 10^{-4} = 1\cdot15 \times 10^6 \text{ V} = 1\cdot15 \text{ MV}.$$

In the air space,

$$E_{max} = \frac{\Phi}{38\cdot3 \times 4\cdot0 \times 10^{-4}}.$$

Therefore

$$\Phi_2 = 3\cdot0 \times 10^6 \times 38\cdot3 \times 4\cdot0 \times 10^{-4} = 4\cdot6 \times 10^4 \text{ V} = 46 \text{ kV}.$$

Therefore the breakdown strength of the air is the governing limitation and
the maximum permissible potential difference is 46 kV.

   If the space is completely filled with air,

$$\Phi = \frac{Q}{4\pi\varepsilon_0}\left(\frac{1}{10^{-2}} - \frac{1}{5\cdot0\times10^{-2}}\right) = \frac{80Q}{4\pi\varepsilon_0}.$$

$$E_{max} = \frac{\Phi}{80 \times 10^{-4}}.$$

Therefore

$$\Phi_3 = 3 \cdot 0 \times 10^6 \times 80 \times 10^{-4} = 2 \cdot 4 \times 10^4 \text{ V} = 24 \text{ kV}.$$

Therefore the percentage increase in maximum permitted potential difference is given by

$$\frac{\Phi_2}{\Phi_3} \times 100 = 192 \text{ per cent.}$$

9.12.   Using the formula derived in the solution to Example 9.2,

$$C = 4\pi\varepsilon_0 r = \frac{0 \cdot 225}{9 \cdot 0 \times 10^9} = 2 \cdot 5 \times 10^{-11} \text{ F} = 25 \text{ pF.}$$

9.13.   Using the formula derived in the solution to Example 9.5,

$$C = \frac{2\pi\varepsilon_0\varepsilon_r d}{\log_e(r_2/r_1)} = \frac{5 \cdot 0 \times 25}{18 \times 10^9 \times \log_e 2 \cdot 5} = 7 \cdot 6 \times 10^{-9} \text{ F} = 7600 \text{ pF.}$$

9.14.   In the solution to Problem 9.4, an expression for the maximum field intensity is derived, therefore

$$E_{\max} = \frac{\Phi}{a \, \log_e(b/a)}.$$

Therefore

$$\Phi = E_{\max} \, a \, \log_e(b/a).$$

9.15.   This problem is similar to Problem 9.5, the potential difference is given by

$$\Phi = \frac{q}{2\pi\varepsilon_0}\left(\frac{1}{\varepsilon_r}\log_e\frac{r_2}{r_1} + \log_e\frac{r_3}{r_2}\right) = \frac{q}{2\pi\varepsilon_0}\left(\frac{1}{5 \cdot 0}\log_e\frac{149}{15} + \log_e\frac{150}{149}\right) = \frac{0 \cdot 466q}{2\pi\varepsilon_0}.$$

In the dielectric,

$$E_{\max} = \frac{\Phi}{5 \cdot 0 \times 0 \cdot 466 \times 15 \times 10^{-3}}.$$

Therefore

$$\Phi_1 = 20 \times 10^6 \times 5 \cdot 0 \times 0 \cdot 466 \times 15 \times 10^{-3} = 6 \cdot 99 \times 10^5 \text{ V} = 699 \text{ kV}$$

In the air gap,

$$E_{\max} = \frac{\Phi}{0 \cdot 466 \times 149 \times 10^{-3}}.$$

Therefore

$$\Phi_2 = 3 \cdot 0 \times 10^6 \times 0 \cdot 466 \times 149 \times 10^{-3} = 208 \times 10^5 \text{ V} = 208 \text{ kV}.$$

Therefore the governing limitation is the maximum field intensity in the air gap. The maximum potential difference is 208 kV.

9.16.   Using the formula derived in the solution to Example 9.6,

$$C = \frac{2\pi\varepsilon_0 l}{\log_e(2d/a)} = \frac{7\cdot0 \times 10^3}{18 \times 10^9 \times \log_e 20000} = 3\cdot9 \times 10^{-8} \text{ F} = 0\cdot039 \text{ } \mu\text{F}.$$

9.17.   This problem is similar to Example 9.6. The geometry of the coordinate system is shown in Fig. 9.7   Then quoting the expression for

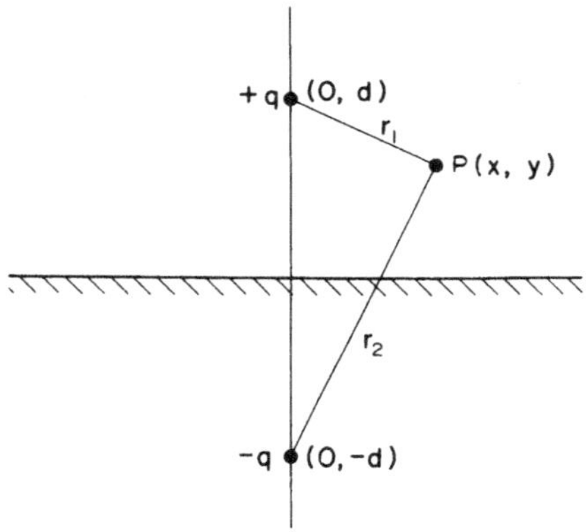

Fig. 9.7. A line source parallel to a plane conducting surface;
illustrating the coordinate system used in the solution to
Problem 9.17.

potential function from the solution to Example 9.6 and substituting for the lengths gives

$$\phi = \frac{q}{2\pi\varepsilon_0} \log_e \frac{r_2}{r_1} = 9 \times 10^9 q \log_e \frac{x^2 + (d + y)^2}{x^2 + (d - y)^2}.$$

9.18.   A formula for the required potential function is given in the solution to Example 9.7, therefore

$$\Phi = \frac{q}{2\pi\varepsilon_0} \log_e \frac{\sqrt{(s + p)} + \sqrt{(s - p)}}{\sqrt{(s + p)} - \sqrt{(s - p)}}.$$

In this problem, $s = 3\cdot0$ m and $p = 1\cdot0$ m, therefore

$$\Phi = 18 \times 10^9 \times 10^{-8} \log_e \frac{2 + \sqrt{2}}{2 - \sqrt{2}} = 180 \log_e 5\cdot83 = 317 \text{ V}.$$

If the approximate formula for a small cylinder from the solution to Example

9.6 is used, this gives

$$\Phi = \frac{q}{2\pi\varepsilon_0} \log_e \frac{2d - a}{a} = 18 \times 10^9 \times 10^{-8} \log_e 5\cdot 0 = 290 \text{ V.}$$

There is an error of about 8 per cent in using the approximate formula.

9.19.   The problem of two parallel wires is similar to that of a wire and a plane earth. Following the derivation used in the solution to Example 9.6, the potential difference between the surfaces of the two wires is given by

$$\Phi = \frac{q}{2\pi\varepsilon_0} \left(\log_e \frac{2d - a}{a} - \log_e \frac{a}{2d - a} \right) = \frac{q}{\pi\varepsilon_0} \log_e \frac{2d - a}{a}.$$

For the sizes in the problem, $d \gg a$, and the capacitance is given by

$$C = \frac{\pi\varepsilon_0}{\log_e(2d/a)} = \frac{1}{36 \times 10^9 \log_e 50} = 7\cdot 1 \times 10^{-12} \text{ F} = 7\cdot 1 \text{ pF/m.}$$

9.20.   The electric field intensity in the uniform field is given by

$$E = \frac{10^8}{10^3} = 10^5 \text{ V/m.}$$

The cylindrical conducting boundary in an otherwise uniform field is the dual of the flux pattern shown in Fig. 3.8, and discussed in the solution to Example 9.10. In that solution it is shown that the maximum electric field intensity in the presence of the cylinder is given by

$$E_{max} = 2E = 2\cdot 0 \times 10^5 \text{ V/m} = 200 \text{ kV/m.}$$

CHAPTER 10

Other Fields

<u>GRAVITATIONAL POTENTIAL</u>

THEORY

The potential values in a gravitational field are defined by analogy with the electrostatic potential given in eqn. (8.1).

$$\text{Potential rise, } A \text{ to } B = \lim_{\delta m \to 0} \left(\frac{\text{work done by external agency in moving } \delta m \text{ from } A \text{ to } B}{\delta m} \right). \tag{10.1}$$

The potential difference is given by a relationship similar to eqn. (8.4),

$$\Phi = \phi_B - \phi_A = \int_A^B - \mathbf{g} \cdot d\mathbf{l}. \tag{10.2}$$

Therefore

$$\mathbf{g} = - \text{grad } \phi. \tag{10.3}$$

The gravitational field is conservative so that

$$\oint \mathbf{g} \cdot d\mathbf{l} = 0. \tag{10.4}$$

The potential energy stored in a gravitational field is given by

$$W = \iiint_{\text{volume}} \tfrac{1}{2} D_g g \, dv \ \text{J}. \tag{10.5}$$

EXAMPLES

10.1.   Find the absolute potential on the surface of a planet of radius R and density ρ.

Answer. The gravitational field intensity has been found in the solution to Example 5.1 (page 80) to be

$$\mathbf{g} = - \frac{4\pi G\rho R^3}{3r^2} \mathbf{U}_r.$$

The potential difference is found by substituting into eqn. (10.2), therefore

$$\Phi = \int_\infty^R - \mathbf{g} \cdot d\mathbf{l} = \tfrac{4}{3}\pi G R^3 \rho \int_\infty^R \frac{1}{r^2} \, dr = - \tfrac{4}{3}\pi G R^2 \rho.$$

10.2.   Obtain an expression for the variation of gravitational potential below the surface of the planet of Example 10.1.

Answer. The gravitational field intensity below the surface of a planet has been found in the solution to Example 5.1 (page 80) to be

$$\mathbf{g} = -\frac{4}{3}\pi G\rho r\,\mathbf{U}_r.$$

Therefore the potential difference is given by

$$\Phi = \int_\infty^r -\mathbf{g}\cdot d\mathbf{l} = \int_\infty^R -\mathbf{g}\cdot d\mathbf{l} + \int_R^r -\mathbf{g}\cdot d\mathbf{l}.$$

The solution to the first part of this integral is the solution to the last example, therefore

$$\Phi = -\frac{4}{3}\pi GR^2\rho + \frac{4}{3}\pi G\rho\int_R^r r\,dr = -\frac{4}{3}\pi GR^2\rho + \frac{2}{3}\pi G\rho(r^2 - R^2).$$

Therefore
$$\Phi = -2\pi G\rho\left(R^2 - \tfrac{1}{3}r^2\right).$$

ELECTRIC CONDUCTION

THEORY

Electric current flows in a manner similar to an ideal fluid so that the principles of field theory can be applied to electric current flow in a homogeneous medium. In the electric current field, the current is the flux.

$$\Psi = I\,A. \tag{10.6}$$

The flux density is the current density, J A/m^2. The field equivalent to *Ohm's law* is

$$\mathbf{J} = \sigma\mathbf{E}, \tag{10.7}$$

where σ is the conductivity measured in S/m and is the conduction current field equivalent to the permittivity. Quoted in its more usual form, Ohm's law is given by

$$V = RI, \tag{10.8}$$

where V is the potential difference and I is the current flux. Therefore, in fields terms, the resistance becomes

$$R = \frac{V}{I} = \frac{\Phi}{\Psi}. \tag{10.9}$$

The potential difference and the electric field intensity are the same in both the electrostatic field and the electric conduction field. By comparing eqn. (10.9) with eqn. (8.8), it is seen that the resistance of a body in the electric conduction field is the inverse of the capacitance of the same shape body in the electrostatic field. When an electric current flows through a material, energy is expended and the material heats up. The rate of power dissipation is given by

$$\frac{\partial W}{\partial t} = \iiint_{volume} JE\, dv \ \text{W}. \tag{10.10}$$

EXAMPLES

10.3.   Find the leakage resistance of 20 m of coaxial cable which is approximately a centre conductor of 2·0 mm diameter and an outer conductor of 5·0 mm inside radius. The insulation of the cable between the two conductors has a conductivity of 10^{-18} S/m.

Answer. The coaxial cable gives a circularly symmetric field system so that it is necessary to specify a line source of conduction current. Let there be an equivalent line source at the centre of the cable of I A/m. Then the flux density in the insulation of the cable is given by

$$J = \frac{I}{2\pi r} \ \text{A/m}^2.$$

The potential difference is obtained by integrating in accordance with eqn. (8.4),

$$\Phi = \int_{r_2}^{r_1} -\mathbf{E} \cdot d\mathbf{l} = \int_{r_2}^{r_1} -\frac{I}{2\pi\sigma r}\, dr = \frac{I}{2\pi\sigma} \log_e \frac{r_2}{r_1}.$$

The total flux is given by

$$\Psi = Id.$$

Therefore the resistance is given by eqn. (10.9),

$$R = \frac{\Phi}{\Psi} = \frac{\log_e(r_2/r_1)}{2\pi\sigma d} = \frac{\log_e 5\cdot 0}{2\pi \times 10^{-18} \times 20} = \frac{1\cdot 6 \times 10^{18}}{2\pi \times 20} = 1\cdot 28 \times 10^{16} \ \Omega.$$

10.4.   Obtain an expression relating the capacitance of a pair of conductors to the resistance between them when the insulation material between them has a conductivity σ and a relative permittivity ε_r.

Answer. The capacitance between any two conductors is given by eqn. (8.8). It is found that eqn. (8.8) reduces to a fields expression which is solely a function of the geometry of the conductors multiplied by the permittivity. For example, in the solution to Example 9.5 (page 173), the capacitance of a coaxial cable is given by

$$C = \frac{\Psi}{\Phi} = \varepsilon_o \varepsilon_r \; \frac{2\pi d}{\log_e(r_2/r_1)} = \varepsilon_o \varepsilon_r K,$$

where K is the factor which is determined by the geometry of the system of conductors. The resistance between any two conductors is given by eqn. (10.9), which again is solely a function of the geometry of the conductors and divided by the conductivity. For the same system of conductors, the geometry of the conductors is the same in both situations. The resistance between the conductors in a coaxial cable is given in the solution to Example 10.3 to be

$$R = \frac{\Phi}{\Psi} = \frac{1}{\sigma} \frac{\log_e(r_2/r_1)}{2\pi d} = \frac{1}{\sigma K},$$

where K is the same as in the expression for the capacitance. Therefore

$$K = \frac{C}{\varepsilon_o \varepsilon_r} = \frac{1}{R\sigma}.$$

Therefore $\qquad\qquad R = \dfrac{\varepsilon_o \varepsilon_r}{\sigma C} \qquad$ and $\qquad C = \dfrac{\varepsilon_o \varepsilon_r}{\sigma R}.$

10.5.   Two similar electrical wires of diameter $2a$ are buried in a uniform ground parallel to one another and a distance $2d$ apart. If the conductivity of the ground is σ and it may be assumed that the surface is so far away that its effect may be ignored, find the resistance between the two wires.

Answer. This is the situation of a line source parallel to a line sink and the flux and equipotential lines for this system are shown in Fig. 10.1. The calculation is similar to the solutions to Example 9.6 (page 174) and Problem 9.19 (page 193), therefore the potential difference between the surfaces of the two wires is given by

$$\Phi = \frac{I}{2\pi\sigma} \left(\log_e \frac{2d-a}{a} - \log_e \frac{a}{2d-a} \right) = \frac{I}{\pi\sigma} \log_e \frac{2d-a}{a},$$

where I is the current out of a unit length of the line source. Then the flux per unit length is given by

$$\Psi = I$$

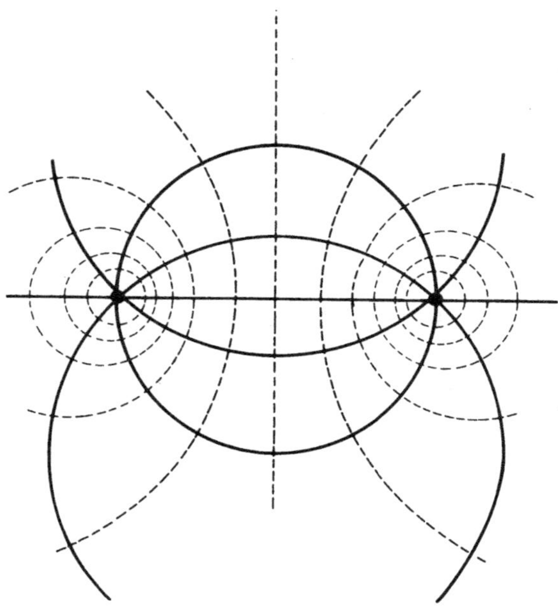

Fig. 10.1. Flux lines and equipotential lines for a line-source
line-sink pair.

and the resistance is given by eqn. (10.9),

$$R = \frac{\Phi}{\Psi} = \frac{1}{\pi\sigma} \log_e \frac{2d - a}{a}.$$

10.6.   A wire is buried at a depth d below the surface of a uniform
ground. It has a leakage current of i A/m; find an expression for the
current flowing in the surface of the ground due to the leakage from the wire.

Answer. The surface of the ground will consist of a flux boundary. This
is an image problem such that there is no current flow across the surface of
the ground. This means that the image will have to be an equal line source
an equal distance above the surface of the ground. The geometry is shown in
Fig. 10.2 and the flux line pattern is the same as Fig. 6.6c (page 109)
bisected down its centre line. The current in the surface of the ground will
be the current flux density on the flux boundary. Taking the dimensions from
the figure, the flux density at P due to one wire is given by

$$J_1 = \frac{i}{2\pi r}.$$

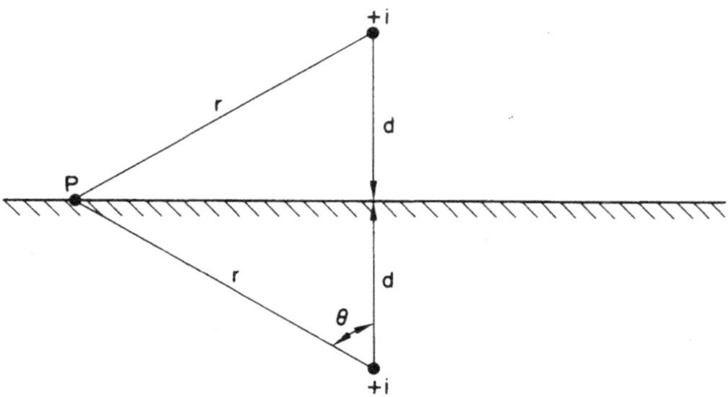

Fig. 10.2. The image of a line source near to a plane flux boundary.

With the contribution from the image source, the component of the current perpendicular to the surface is zero and the combined component parallel to the surface is given by

$$J = 2J_1 \sin \theta = 2\frac{i}{2\pi} \frac{\cos \theta}{d} \sin \theta = \frac{i}{2\pi d} \sin 2\theta.$$

FLUID FLOW THROUGH PERMEABLE MEDIA

THEORY

The flow of liquids through permeable media is found to be analogous to electrical conduction. Fluid flow through permeable media is most frequently met in connection with water seepage through the ground. Part of the fields analogy is the same as for ideal fluid flow. Flux is the rate of volume flow of the fluid, and flux density is the velocity of flow of the liquid. It is necessary to assume that the fluid is incompressible and of constant density. The presence of the porous medium causes a resistance to flow. *Darcy's law* for fluid flow is analogous to Ohm's law. The rate of flow of the fluid is proportional to the difference of pressure in the direction of flow. Therefore

$$\mathbf{v} = - k \text{ grad } p. \tag{10.11}$$

The velocity is the vector flux density. There is no quantity equivalent to the field intensity; it is the gradient of the potential which is the pressure. The constant of proportionality k is called the permeability of the medium, which is the fields equivalent to the permittivity. The pressure difference is given by the potential difference

$$\Phi = p_B - p_A = -\frac{1}{k} \int_A^B \mathbf{v} \cdot d\mathbf{l}. \qquad (10.12)$$

In any fluid the pressure is given by Bernoulli's equation,

$$p + \rho g z + \tfrac{1}{2}\rho v^2 = \text{constant}. \qquad (10.13)$$

It will be noticed that this form of Bernoulli's equation differs from eqn. (6.1) (page 95) in that there is now a term included taking into consideration the pressure due to the height of the fluid. In permeable fluid flow the velocity is small, and the term in Bernoulli's equation involving the velocity may be ignored. Then the total pressure has two components: the first term in the equation is called the excess pore water pressure and the second term is the pressure due to the head of the fluid. The pressure at any point can be indicated by the height a liquid rises up a manometer tube at that point. The excess pore water pressure is the manometric height relative to the height of the point, and the total pressure is the manometric height relative to the datum for the dimension z. The potential difference between two points is the same as the height difference between the free liquid surface of manometers at these points.

Energy is expended in forcing the fluid through the permeable medium which goes towards heating up the fluid and the medium. The rate of dissipation of energy is given by

$$\frac{\partial w}{\partial t} = v \text{ grad } p \text{ W/m}^3. \qquad (10.14)$$

EXAMPLES

10.7.   A vertical rainwater pipe becomes blocked with 45 mm of earth of permeability 10^{-5} m³ s/kg. If the blockage is at the bottom of the pipe which is 10 m high, what is the smallest velocity of water flowing into the pipe that will cause it to overflow?

Answer.   The flow can be assumed to be uniform through the blockage.

Therefore
$$\text{grad } p = \frac{p_1 - p_2}{d} = \frac{\rho g h}{d}.$$

Therefore the velocity can be found by application of eqn. (10.11), which gives

$$v = \frac{k\rho g h}{d} = \frac{10^{-5} \times 10^3 \times 9\cdot81 \times 10}{45 \times 10^{-3}} = 21\cdot8 \text{ m/s}.$$

10.8.   A flower-pot with a drain-hole in the bottom is filled to the brim with earth and a drip maintains the top with a covering of water. The pot is in the shape of a cylindrical taper having a diameter of 160 mm at the top and a diameter of 120 mm at the bottom. Its height is 100 mm. Assuming that the liquid flows uniformly through any horizontal cross-section of the pot, find the volume rate of flow through the pot when the permeability of the earth is 10^{-4} m³ s/kg.

Answer. In order to show the method of derivation, the result will be found in symbols and then substitution will be made for the dimensions given in the question. Let the flower-pot have the dimensions shown in Fig. 10.3.

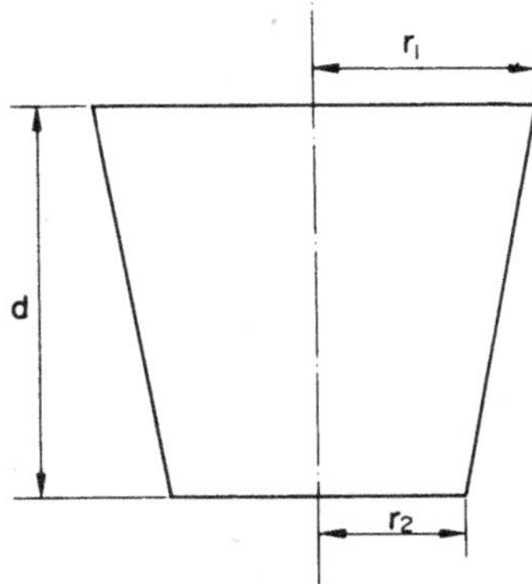

Fig. 10.3. The dimensions of the flower-pot of Example 10.8.

The earth filling will have a permeability k. The pressure head difference between the top and bottom of the pot will cause the water to permeate through the earth and out of the base. The assumption given in the question means that the whole of the bottom layer of earth in the pot is at the same pressure. At any height x above the base of the pot, the volume flow will be constant. Therefore

$$Q = v_x \pi \left(r_2 + \frac{r_1 - r_2}{d} x \right)^2.$$

From eqn. (10.11),

$$v_x = - k \frac{\partial p}{\partial x}.$$

Therefore

$$\frac{\partial p}{\partial x} = -\frac{Q}{k\pi\left(r_2 + \dfrac{r_1 - r_2}{d}\,x\right)^2}$$

and

$$p_1 - p_2 = -\int_0^d \frac{Q\,dx}{k\pi\left(r_2 + \dfrac{r_1 - r_2}{d}\,x\right)^2} = \frac{Qd}{k\pi r_1 r_2}.$$

But

$$p_1 - p_2 = \rho g d.$$

Therefore

$$Q = \rho g k \pi r_1 r_2.$$

Inserting numbers from the problem,

$$Q = 10^3 \times 9{\cdot}81 \times 10^{-4} \times \pi \times 80 \times 60 \times 10^{-6} = 1{\cdot}48 \times 10^{-2}\ \mathrm{m^3/s}.$$

10.9.   A pipe is inserted a great depth below the surface of a uniform ground and used to discharge fluid into the ground. If the pipe orifice is of radius R and can be assumed to be equivalent to a point source of fluid of that radius, obtain an expression for the rate of discharge of the fluid in terms of the depth d of the end of the pipe below the free surface of the liquid.

   Answer. Assume a point source of fluid flux of Q m^3/s. Therefore at any radius r in the earth the flux density is given by

$$v = \frac{Q}{4\pi r^2}.$$

The pressure difference is given by the potential difference,

$$p_1 - p_2 = -\frac{1}{k}\int_{r_2}^{r_1} \frac{Q}{4\pi r^2}\,dr = \frac{Q}{4\pi k}\left(\frac{1}{r_1} - \frac{1}{r_2}\right).$$

If the surface of the ground is supposed to be a long distance away, $r_2 = \infty$ and r_1 is given by the radius of the source. Therefore

$$p_1 - p_2 = \frac{Q}{4\pi k R}.$$

The potential difference is given by the pressure head at the end of the pipe. The pipe must be assumed to be full to provide maximum flow then

$$p_1 - p_2 = \rho g d.$$

Therefore $\qquad\qquad\qquad\qquad Q = 4\pi k\rho g dR$ m³/s.

10.10. A water channel of rectangular cross-section, 200 m wide by 10 m deep, has impermeable walls of base width 10 m resting on the plane surface of the ground. If the permeability of the soil of the ground is 10^{-5} m³ s/kg, estimate the seepage losses per unit length of channel through the ground.

Answer. It is necessary to assume that the channel is resting on the surface of a uniform ground of infinite extent. Then the flux and potential lines will be the dual of those due to a line source and a line sink. A section through the channel, together with the estimated flux and potential lines, is shown in Fig. 10.4. The field pattern drawn is that of a line

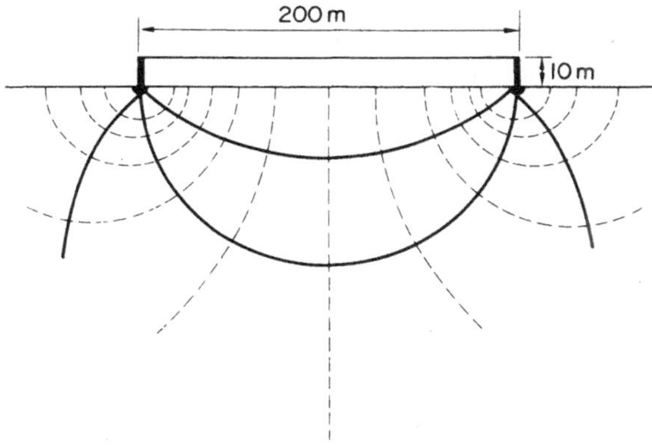

Fig. 10.4. Seepage flux and potential lines through a uniform
ground below a water channel.

source or a line sink coincident with the base of each of the walls. As this is the dual of the problem, the flux function of the line source and line sink is the potential function of this groundwater problem. The assumption needs to be made that the base of the walls are semicircular in section. The total potential drop from inside the channel to the surface of the ground outside is $\frac{1}{2}q$, where q C/m is the strength of the equivalent line source. The flux function in the groundwater problem is the potential function of the line source and line sink. The problem is similar to Example 10.5. The dual of the potential difference given in the solution to that problem is

$$\Psi = \frac{q}{\pi}\,\log_e\frac{d-a}{a}.$$

where a is the radius of the circular base of the walls and d is the width of the channel. If the potential at the free surface of the ground is taken to be zero, then this is also the potential at the free surface of the water in the channel, and the potential at the bottom of the channel will be equal to the pressure due to the head of water in the channel. Therefore

$$\Phi = \rho g h,$$

but this potential difference is equal to $\frac{1}{2}q/k$, where k is the permeability of the ground. Therefore

$$q = 2\rho g h k$$

and

$$\psi = \frac{2\rho g h k}{\pi} \log_e \frac{d - a}{a} = \frac{2 \times 10^3 \times 9 \cdot 81 \times 10 \times 10^{-5}}{\pi} \log_e \frac{195}{5} = 2 \cdot 3 \text{ m}^2/\text{s}.$$

If the water in the channel is flowing at $1 \cdot 0$ m/s, this seepage loss is $0 \cdot 1$ per cent per metre of channel. It is negligible provided that the channel is not too long, i.e. it becomes 10 per cent of the flow if the channel is 100 m long.

PROBLEMS ON GRAVITATION

10.1.   Find an expression for the variation of gravitational potential inside a thin spherical shell of surface density σ kg/m^2. (0.)

10.2.   A thick spherical shell is made of iron of density 8000 kg/m^3 and has an inner diameter of $6 \cdot 0$ m and an outer diameter of $10 \cdot 0$ m. Plot the variation of gravitational absolute potential with radius both inside and outside the shell.

10.3.   Obtain an expression for the variation of gravitational potential with radius inside an infinitely long solid circular cylinder of density ρ and diameter $2R$. $(\pi G \rho (r^2 - R^2).)$

10.4.   Find an expression for the gravitational intensity inside a sphere whose density is proportional to the distance from the outside of the sphere to the point in question. $(\pi G \rho_m r(4R - 3r)/3R.)$

PROBLEMS ON ELECTRICAL CONDUCTION

10.5.   Calculate both the capacitance and resistance per metre of a coaxial cable whose inner conductor is 2·5 mm radius and outer conductor is 10·0 mm internal radius. The insulation material which fills the space between the conductors has a relative permittivity of 2·5 and a conductivity of 10^{-12} S/m.   (100 pF/m; $2\cdot2 \times 10^{11}$ Ω/m.)

10.6.   A capacitor utilizes an insulating material between its plates having a relative permittivity of 9·0 and a resistivity of 10^{13} Ω/m. If its capacitance is 100 μF, find its leakage resistance.   (8·0 MΩ.)

10.7.   An earthing electrode consists of a highly conducting sphere buried in a ground of uniform conductivity σ. Prove that the resistance between the sphere and earth is $1/(4\pi\sigma R)$, where R is the radius of the sphere.

PROBLEMS ON FLUID FLOW THROUGH PERMEABLE MEDIA

10.8.   A cylindrical flower-pot, 0·1 m diameter, is filled to within 10 mm of the rim with soil of permeability 10^{-5} m³ s/kg. If it can be assumed that there is uniform fluid flow through the soil which is 0·1 m deep, find the greatest flow of water into the top of the flower-pot without it running over.   ($8\cdot5 \times 10^{-4}$ m³/s.)

10.9.   A bore pipe is sunk into a gravel bed to extract water. The perforated end of the pipe may be assumed to approximate to a point sink of fluid flow. Calculate the relationship between the discharge of water and the negative pressure head required at the bottom of the pipe to provide this discharge if the permeability of gravel is 10 m³ s/kg and the diameter of the pipe is 200 mm.   ($Q = 12\cdot6p$.)

10.10.   A rainwater pipe discharges the rain off a roof of area 100 m². It is 20 m high and 100 mm in diameter and enters straight into clay soil of permeability $4\cdot0 \times 10^{-6}$ m³ s/kg. Find the maximum rate of rainfall which will cause the pipe to overflow.   ($4\cdot93 \times 10^{-3}$ m/s.)

SOLUTIONS

10.1.   According to Gauss's law it is possible to take a spherical Gaussian surface anywhere inside the shell which will enclose no mass. Because the surface does not enclose any gravitational flux sources, there will be no net flux flow through the Gaussian surface. Therefore there will be no flux density and consequently no field intensity anywhere inside the thin spherical shell. The gravitational potential difference is given by eqn. (10.2), but as $g = 0$ there is *no* variation of gravitational potential inside the thin spherical shell.

10.2.   We will first work out the various values of potential in symbols. Let the dimensions of the sphere be: inner radius a, outer radius b. Therefore, applying Gauss's law, the field intensity can be obtained in different parts of the field:

for $a < r < b$,

$$g = - \frac{4\pi G \frac{4}{3}\pi (r^3 - a^3)\rho}{4\pi r^2} = - \frac{4}{3}\pi G\rho\left(r - \frac{a^3}{r^2}\right);$$

for $r > b$,

$$g = - \frac{4\pi G \frac{4}{3}\pi (b^3 - a^3)\rho}{4\pi r^2} = - \frac{4}{3}\pi G\rho(b^3 - a^3)\frac{1}{r^2}.$$

The absolute potential is obtained by taking $\phi = 0$ when $r = \infty$. Therefore, integrating in accordance with eqn. (10.2):

for $r > b$,

$$\phi_1 = \int_{\infty}^{r_1} -\mathbf{g}\cdot d\mathbf{l} = \int_{\infty}^{r_1} \frac{4}{3}\pi G\rho(b^3 - a^3)\frac{dr}{r^2} = - \frac{4}{3}\pi G\rho(b^3 - a^3)\frac{1}{r_1};$$

when $r = b$,

$$\phi_b = - \frac{4}{3}\pi G\rho(b^3 - a^3)\frac{1}{b};$$

for $a < r < b$,

$$\phi_2 = \phi_b + \int_{b}^{r_2} -\mathbf{g}\cdot d\mathbf{l} = \phi_b + \int_{b}^{r_2} \frac{4}{3}\pi G\rho\left(r - \frac{a^3}{r^2}\right)dr$$

$$= \phi_b + \frac{4}{3}\pi G\rho\left\{\tfrac{1}{2}(r_2^2 - b^2) + a^3\left(\frac{1}{r_2} - \frac{1}{b}\right)\right\} = - \frac{4}{3}\pi G\rho\left(\frac{3}{2}b^2 - \frac{1}{2}r_2^2 - \frac{a^3}{r_2}\right);$$

when $r_2 = a$,

$$\phi_a = - 2\pi G\rho(b^2 - a^2).$$

A graph of the variation of field intensity with radius is similar to Fig. 2.11 (page 32). Inserting numbers from the problem into the expressions for the potential gives:

$$\phi_b = - \frac{4}{3}\pi \times 6 \cdot 67 \times 10^{-11} \times 8000 \times (25 - \frac{27}{5}) = 4 \cdot 4 \times 10^{-5} \ \text{m}^2/\text{s}^2,$$

$$\phi_a = - 2\pi \times 6 \cdot 67 \times 10^{-11} \times 8000 \times (25 - 9) = 5 \cdot 4 \times 10^{-5} \ \text{m}^2/\text{s}^2.$$

Inside the inner radius there is no variation of potential. The required graph of variation of potential is given in Fig. 10.5.

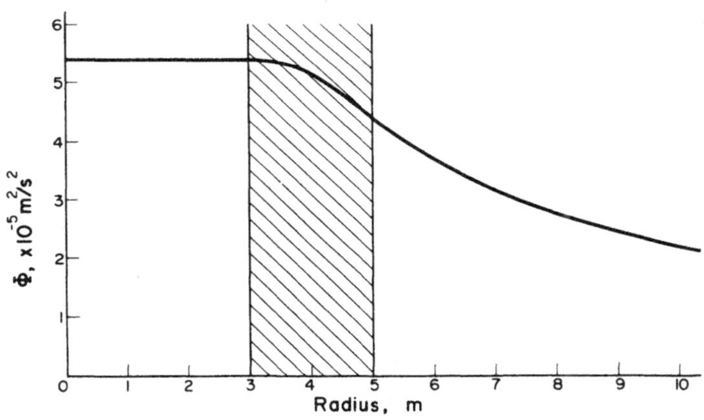

Fig. 10.5. The solution to Problem 10.2. The variation of gravitational potential in the region of a thick spherical shell.

10.3.   For the cylindrically symmetric system, using Gauss's law, the field intensity inside the solid cylinder is given by

$$g = - \frac{4\pi G\pi r^2 \rho l}{2\pi r l} = - 2\pi G r \rho.$$

The potential difference between two points inside the cylinder is given by

$$\phi_2 - \phi_1 = \int_1^2 - \mathbf{g} \cdot d\mathbf{l} = \int_{r_1}^{r_2} 2\pi G r \rho \, dr = \pi G\rho(r_2^2 - r_1^2).$$

If $r_2 = r$ is a radius inside the cylinder and $r_1 = R$ is the outside radius of the cylinder, the potential difference between the outside surface of the cylinder and any point inside is negative and is given by

$$\Phi = \pi G \rho (r^2 - R^2).$$

10.4.   Let the outside radius of the sphere be R. Then the density at any radius r inside the sphere will be given by

$$\rho = \rho_m \frac{R - r}{R},$$

where ρ_m is the density at the outside radius of the sphere. Then at any radius r, the total enclosed mass is given by

$$M = \int_0^r \rho_m \frac{R - r}{R} 4\pi r^2 \, dr = 4\pi \rho_m \left(\frac{r^3}{3} - \frac{r^4}{4R}\right).$$

The flux density is given by

$$D_g = \frac{M}{4\pi r^2} = \rho_m \left(\frac{r}{3} - \frac{r^2}{4R}\right).$$

Therefore the gravitational field intensity is given by

$$g = - 4\pi G D_g = - \pi G \rho_m r(4R - 3r)/3R.$$

10.5.   For a cylindrically symmetric system, assume a line source flux of q C/m or i A/m. The potential difference is given in eqn. (9.3)(page 167), therefore

$$\Phi = \frac{q}{2\pi \varepsilon_0 \varepsilon_r} \log_e \frac{r_2}{r_1}$$

or

$$\Phi = \frac{i}{2\pi \sigma} \log_e \frac{r_2}{r_1}.$$

The capacitance is given by eqn. (8.8) which gives

$$C = \frac{\Psi}{\Phi} = \frac{2\pi \varepsilon_0 \varepsilon_r}{\log_e(r_2/r_1)}.$$

The resistance is given by eqn. (10.9) which gives

$$R = \frac{\Phi}{\Psi} = \frac{\log_e(r_2/r_1)}{2\pi \sigma}.$$

Inserting numbers from the problem gives,

$$C = \frac{2\pi \times 2 \cdot 5}{36\pi \times 10^9 \times \log_e 4} = 1 \cdot 0 \times 10^{-10} \text{ F/m} = 100 \text{ pF/m},$$

$$R = \frac{\log_e 4}{2\pi \times 10^{-12}} = 2\cdot2 \times 10^{11} \ \Omega/\text{m}.$$

10.6.   This problem is similar to Example 10.4. Using the formula derived in the solution to that example,

$$R = \frac{\varepsilon_0 \varepsilon_r}{\sigma C} = \frac{9\cdot0}{36\pi \times 10^9 \times 10^{-13} \times 100 \times 10^{-6}} = 8\cdot0 \times 10^6 \ \Omega = 8\cdot0 \ \text{M}\Omega.$$

10.7.   This is a spherically symmetrical problem. Assume that there is a point source of flux of I A at the centre of the sphere. Then the field intensity is given by

$$E = \frac{I}{4\pi r^2 \sigma} \ \text{V/m}.$$

Then the absolute value of the potential of the surface of the sphere is given by

$$\Phi = \int_\infty^R - \mathbf{E} \cdot d\mathbf{l} = \int_\infty^R - \frac{I \ dr}{4\pi r^2 \sigma} = \frac{I}{4\pi \sigma R}.$$

The resistance is obtained from eqn. (10.9), therefore

$$R = \frac{\Phi}{\Psi} = \frac{\Phi}{I} = \frac{1}{4\pi \sigma R}.$$

10.8.   There will be a uniform fluid flow field through the soil in the flower-pot. If the fluid velocity is U m/s, the field intensity in the soil is U/k N/m^4. The limiting condition is given by

$$\text{grad } p = \frac{U}{k}.$$

The pressure gradient is given by the total pressure head between the top and bottom of the flower-pot. Therefore

$$\text{grad } p = \frac{\rho g h}{d} = \frac{10^3 \times 9\cdot81 \times 0\cdot11}{0\cdot1} = 1\cdot08 \times 10^4 \ \text{N/m}^4.$$

The total flow of water is given by

$$Q = U(\text{area}) = 1\cdot08 \times 10^4 \times 10^{-5} \times \pi \times (0\cdot05)^2 = 8\cdot48 \times 10^{-4} \ \text{m}^3/\text{s}.$$

10.9.   This problem is similar to Example 10.9 except that in this problem the pressure difference is being used to extract water from the ground. Using the relationship derived in the solution to that Example, gives

$$p = \Phi = \frac{Q}{4\pi kR}.$$

Therefore $\qquad Q = 4\pi kRp = 4\pi \times 10 \times 0\cdot 1p = 12\cdot 57p.$

10.10.　　　For the simplest approximation, assume that the pipe end acts as a point source of fluid below the surface of the soil and that it is sufficiently deep for the surface to be ignored. Assume that the end of the pipe forms a spherical surface of the same diameter as the pipe. Then the absolute potential of the surface of the sphere has been derived in the soqution to Example 10.9 and is given by

$$\Phi = p = \frac{Q}{4\pi kR}.$$

The maximum pressure is given by the head of the rainwater pipe when it is full, which is

$$p = \rho gh = 1000 \times 9\cdot 81 \times 20 = 1\cdot 96 \times 10^5 \text{ N/m}^2.$$

The total discharge of water is given by

$$Q = (\text{rainfall}) \times (\text{area of roof}) = 4\pi kRp.$$

Therefore

$$\text{rainfall} = \frac{Q}{\text{area}} = \frac{4\pi \times 4\cdot 0 \times 10^{-6} \times 0\cdot 05 \times 1\cdot 96 \times 10^5}{100} = 4\cdot 93 \times 10^{-3} \text{ m/s}.$$

FURTHER PROBLEMS ON GRAVITATION

10.11.　　　Find the value of the absolute gravitational potential on the surface of the earth whose mean density is 5500 kg/m^3. ($-6\cdot 2 \times 10^7$ m^2/s^2.)

10.12.　　　Find the value of the gravitational potential difference between the surface of the earth and a point at a depth of $1\cdot 0$ km below the surface of the earth. ($-9\cdot 8 \times 10^3$ m^2/s^2.)

10.13.　　　Find the gravitational potential difference between the centre and outside of a solid, long, right circular cylinder of iron of density 8000 kg/m^3 and diameter $0\cdot 2$ m. ($-1\cdot 67 \times 10^{-8}$ m^2/s^2.)

FURTHER PROBLEMS ON ELECTRICAL CONDUCTION

10.14.   Find the resistance of a cylindrical block of composition material having electrical connections made to conducting surfaces across each of the plane ends. The cylinder is 4·0 mm diameter and 20 mm long and is made of material of resistivity 6.0×10^{-2} Ω m.   (95 Ω.)

10.15.   A coaxial cable consists of an inner conductor of 2·0 mm radius surrounded by an insulating material of relative permittivity 5·0 and a conductivity of 10^{-14} S/m and is surrounded by an earthed conducting sheath of inside radius 10 mm. If the potential difference of the inner conductor relative to earth is 100 V, find the leakage current per unit length of cable. (3·9 pA/m.)

10.16.   An earthing electrode consists of a highly conducting sphere of radius 0·4 m buried in a uniform conducting ground of conductivity 0·6 S/m. Assuming that the sphere is buried at such a depth that the effect of the surface of the ground may be ignored, find the resistance between the sphere and earth.   (0·33 Ω.)

10.17.   Two parallel bare electric wire conductors pass through a tank of liquid of conductivity 10^{-8} S/m. Find the resistance between the wires if they are of length 1·0 m, 3·0 mm diameter, and 10 mm apart.   (55 MΩ.)

FURTHER PROBLEMS ON FLUID FLOW THROUGH PERMEABLE MEDIA

10.18.   A porous filter in a pipe consists of 0·2 m length of a material of permeability 0·01 m^3 s/kg. Find the pressure difference needed to maintain a flow rate of 1·0 m/s through the filter.   (20 N/m^2.)

10.19.   A rainwater pipe, 8·0 m high, becomes blocked at its bottom with 0·2 m of earth of permeability 4.0×10^{-6} m^3 s/kg. What is the smallest velocity of water flowing into the top of the pipe that will cause it to overflow?   (1·57 m/s.)

10.20.   A porous filter consists of a circular tube of material of permeability k having its inside diameter $2a$ and an outside diameter of $2b$. If the tube is of length d and is horizontal, find the relationship between the rate of discharge radially through the tube and the pressure drop across it.   ($\Psi = 2\pi kpd/\log_e(b/a)$.)

FURTHER SOLUTIONS

10.11.   This problem is the same as Example 10.1. Using the formula derived in the solution to that example,

$$\Phi = -\tfrac{4}{3}\pi G R^2 \rho = -\tfrac{4}{3}\pi \times 6\cdot67 \times 10^{-11} \times (6\cdot37)^2 \times 10^{12} \times 5500 = -6\cdot2 \times 10^7 \ \mathrm{m^2/s^2}.$$

10.12.   This problem is the same as Example 10.2. Using the formula derived in the solution to that example and simplifying because depth $d \ll R$ gives

$$\Phi = -\tfrac{2}{3}\pi G \rho d(2R - d) = -\tfrac{2}{3}\pi \times 6\cdot67 \times 10^{-11} \times 5500 \times 10^6 \times 2 \times 6\cdot37 \times 1\cdot0 \times 10^3$$

$$= -9\cdot8 \times 10^3 \ \mathrm{m^2/s^2}.$$

10.13.   This problem is the same as Problem 10.3. Using tne formula derived in the solution to that problem when $r = 0$ at the centre of the cylinder,

$$\phi_c - \phi_0 = -\pi G \rho R^2 = -\pi \times 6\cdot67 \times 10^{-11} \times 8000 \times 0\cdot01 = -1\cdot67 \times 10^{-8} \ \mathrm{m^2/s^2}.$$

When $r = R$ at the outside of the cylinder, $\phi = \phi_0$. Therefore

$$\Phi = -1\cdot67 \times 10^{-8} \ \mathrm{m^2/s^2}.$$

10.14.   In this problem there is a plane electric current field, therefore

$$R = \frac{(\text{length})}{\sigma(\text{area})} = \frac{\rho(\text{length})}{(\text{area})} = \frac{6\cdot0 \times 10^{-2} \times 20 \times 10^{-3}}{\pi \times 4\cdot0 \times 10^{-6}} = 95 \ \Omega.$$

10.15.   This is a circularly symmetrical problem in two dimensions, similar to Example 10.3. If i is the strength of the equivalent line current source, the potential difference is given in the solution to that example to be

$$\Phi = \frac{i}{2\pi\sigma} \log_e \frac{r_2}{r_1}.$$

Therefore

$$i = \frac{\Phi 2\pi\sigma}{\log_e(r_2/r_1)} = \frac{100 \times 2\pi \times 10^{-14}}{\log_e 5\cdot0} = 3\cdot9 \times 10^{-12} \ \mathrm{A/m} = 3\cdot9 \ \mathrm{pA/m}.$$

10.16.   This problem is the same as Problem 10.7. Using the formula derived in the solution to that problem,

$$R = \frac{1}{4\pi\sigma r} = \frac{1}{4\pi \times 0\cdot6 \times 0\cdot4} = 0\cdot33 \ \Omega.$$

10.17.   This problem is the same as Example 10.5. Using the formula derived in the solution to that problem,

$$R = \frac{1}{\pi\sigma} \log_e \frac{d - a}{a} = \frac{1}{\pi \times 10^{-8}} \log_e \frac{8\cdot5}{1\cdot5} = 5\cdot5 \times 10^7 \ \Omega = 55 \ \text{M}\Omega.$$

10.18.   Assuming that there is uniform flow through the plug,

$$v = k \ \text{grad} \ p = \frac{k(p_2 - p_1)}{d}.$$

Therefore

$$p_2 - p_1 = \frac{vd}{k} = \frac{1\cdot0 \times 0\cdot2}{0\cdot01} = 20 \ \text{N/m}^2.$$

10.19.   This problem is similar to Example 10.7. Using the formula derived in the solution to that example,

$$v = \frac{k\rho gh}{d} = \frac{4\cdot0 \times 10^{-6} \times 10^3 \times 9\cdot81 \times 8\cdot0}{0\cdot2} = 1\cdot57 \ \text{m/s}.$$

10.20.   The flow through the porous tube will be radially outwards so that this is a two-dimensional circularly symmetrical field. Therefore assuming an equivalent line source of q, the potential difference is given by

$$\Phi = \frac{q}{2\pi k} \log_e \frac{b}{a} = p.$$

Therefore

$$\text{discharge} = \Psi = qd = \frac{2\pi kpd}{\log_e (b/a)}.$$

CHAPTER 11

Conductive Heat Transfer

THEORY

Heat energy may be transferred by radiation, convection, or conduction. In conduction, the heat energy appears to flow in a manner analogous to the flow of a fluid or an electric current. The rate of heat flow through a body is porportional to the temperature difference across the ends of the body. The relationship is directly analogous to Ohm's law or Darcy's law,

$$T = RQ, \tag{11.1}$$

where T is the temperature difference, Q is the rate of heat flow, and R is the thermal resistance. Heat is energy, so that the flow of heat energy is measured in units of W. Temperature is the potential and heat flow is the flux. Therefore

$$\Phi = T \text{ K} \tag{11.2}$$

and

$$\Psi = Q \text{ W.} \tag{11.3}$$

The field equivalent to eqn. (11.1) is

$$\mathbf{q} = - k \text{ grad } T, \tag{11.4}$$

where $\mathbf{q}$ is the flux density measured in units of W/m^2 and k is the thermal conductivity measured in units of W/m K. Therefore the potential difference is given by an equation similar to eqn. (8.4)

$$\Phi = \phi_B - \phi_A = T_B - T_A = - \frac{1}{k} \int_A^B \mathbf{q} \cdot d\mathbf{l} . \tag{11.5}$$

At a surface which is the boundary between a solid and a fluid, there may be an abrupt temperature difference between the temperature of the fluid and the temperature of the surface of the solid. The temperature change occurring between a solid and a fluid is shown in Fig. 11.1. The instantaneous change of temperature on passing through the boundary between the solid and the bulk of the fluid is a property of the particular solid and fluid under consideration and is proportional to the rate of heat flow across the surface. Mathematically the heat transfer across the surface is described by the introduction of a *surface heat transfer coefficient*. Then the rate of heat transfer across the surface is given by

214

$$q = h(T_1 - T_2),\tag{11.6}$$

where h is the surface heat transfer coefficient and the temperatures are of
the surface and the fluid as shown in Fig. 11.1.

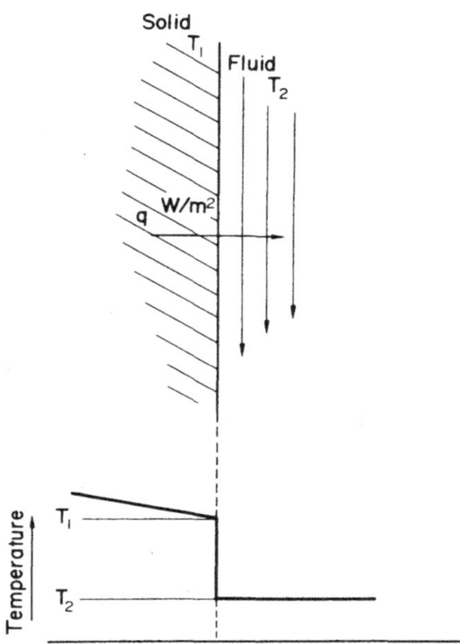

Fig. 11.1.   The abrupt temperature change across a boundary
between a solid and a fluid.

EXAMPLES

11.1.   A room has an internal surface area of wall of 60 m². The inside
surface temperature of the walls is found to be 15°C and the outside surface
temperature 5°C. If the walls are 200 mm thick and are made of a material
having a thermal conductivity of 2·0 W/m K, find the rate of heat loss through
the walls.

Answer. It must be assumed that the heat flow through the walls will be
uniform and the same as that through a plane wall of the same area and the
same thickness. To generate the uniform field we need a plane source parallel
to a similar plane sink. A uniform plane source of strength $+Q$ W is shown in
Fig. 11.2 a distance d from a uniform plane sink of strength $-Q$ W. Therefore
the flux density is given by

$$q = \frac{Q}{A} \text{ W/m}^2.$$

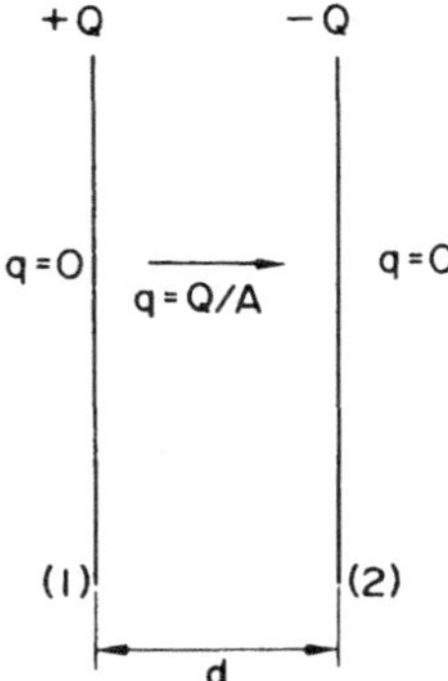

Fig. 11.2. Uniform thermal conduction between parallel plane surfaces.

The potential difference is given by eqn. (11.5), therefore

$$\phi_2 - \phi_1 = -T = \frac{1}{k} \int_0^d - q\, dx = - \frac{Qd}{kA}.$$

Therefore the total rate of heat flow is given by

$$Q = \frac{TkA}{d}\ \text{W.}$$

Inserting numbers from the problem gives

$$Q = \frac{10 \times 2{\cdot}0 \times 60}{0{\cdot}2} = 6{\cdot}0 \times 10^3\ \text{W} = 6{\cdot}0\ \text{kW.}$$

11.2.   A spherical concrete pressure vessel has an internal diameter of
10 m and an external diameter of 12 m. The thermal conductivity of concrete
is 1·0 W/m K. If the inside wall of the vessel is at 100^0C and the external
wall is at 0^0C, find the rate of heat loss through the walls of the vessel.

   Answer. This is a spherically symmetrical problem. It is necessary to
specify a point source of strength Q at the centre of the system, and this
quantity will be the required heat loss. Then at any radius r the flux density
will be given by

$$q = \frac{Q}{4\pi r^2}.$$

The potential difference between any two radii r_1 and r_2 is given by

$$T = \frac{1}{k} \int_{r_2}^{r_1} - \mathbf{q} \cdot d\mathbf{l} = \frac{1}{k} \int_{r_2}^{r_1} - \frac{Q}{4\pi r^2}\, dr = \frac{Q}{4\pi k} \left(\frac{1}{r_1} - \frac{1}{r_2} \right).$$

Therefore, inserting numbers from the question,

$$Q = \frac{4\pi kT}{\left(\dfrac{1}{r_1} - \dfrac{1}{r_2}\right)} = \frac{4\pi \times 1{\cdot}0 \times 100}{(0{\cdot}20 - 0{\cdot}167)} = 3{\cdot}8 \times 10^4 \ \text{W} = 38 \ \text{kW}.$$

11.3.   A steam-pipe has a bore of 50 mm and an outside diameter of 70 mm. It is made of steel of thermal conductivity 50 W/m K. If the temperature of the inside surface of the pipe is 100°C and that of the outside surface is 90°C, find the rate of heat loss through the pipe per unit length.

Answer. This is a circularly symmetrical problem in two dimensions. Assume a line source of strength Q W/m at the axis of the pipe. Then the flux density at any radius r is given by

$$q = \frac{Q}{2\pi r} \ \text{W/m}^2.$$

The potential difference is obtained from eqn. (11.5), therefore

$$T = \frac{1}{k} \int_{r_2}^{r_1} - \mathbf{q} \cdot d\mathbf{l} = \frac{1}{k} \int_{r_2}^{r_1} - \frac{Q}{2\pi r} \, dr = \frac{Q}{2\pi k} \log_e \frac{r_2}{r_1}.$$

Therefore, inserting numbers from the problem gives

$$Q = \frac{2\pi kT}{\log_e(r_2/r_1)} = \frac{2\pi \times 50 \times 10}{\log_e 1{\cdot}4} = 9{\cdot}36 \times 10^3 \ \text{W/m} = 9{\cdot}36 \ \text{kW/m}.$$

11.4.   A spherical hot water tank is 1·0 m inside diameter and is made of copper 10 mm thick which has a thermal conductivity of 400 W/m K and is lagged with 25 mm thickness of insulating material of thermal conductivity 0·1 W/m K. If the water temperature is 60°C, which may be assumed to be the temperature of the inside wall of the tank, and the outside wall temperature of the insulation is 30°C, find the rate of heat loss from the tank and the temperature of the junction between the tank and the insulation.

Answer. Again, this is a problem where the potential difference is given and it is required to find the flux. However, it is necessary to know the flux in order to calculate the potential difference from eqn. (11.5). Therefore the complete problem will be calculated in symbols so that an analytical expression can be obtained for the required answer and then numbers will be inserted into that expression. The various dimensions and properties are specified in Fig. 11.3, which is not drawn to scale according to the dimensions given in the question. As this is a spherically symmetrical problem, we will specify a point source Q at the centre of the system. Then at any

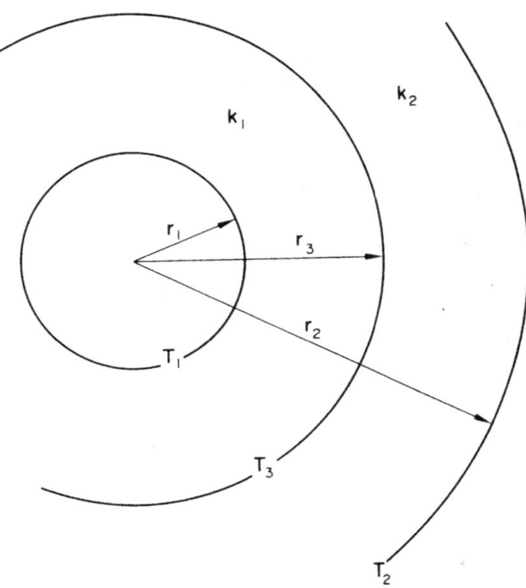

Fig. 11.3. Specifying temperatures and radii for a spherically
   symmetrical heat conduction problem, as used in Example 11.4.

radius r the flux density is given by

$$q = \frac{Q}{4\pi r^2} \ \text{W/m}^2 .$$

Then the potential difference between any two radii in the same material is
given by application of eqn. (11.5),

$$T_B - T_A = \frac{1}{k} \int_A^B - \mathbf{q} \cdot d\mathbf{l} = \frac{1}{k} \int_{r_A}^{r_B} - \frac{Q}{4\pi r^2} \, dr = \frac{Q}{4\pi k}\left(\frac{1}{r_B} - \frac{1}{r_A}\right).$$

Inserting the symbols from Fig. 11.3 gives

$$T_1 - T_3 = \frac{Q}{4\pi k_1}\left(\frac{1}{r_1} - \frac{1}{r_3}\right),$$

$$T_3 - T_2 = \frac{Q}{4\pi k_2}\left(\frac{1}{r_3} - \frac{1}{r_2}\right),$$

and $$T_1 - T_2 = (T_1 - T_3) + (T_3 - T_2).$$

Inserting some numbers from the problem gives

$$T_1 - T_3 = \frac{Q}{4\pi \times 400}\left(\frac{1}{0\cdot50} - \frac{1}{0\cdot51}\right) = \frac{Q}{4\pi} \times 9\cdot80 \times 10^{-5}$$

and
$$T_3 - T_2 = \frac{Q}{4\pi \times 0\cdot1}\left(\frac{1}{0\cdot51} - \frac{1}{0\cdot535}\right) = \frac{Q}{4\pi} \times 9\cdot16 \times 10^{-1}.$$

Therefore
$$T_1 - T_2 = \frac{Q}{4} \times 0\cdot916 = 60 - 30 = 30.$$

Therefore
$$Q = 412 \text{ W.}$$

T_3 must be found by subtracting a small temperature difference from T_1: otherwise it will be very inaccurate. Therefore

$$T_3 = 60\cdot0 - \frac{412 \times 9\cdot80 \times 10^{-5}}{4\pi} = 60\cdot0 - 0\cdot0032 = 59\cdot9968^{\circ}\text{C.}$$

11.5.   The walls of a cold room consist of 200 mm of thermal insulation inside 200 mm of brickwork. The thermal conductivity of the insulation is 0·2 W/m K and its surface heat transfer coefficient into the room is 6·0 W/m² K. The thermal conductivity of the brickwork is 1·0 W/m K and its surface heat transfer coefficient to the surroundings is 20 W/m² K. Find the heat flow into the room per unit area of wall and the surface temperatures of the wall and the insulation when the inside air temperature is -10°C and the outside air temperature is 20°C.

   Answer. The geometry of a section of the wall is shown in Fig. 11.4, where all the dimensions and the properties of the wall are shown in symbols. It is possible to write down the expressions for the heat flux density in terms of the temperature drop across each part of the wall. Therefore,

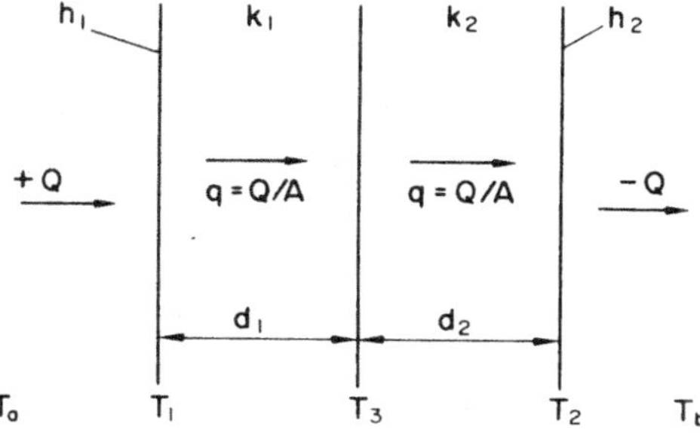

Fig. 11.4. Uniform thermal conduction through plane sheets of material having abrupt temperature changes at the surfaces.

$$q = h_1(T_a - T_1) \quad \text{across boundary 1,}$$

$$q = \frac{k_1}{d_1}(T_1 - T_3) \quad \text{through medium 1,}$$

$$q = \frac{k_2}{d_2}(T_3 - T_2) \quad \text{through medium 2,}$$

$$q = h_2(T_2 - T_b) \quad \text{across boundary 2.}$$

Inserting numbers from the problem gives

$$T_a - T_1 = \frac{q}{h_1} = \frac{q}{20} = 0\cdot05\ q,$$

$$T_1 - T_3 = \frac{qd_1}{k_1} = \frac{q\ 0\cdot2}{1\cdot0} = 0\cdot20\ q,$$

$$T_3 - T_2 = \frac{qd_2}{k_2} = \frac{q\ 0\cdot2}{0\cdot2} = 1\cdot00\ q,$$

$$T_2 - T_b = \frac{q}{h_2} = \frac{q}{6\cdot0} = 0\cdot167\ q.$$

Therefore

$$T_a - T_b = (T_a - T_1) + (T_1 - T_3) + (T_3 - T_2) + (T_2 - T_b)$$

$$= q\ (0\cdot05 + 0\cdot20 + 1\cdot00 + 0\cdot17) = 1\cdot42\ q.$$

Therefore
$$q = \frac{1}{1\cdot42}(T_a - T_b) = \frac{30}{1\cdot42} = 21\ \text{W/m}^2.$$

To find the temperatures, look for the smallest temperature differences
to achieve the greatest accuracy. Therefore

$$T_1 = T_a - 0\cdot05q = 20 - 1\cdot05 = 18\cdot95^{\circ}\text{C},$$

$$T_3 = T_1 - 0\cdot20q = 18\cdot95 - 4\cdot20 = 14\cdot75^{\circ}\text{C},$$

$$T_2 = T_b + 0\cdot167q = -10 + 3\cdot5 = -6\cdot5^{\circ}\text{C}.$$

Therefore the temperatures through the insulation of the wall and the
surrounding temperatures are:

$$T_a = 20\cdot0^{\circ}\text{C}, \qquad T_2 = -6\cdot5^{\circ}\text{C},$$

$$T_1 = 19\cdot0^{\circ}\text{C}, \qquad T_b = -10\cdot0^{\circ}\text{C}.$$

$$T_3 = 14\cdot7^{\circ}\text{C},$$

As a check, the final temperature difference is calculated:

$$T_3 - T_2 = 1\cdot00q = 21 = 14\cdot7 + 6\cdot5 \quad \text{checked}$$

11.6.   A lagged hot-water pipe consists of a pipe of 20 mm inside radius,
5·0 mm thick made of steel of thermal conductivity 50 W/m K and surface heat
transfer coefficient to water of 400 W/m² K, surrounded by insulation 25 mm
thick having a thermal conductivity of 0·25 W/m K and a surface heat transfer
coefficient to air of 10 W/m² K. If the water temperature is 60°C and the
air outside is 20°C, find the rate of heat loss from the water to the
surroundings and the surface temperatures of the pipe and insulation.

Answer. Again it is necessary to specify a heat flux source strength
and to calculate the temperature drop in terms of it. It will be easier to
follow the process of calculation of the potential differences if the
calculation initially is performed in symbols. The geometry of the system
and the material properties are shown symbolically in Fig. 11.5. As this is a

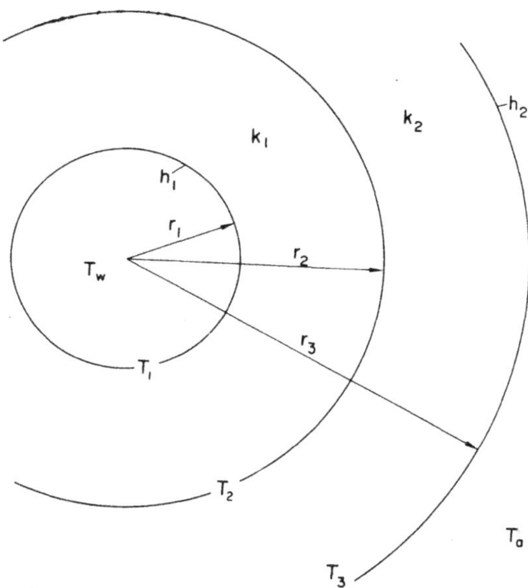

Fig. 11.5. Illustrating thermal conduction through the walls of
a lagged hot-water pipe.

cylindrically symmetrical system, we will specify a line source of flux of Q
at the axis of the pipe. Then the flux density at any radius r is given by

$$q = \frac{Q}{2\pi r}.$$

At the inside surface, the heat transfer is given by

$$q = h_1(T_w - T_1) \qquad \text{or} \qquad T_w - T_1 = \frac{Q}{2\pi r_1 h_1}.$$

The temperature drop across the pipe is given by

$$T_1 - T_2 = \frac{1}{k_1} \int_{r_1}^{r_2} \frac{Q}{2\pi r}\, dr = \frac{Q}{2\pi k_1} \log_e \frac{r_2}{r_1}$$

and that across the lagging by

$$T_2 - T_3 = \frac{Q}{2\pi k_2} \log_e \frac{r_3}{r_2}.$$

At the outside surface, the temperature drop is given by

$$T_3 - T_a = \frac{Q}{2\pi r_3 h_2}.$$

Inserting numbers into these equations gives

$$T_w - T_1 = \frac{Q}{2\pi \times 0\cdot02 \times 400} = \frac{0\cdot125\,Q}{2\pi},$$

$$T_1 - T_2 = \frac{Q}{2\pi \times 50} \log_e \frac{25}{20} = \frac{0\cdot004\,Q}{2\pi},$$

$$T_2 - T_3 = \frac{Q}{2\pi \times 0\cdot25} \log_e \frac{50}{25} = \frac{2\cdot773\,Q}{2\pi}$$

$$T_3 - T_a = \frac{Q}{2\pi \times 10 \times 0\cdot05} = \frac{2\cdot00\,Q}{2\pi}.$$

Therefore

$$T_w - T_a = (T_w - T_1) + (T_1 - T_2) + (T_2 - T_3) + (T_3 - T_a)$$
$$= \frac{Q}{2\pi}(0\cdot125 + 0\cdot004 + 2\cdot773 + 2\cdot000) = \frac{4\cdot90 Q}{2\pi}.$$

Therefore

$$Q = \frac{40 \times 2\pi}{4\cdot9} = 51 \text{ W/m.}$$

To find the temperatures, we have to substitute into the preceding equations, therefore

$$T_1 = T_w - \frac{0\cdot125\,Q}{2\pi} = 60 - 1\cdot02 = 58\cdot98\,^{\circ}\text{C},$$

$$T_2 = T_1 - \frac{0.004 Q}{2\pi} = 58\cdot98 - 0\cdot032 = 58\cdot95\,^{\circ}\text{C},$$

$$T_3 = T_a + \frac{2\cdot0 Q}{2\pi} = 20 + 16\cdot3 = 36\cdot3\,^{\circ}\text{C}.$$

As a check, we calculate the final temperature difference,

$$T_2 - T_3 = \frac{2\cdot773\,Q}{2\pi} = 22\cdot6 = 58\cdot9 - 36\cdot3.$$

Again there is one small temperature difference which is difficult to calculate as the small difference between two large numbers.

11.7.   A spherical volume of nuclear fuel is generating heat at the rate
of $5 \cdot 0$ MW/m^3 uniformly throughout its volume. It is a solid sphere of 20 mm
diameter.   If the fuel has a thermal conductivity of 40 W/m K and a surface
heat transfer coefficient of 400 W/m^2 K to the surroundings, find the surface
temperature of the sphere and the maximum temperature inside the sphere when
the surroundings are at 10°C.

Answer. The flux density inside a spherical volume having a uniformly
distributed source is given in eqn. (2.14)(page 14), therefore inside the
sphere,

$$q = \tfrac{1}{3}\rho r,$$

where ρ is the rate of heat generation. The maximum temperature occurs at the
centre of the sphere. The potential difference between the centre and the
surface of the sphere is given by

$$T_{\max} - T_s = \frac{1}{k} \int_R^O - \mathbf{q} \cdot d\mathbf{l} = \frac{1}{k} \int_R^O - \tfrac{1}{3}\rho r \, dr = \frac{\rho R^2}{6k}.$$

For the heat transfer at the surface,

$$q = \tfrac{1}{3}\rho R = h(T_s - T_a).$$

Therefore

$$T_s = T_a + \frac{\rho R}{3h} = 10 + \frac{5 \cdot 0 \times 10^6 \times 0 \cdot 01}{3 \times 400} = 10 + 41 \cdot 7 = 51 \cdot 7\,^{\mathrm{o}}\mathrm{C}.$$

Therefore

$$T_{\max} = T_s + \frac{\rho R^2}{6k} = 51 \cdot 7 + \frac{5 \cdot 0 \times 10^6 \times 10^{-4}}{6 \times 40} = 51 \cdot 7 + 2 \cdot 1 = 53 \cdot 8\,^{\mathrm{o}}\mathrm{C}.$$

11.8.   A solid copper bar of diameter 120 mm carries an electric current
density of 20 MA/m^2. The electrical conductivity of copper is $5 \cdot 0 \times 10^7$ S/m
and its thermal conductivity is 400 W/m K. It is surrounded by a cooling
fluid at 10°C and the surface heat transfer coefficient to the cooling fluid
is 1000 W/m^2 K. Find the surface temperature of the bar and the maximum temperature inside the bar.

Answer. The rate of heat generation inside the bar is determined from
eqn. (10.10)(page 196). Therefore

$$\rho = \frac{J^2}{\sigma} = \frac{400 \times 10^{12}}{5 \cdot 0 \times 10^7} = 8 \cdot 0 \times 10^6 \text{ W/m}^3.$$

For the cylindrically symmetrical system inside the bar, the flux density at
any radius r is given by

$$q = \frac{\pi r^2 \rho}{2\pi r} = \tfrac{1}{2}r\rho.$$

The potential difference between the surface and the centre of the bar is
given by

$$T_{max} - T_s = \frac{1}{k} \int_R^0 - \mathbf{q} \cdot d\mathbf{l} = \frac{1}{k} \int_R^0 - \tfrac{1}{2} r\rho \, dr = \frac{\rho R^2}{4k}.$$

The potential difference across the surface is given by

$$q = \tfrac{1}{2} R\rho = h(T_s - T_a).$$

Therefore

$$T_s = T_a + \frac{R\rho}{2h} = 10 + \frac{6 \cdot 0 \times 10^{-2} \times 8 \cdot 0 \times 10^6}{2 \times 1000} = 10 + 240 = 250^{\circ}\text{C}$$

and the maximum temperature is given by

$$T_{max} = T_s + \frac{\rho R^2}{4k} = 250 + \frac{8 \cdot 0 \times 10^6 \times 36 \times 10^{-4}}{4 \times 400} = 250 + 18 = 268^{\circ}\text{C}.$$

PROBLEMS

11.1.   The inner surface of a 230 mm brick wall of a furnace is kept at
800°C and it is found that the outer surface temperature is 150°C. Calculate
the heat loss per square metre of wall area given that the thermal conductivity
is $1 \cdot 0$ W/m K.   ($2 \cdot 83$ kW/m^2.)

11.2.   An insulating brick wall 230 mm thick and of conductivity $0 \cdot 25$
W/m K is added to the outside of the furnace wall of Problem 11.1.   Calculate
the reduction in heat loss and the brick interface and outer surface temperature.
Assume that the inner surface temperature is unchanged, that the surroundings
are at 20°C and that the surface heat transfer coefficient for the outer surface
is 12 W/m^2 K.   (78 per cent; 655°C; 72°C.)

11.3.   Water at 60°C flows through a 50 mm bore steel pipe, $6 \cdot 0$ mm thick.
The surroundings are at 20°C.   The pipe conductivity is 50 W/m K and the surface
heat transfer coefficients inside and outside are 1000 and 13 W/m^2 K respectively.
Calculate the heat loss per unit length of pipe and the temperature of the
surfaces.   (100 W/m; $59 \cdot 36^{\circ}$C; $59 \cdot 29^{\circ}$C.)

11.4.   An insulating layer 25 mm thick and conductivity $0 \cdot 2$ W/m K is added
to the outside of the pipe of Problem 11.3.   Assuming the same inside heat
transfer coefficient and an outside one of 10 W/m K, calculate the heat loss
per unit length and the surface and interface temperatures.   (52 W/m; $59 \cdot 67^{\circ}$C;
$59 \cdot 63^{\circ}$C; $34 \cdot 8^{\circ}$C.)

11.5. A pipe of outside diameter 75 mm is covered with two layers of insulating material each 25 mm thick. The conductivity of one material is four times that of the other. Show that the combined conductivity of the two layers is 28 per cent more when the better insulating material is put on the outside than when it is put on the inside.

11.6. A hollow cylindrical copper bar, having internal and external diameters of 12 mm and 50 mm respectively, carries an electric current density of 50 MA/m^2. The electrical resistivity is $2 \cdot 0 \times 10^{-8}$ Ω m and the thermal conductivity is 400 W/m K. When the outer surface is maintained at 40°C and no heat is removed through the central hole, find the position and value of the maximum temperature. (55$\cdot$2^{o}C at 6$\cdot$0 mm radius.)

11.7. The hollow copper bar of Problem 11.6, carrying the same current density, now has the inner surface cooled to 25°C, with the outer still at 40°C. Find the position and value of the maximum temperature and the heat removed internally and externally. (41$\cdot$9^{o}C at 19$\cdot$4 mm radius; 53$\cdot$2 kW/m; 39$\cdot$3 kW/m.)

11.8. A long length of hot-water pipe 100 mm in diameter is to be lagged in the most efficient way. Two different thermal insulating materials are available, but sufficient only to provide one complete layer of each. One material is 10 mm thick but has a thermal conductivity which is only half of that of the other which is 20 mm thick. In what order should they be placed? What is the percentage increase in conductivity resulting from placing the materials in the wrong order? (8$\cdot$1 per cent.)

11.9. Gas at a temperature of 400°C is contained in a spherical concrete pressure vessel of internal diameter 3$\cdot$0 m and external diameter 4$\cdot$0 m. Calculate the rate of heat loss from the vessel when the surrounding air is at 20^{o}C. The thermal conductivity of concrete is 1$\cdot$0 W/m K and the internal and external heat transfer coefficients are 10 and 6$\cdot$0 W/m^2 K respectively. (18$\cdot$9 kW.)

11.10. A nuclear reactor is fuelled with uranium in the form of tubular elements of 50 mm inside diameter and 100 mm outside diameter. It is cooled by a gas which maintains the temperature of both the internal and external faces of the elements constant at 400°C. If the rate of heat generation is constant throughout the element at 30 MW/m^3 and the thermal conductivity of uranium is 35 W/m K, find the position and magnitude of the maximum uranium temperature. (468°C at 36$\cdot$8 mm radius.)

SOLUTIONS

11.1. As no other information is given, it must be assumed that the furnace is large enough to be considered to have walls which are plane surfaces. Then the thermal conduction field is a uniform field and the problem is similar to Example 11.1. Using the formula derived in the solution to that example gives

$$q = \frac{k(T_1 - T_2)}{d} = \frac{1 \cdot 0(800 - 150)}{230 \times 10^{-3}} = 2 \cdot 83 \times 10^3 \text{ W/m}^2 = 2 \cdot 83 \text{ kW/m}^2.$$

11.2. Continuing the calculation from the solution to the last problem, the potential difference across the wall is given by

$$\Phi_1 = T_1 - T_2 = \frac{q d_1}{k_1} = \frac{230 q}{10^3}.$$

That across the insulation is given by

$$\Phi_2 = T_2 - T_3 = \frac{q d_2}{k_2} = \frac{230 \times 4 q}{10^3}.$$

The potential difference at the outer surface is given by eqn. (11.6) which becomes

$$\Phi_3 = T_3 - T_a = \frac{q}{h} = \frac{q}{12}.$$

Therefore the total potential difference is given by

$$\Phi = \Phi_1 + \Phi_2 + \Phi_3 = T_1 - T_a = q(0 \cdot 230 + 0 \cdot 920 + 0 \cdot 0835) = 1 \cdot 233 q.$$

Therefore

$$q = \frac{800 - 20}{1 \cdot 233} = \frac{780}{1 \cdot 233} = 633 \text{ W/m}^2.$$

Percentage change in heat loss is given by

$$\left(\frac{2830 - 633}{2830} \right) \times 100 = 78 \text{ per cent.}$$

To find the intermediate temperatures,

$$T_2 = 800 - 633 \times 0 \cdot 230 = 655^{\circ}\text{C},$$

$$T_3 = 20 + \frac{633}{12} = 72 \cdot 7^{\circ}\text{C}.$$

The check calculation,

$$T_2 - T_3 = 633 \times 0 \cdot 920 = 582 = 655 - 73.$$

11.3.   This is a cylindrically symmetrical problem with potential differences across the surfaces. It is similar to Example 11.6. Using the formulae derived in the solution to that example,

$$T_w - T_1 = \frac{Q}{2\pi r_1 h_1} = \frac{Q}{2\pi}\left(\frac{10^3}{25 \times 1000}\right) = \frac{0\cdot040\,Q}{2\pi},$$

$$T_1 - T_2 = \frac{Q}{2\pi k}\log_e \frac{r_2}{r_1} = \frac{Q}{2\pi} \times \frac{1}{50}\log_e\frac{31}{25} = \frac{0\cdot0043\,Q}{2\pi},$$

$$T_2 - T_a = \frac{Q}{2\pi r_2 h_2} = \frac{Q}{2\pi}\left(\frac{10^3}{31 \times 13}\right) = \frac{2\cdot481\,Q}{2\pi}.$$

Therefore

$$T_w - T_a = (T_w - T_1) + (T_1 - T_2) + (T_2 - T_a)$$

$$= \frac{Q}{2\pi}(0\cdot040 + 0\cdot0043 + 2\cdot481) = \frac{2\cdot525\,Q}{2\pi}.$$

Therefore
$$Q = \frac{2\pi(60 - 20)}{2\cdot525} = 99\cdot6 \text{ W/m}.$$

To find the pipe temperatures,

$$T_1 = T_w - \frac{99\cdot6 \times 0\cdot040}{2\pi} = 60 - 0\cdot64 = 59.36^{\circ}\text{C},$$

$$T_2 = T_1 - \frac{99\cdot6 \times 0\cdot0043}{2\pi} = 59\cdot36 - 0\cdot07 = 59.29^{\circ}\text{C}.$$

11.4.   This problem continues from the last problem, therefore

$$T_w - T_1 = \frac{Q}{2\pi r_1 h_1} = \frac{0\cdot040\,Q}{2\pi},$$

$$T_1 - T_2 = \frac{Q}{2\pi k_1}\log_e \frac{r_2}{r_1} = \frac{0\cdot0043\,Q}{2\pi},$$

$$T_2 - T_3 = \frac{Q}{2\pi k_2}\log_e \frac{r_3}{r_2} = \frac{Q}{2\pi} \times \frac{1}{0\cdot2}\log_e\frac{56}{31} = \frac{2\cdot9568\,Q}{2\pi},$$

$$T_3 - T_a = \frac{Q}{2\pi r_3 h_2} = \frac{Q}{2\pi}\left(\frac{10^3}{56 \times 10}\right) = \frac{1\cdot7857\,Q}{2\pi}.$$

Therefore

$$T_w - T_a = \frac{Q}{2\pi}(0\cdot040 + 0\cdot0043 + 2\cdot9568 + 1\cdot7857) = \frac{4\cdot7868\,Q}{2\pi}.$$

Therefore
$$Q = \frac{2\pi \times 40}{4\cdot7868} = 52\cdot5 \text{ W/m}.$$

To find the various surface temperatures,

$$T_1 = T_w - \frac{52\cdot5 \times 0\cdot040}{2\pi} = 60 - 0\cdot333 = 59.67^{\circ}\text{C},$$

$$T_2 = T_1 - \frac{52 \cdot 5 \times 0 \cdot 0043}{2\pi} = 59 \cdot 67 - 0 \cdot 036 = 59 \cdot 63^{\circ}C,$$

$$T_3 = T_a + \frac{52 \cdot 5 \times 1 \cdot 7857}{2\pi} = 20 + 14 \cdot 92 = 34 \cdot 92^{\circ}C.$$

The check calculation,

$$T_2 - T_3 = \frac{52 \cdot 5 \times 2 \cdot 9568}{2\pi} = 24 \cdot 7 = 59 \cdot 6 - 34 \cdot 9.$$

11.5.   Quoting the formulae that were used in the solutions to the last two problems,

$$T_1 - T_2 = \frac{Q}{2\pi k_1} \log_e \frac{r_2}{r_1} = \frac{Q}{2\pi k_1} \log_e \frac{62 \cdot 5}{37 \cdot 5} = \frac{0 \cdot 5111 Q}{2\pi k_1},$$

$$T_2 - T_3 = \frac{Q}{2\pi k_2} \log_e \frac{r_3}{r_2} = \frac{Q}{2\pi k_2} \log_e \frac{87 \cdot 5}{62 \cdot 5} = \frac{0 \cdot 3365 Q}{2\pi k_2}.$$

When the better insulating material is on the outside,

$$T_1 - T_3 = \frac{Q}{2\pi k} \left(\frac{0 \cdot 5111}{4} + 0 \cdot 3365 \right) = \frac{0 \cdot 464 \, Q}{2\pi k}.$$

When the better insulating material is on the inside,

$$T_1 - T_3 = \frac{Q}{2\pi k} \left(0 \cdot 5111 + \frac{0 \cdot 3365}{4} \right) = \frac{0 \cdot 595 Q}{2\pi k}.$$

Therefore the percentage increase in conductivity is given by

$$\left(\frac{1}{0 \cdot 464} - \frac{1}{0 \cdot 595} \right) \times 0 \cdot 595 \times 100 = 28 \text{ per cent.}$$

11.6.   The rate of heat generation inside the bar is given by

$$\rho = \frac{J^2}{\sigma} = 2 \cdot 0 \times 10^{-8} \times 25 \times 10^{14} = 5 \cdot 0 \times 10^7 \text{ W/m}^3.$$

The heat flux density at any radius r inside the bar is given by

$$q = \frac{\rho \pi (r^2 - r_1^2)}{2\pi r} = \tfrac{1}{2}\rho \left(r - \frac{r_1^2}{r} \right) \text{ W/m}^2.$$

As no heat is removed through the central hole, the flux density at the inside radius will be zero and the maximum temperature will be at the inner radius of the bar. Calculating the potential difference across the thickness of the bar gives

$$T_1 - T_2 = \frac{1}{k} \int_{r_2}^{r_1} - \mathbf{q} \cdot d\mathbf{l} = \frac{1}{k} \int_{r_2}^{r_1} - \tfrac{1}{2}\rho \left(r - \frac{r_1^2}{r} \right) dr$$

$$= \frac{\rho}{2k}\left\{\tfrac{1}{2}r_2^2 - r_1^2\left(\tfrac{1}{2} + \log_e \frac{r_2}{r_1}\right)\right\}.$$

Inserting numbers from the problem gives

$$T_1 = 40 + \frac{5\cdot0 \times 10^7 \times 10^{-6}}{2 \times 400}\left\{312 - 36\left(0\cdot5 + 1\cdot427\right)\right\} = 40 + 15\cdot2 = 55\cdot2^0\text{C}.$$

11.7.   In order to make allowance for the flow of heat into the centre of the bar it is necessary to postulate a negative heat flux source along the axis of the bar. Let this source be of strength $-Q$. Then using some results from the solution to the last problem, the heat flux density at radius r inside the bar is given by

$$q = -\frac{Q}{2\pi r} + \tfrac{1}{2}\rho\left(r - \frac{r_1^2}{r}\right).$$

The potential difference across the thickness of the bar is given by

$$T_1 - T_2 = \frac{1}{k}\int_{r_2}^{r_1} -\mathbf{q} \cdot d\mathbf{l} = -\frac{Q}{2\pi k}\log_e \frac{r_2}{r_1} + \frac{\rho}{2k}\left\{\tfrac{1}{2}r_2^2 - r_1^2\left(\tfrac{1}{2} + \log_e \frac{r_2}{r_1}\right)\right\}.$$

Inserting numbers gives

$$25 - 40 = \frac{1\cdot427\,Q}{2\pi \times 400} + 15\cdot2.$$

Therefore $$Q = \frac{30\cdot2 \times 2\pi \times 400}{1\cdot427} = 5\cdot32 \times 10^4 \text{ W/m} = 53\cdot2 \text{ kW/m}.$$

The total heat generated inside the bar is given by

$$\rho\pi(r_2^2 - r_1^2) = 5\cdot0 \times 10^7 \times \pi \times 10^{-6} \times (625 - 36) = 9\cdot25 \times 10^4 \text{ W/m} = 92\cdot5 \text{ kW/m}.$$

Therefore the heat flow to the external surface = $92\cdot5 - 53\cdot2 = 39\cdot3$ kW/m.

The maximum temperature occurs where the potential gradient is zero. Therefore

$$0 = -\frac{Q}{2\pi r} + \tfrac{1}{2}\rho\left(r - \frac{r_1^2}{r}\right).$$

Therefore

$$r^2 = r_1^2 + \frac{Q}{\pi\rho} = (36 + 339) \times 10^{-6} = 375 \times 10^{-6} \text{ m}^2.$$

Therefore $$r = 19\cdot4 \text{ mm}.$$

The maximum temperature is given by

$$T_1 - T_3 = -\frac{Q}{2\pi k}\log_e \frac{r_3}{r_1} + \frac{\rho}{2k}\left\{\tfrac{1}{2}r_3^2 - r_1^2\left(\tfrac{1}{2} + \log_e \frac{r_3}{r_1}\right)\right\}$$

or, integrating for the smaller temperature drop,

$$T_3 - T_2 = -\frac{Q}{2\pi k}\log_e\frac{r_2}{r_3} + \frac{\rho}{2k}\left(\tfrac{1}{2}r_2^2 - \tfrac{1}{2}r_3^2 - r_1^2\log_e\frac{r_2}{r_3}\right)$$

Therefore

$$T_3 = T_2 - \frac{1}{2k}\left(\frac{Q}{\pi} + \rho r_1^2\right)\log_e\frac{r_2}{r_3} + \frac{\rho}{4k}(r_2^2 - r_3^2).$$

Inserting numbers gives

$$T_3 = 40 - \frac{1}{800}\left(\frac{53\cdot2\times10^3}{\pi} + 50\times36\right)\log_e\frac{25}{19\cdot4} + \frac{50}{1600}(625 - 375)$$

$$= 40 - 5\cdot9 + 7\cdot8 = 41\cdot9^{\circ}\text{C}.$$

11.8.   Comparison with the solution to Problem 11.4 will show that the total heat loss per unit length of pipe is given by

$$Q = \frac{2\pi(T_1 - T_2)}{\dfrac{1}{k_1}\log_e\dfrac{r_3}{r_1} + \dfrac{1}{k_2}\log_e\dfrac{r_2}{r_3}}.$$

For the first condition, r_1 = 50 mm, r_3 = 60 mm, r_2 = 80 mm, k_1 = k, and k_2 = $2k$. Therefore

$$Q = \frac{2\pi(T_1 - T_2)}{\dfrac{1}{k}\log_e\dfrac{60}{50} + \dfrac{1}{2k}\log_e\dfrac{80}{60}} = \frac{2\pi k(T_1 - T_2)}{0\cdot1823 + \tfrac{1}{2}\times0\cdot2877} = \frac{2\pi k(T_1 - T_2)}{0\cdot3262}.$$

For the second condition, r_1 = 50 mm, r_3 = 70 mm, r_2 = 80 mm, k_1 = $2k$, and k_2 = k. Therefore

$$Q = \frac{2\pi(T_1 - T_2)}{\dfrac{1}{2k}\log_e\dfrac{70}{50} + \dfrac{1}{k}\log_e\dfrac{80}{70}} = \frac{2\pi k(T_1 - T_2)}{\tfrac{1}{2}\times0\cdot3365 + 0\cdot1336} = \frac{2\pi k(T_1 - T_2)}{0\cdot3018}.$$

Therefore the first condition with the thicker material on the outside leads to the smallest heat loss. The percentage increase in conductivity due to putting the insulation layers in the wrong order is

$$\left(\frac{1}{0\cdot3018} - \frac{1}{0\cdot3262}\right)\times0\cdot3262\times100 = 8\cdot1 \text{ per cent.}$$

11.9.   This is a spherically symmetrical system. The potential difference across two surfaces in a spherically symmetrical system has been derived in the solution to Example 11.2. The temperature drop across the surface is given by eqn. (11.6). Therefore the relevant equations for this problem are:

$$T_1 - T_2 = \frac{Q}{4\pi k}\left(\frac{1}{r_1} - \frac{1}{r_2}\right) = \frac{Q}{4\pi \times 1.0}\left(\frac{1}{1.5} - \frac{1}{2.0}\right) = \frac{0.1667Q}{4\pi},$$

$$400 - T_1 = \frac{Q}{4\pi r_1^2 h_1} = \frac{Q}{4\pi \times 1.5^2 \times 10} = \frac{0.0444Q}{4\pi},$$

$$T_2 - 20 = \frac{Q}{4\pi r_2^2 h_2} = \frac{Q}{4\pi \times 2.0^2 \times 6.0} = \frac{0.0417Q}{4\pi}.$$

Therefore

$$400 - 20 = \frac{Q}{4\pi}(0.1667 + 0.0444 + 0.0417) = \frac{0.2528Q}{4\pi}.$$

Therefore

$$Q = \frac{4\pi \times 380}{0.253} = 1.89 \times 10^4 \ \text{W} = 18.9 \ \text{kW}.$$

11.10.   This problem is similar to Problem 11.7. Using the formula derived in the solution to that problem for the temperature difference between the two surfaces gives

$$0 = -\frac{Q}{2\pi k}\log_e\frac{r_2}{r_1} + \frac{\rho}{2k}\left\{\tfrac{1}{2}r_2^2 - r_1^2\left(\tfrac{1}{2} + \log_e\frac{r_2}{r_1}\right)\right\}.$$

Therefore

$$Q = \pi\rho\left(\frac{r_2^2 - r_1^2}{2\log_e\dfrac{r_2}{r_1}} - r_1^2\right) = \pi \times 30 \times 10^6 \times\left(\frac{2500 - 625}{2 \times 0.6931} - 625\right) \times 10^{-6}$$

$$= \pi \times 30 \times 727.6 = 6.857 \times 10^4 \ \text{W/m} = 68.57 \ \text{kW/m}.$$

The position of the maximum temperature is given by

$$r^2 = r_1^2 + \frac{Q}{\pi\rho} = 625 \times 10^{-6} + \frac{6.857 \times 10^4}{\pi \times 30 \times 10^6} = 1352.6 \times 10^{-6} \ \text{m}^2.$$

Therefore $r = 36.78$ mm.

The maximum temperature is given by

$$T_m = T_1 + \frac{1}{2k}\left(\frac{Q}{\pi} + \rho r_1^2\right)\log_e\frac{r_3}{r_1} - \frac{\rho}{2k}\left(r_3^2 - r_1^2\right)$$

$$= 400 + \frac{1}{70}\left(\frac{68570}{\pi} + 30 \times 10^6 \times 625 \times 10^{-6}\right)0.3839$$

$$- \frac{30 \times 10^6}{2 \times 70}(1352.6 - 625) \times 10^{-6}$$

$$= 400 + 67.9 = 467.9\,^\circ\text{C}.$$

FURTHER PROBLEMS

11.11.   The outside wall of a room has an average thermal conductivity of $1 \cdot 0$ W/m K. The area of the wall is 25 m^2 and its thickness is 250 mm. If $1 \cdot 0$ kW heating in the room is sufficient to keep the inside surface temperature of the wall at 20°C, find the temperature of the outside surface of the wall assuming no heat loss elsewhere. (10°C.)

11.12.   Find the rate of heat input required to keep the inside of a house at a temperature of 25°C when the air temperature outside is 0°C. Assume that the house has a uniform heat loss from all its external surfaces except the floor through which no heat is lost. The house is $10 \cdot 0$ m by $10 \cdot 0$ m by $8 \cdot 0$ m high with a flat roof and the walls (and roof) are made of brick, 230 mm thick, having a thermal conductivity of $2 \cdot 0$ W/m K and a surface heat transfer coefficient of 20 W/m^2 K.   ($63 \cdot 6$ kW.)

11.13.   A spherical pressure vessel is made of concrete with a thermal conductivity of $1 \cdot 5$ W/m K. It has an external diameter of 10 m and a wall thickness of $0 \cdot 5$ m. If the temperature of the outside surface of the wall is 30°C and 25 kW heating is required to maintain the gas inside at a constant temperature, find the temperature of the inside surface of the wall. ($59 \cdot 5^{\circ}$C.)

11.14.   A copper hot-water pipe 24 mm diameter is insulated with 12 mm of insulation having a thermal conductivity of $0 \cdot 15$ W/m K. If there is negligible temperature difference between the outside of the pipe and the water at 60°C, and if the temperature of the outside of the insulation is 30°C, find the rate of heat loss from the pipe. ($40 \cdot 8$ W/m.)

11.15.   An approximation for an electric light bulb consists of a sphere of 50 mm diameter. If the surface heat transfer coefficient of the material of the bulb envelope is 120 W/m^2 K, find the surface temperature of the outside of a 60 W bulb when the air temperature is 20°C.   ($83 \cdot 7^{\circ}$C.)

11.16.   A steel steam-pipe carries water at 100°C. The pipe is $0 \cdot 2$ m outside diameter, 20 mm thick, and has a thermal conductivity of 30 W/m K and a surface heat transfer coefficient of 600 W/m^2 K. It is surrounded by 100 mm thick layer of insulation having a conductivity of $0 \cdot 25$ W/m K and a surface heat transfer coefficient of $6 \cdot 0$ W/m^2 K to the surrounding air at 25°C. Find the rate of heat loss through the insulation and all the surface temperatures of the pipe and insulation. (133 W/m; $99 \cdot 56^{\circ}$C; $99 \cdot 40^{\circ}$C; $42 \cdot 8^{\circ}$C.)

11.17.   A solid copper bar of diameter 200 mm carries a uniformly distributed electric current of 50 kA. The electrical conductivity of copper is 5.0×10^7 S/m and its thermal conductivity is 400 W/m K. It is surrounded by air at 30°C and the surface heat transfer coefficient to the air is 20 W/m^2 K. Find the surface temperature of the bar and the maximum temperature inside the bar.   ($156 \cdot 6^{\circ}$C; $156 \cdot 9^{\circ}$C.)

11.18.   The solid copper conductor of Problem 11.17 is replaced by a hollow copper tube of the same outside diameter and inside diameter 20 mm carrying cooling fluid at 40°C. Find the surface temperature of the bar if the inside surface heat transfer coefficient to the fluid is 1000 W/m^2 K and the other conditions are the same as in Problem 11.17.   ($60 \cdot 3^{\circ}$C.)

11.19.   A long hollow concrete tube of internal radius 150 mm and external radius 300 mm has a large number of electric heating wires buried in it, uniformly distributed across the cross-section. Heat is supplied to the block at a rate of $8 \cdot 0$ kW/m and the inner and outer surfaces of the tube are maintained at a temperature of 30°C. Find the magnitude and position of the maximum temperature reached. The thermal conductivity of concrete is $1 \cdot 0$ W/m K.   (136°C at 221 mm radius.)

11.20.   A radioactive metal sphere of diameter 20 mm is surrounded by a layer of lead 50 mm thick. The temperature of the surrounding air is 25°C. If the rate of heat generation in the radioactive material is uniform at 30 MW/m^3, find the temperature of the two surfaces of the lead jacket and the maximum temperature inside the radioactive material. The thermal conductivity of the radioactive material is 30 W/m K, the thermal conductivity of lead is 35 W/m K, and the surface heat transfer coefficient to the surrounding air is 50 W/m^2 K.   (81°C; 104°C; 121°C.)

FURTHER SOLUTIONS

11.11. The heat flux density relationship is

$$q = \frac{Q}{A} = \frac{k}{d}(T_1 - T_2).$$

Therefore

$$T_2 = 20 - \frac{1 \cdot 0 \times 10^3 \times 0 \cdot 25}{25 \times 1 \cdot 0} = 10^{\mathrm{o}}\mathrm{C}.$$

11.12. The total wall area is given by

$$A = 2(10 + 10) \times 8 \cdot 0 + 10 \times 10 = 420\ \mathrm{m}^2.$$

The heat flux density is given by

$$q = \frac{Q}{A} = \frac{k}{d}(T_1 - T_2) = h(T_2 - T_a).$$

Therefore

$$Q = \frac{A(T_1 - T_a)}{\frac{d}{k} + \frac{1}{h}} = \frac{420 \times 25}{\frac{0 \cdot 23}{2 \cdot 0} + \frac{1}{20}} = \frac{420 \times 25}{0 \cdot 165} = 6 \cdot 36 \times 10^4\ \mathrm{W} = 63 \cdot 6\ \mathrm{kW}.$$

11.13. Using the formula derived in the solution to Example 11.2,

$$T_1 - T_2 = \frac{Q}{4\pi k}\left(\frac{1}{r_1} - \frac{1}{r_2}\right).$$

Therefore

$$T = 30 + \frac{25 \times 10^3}{4\pi \times 1 \cdot 5} \times \left(\frac{1}{4 \cdot 5} - \frac{1}{5 \cdot 0}\right) = 30 + 29 \cdot 5 = 59 \cdot 5^{\mathrm{o}}\mathrm{C}.$$

11.14. Using the formula derived in the solution to Example 11.3,

$$Q = \frac{2\pi k(T_1 - T_2)}{\log_e(r_2/r_1)} = \frac{2\pi \times 0 \cdot 15 \times 30}{\log_e(24/12)} = 40 \cdot 8\ \mathrm{W/m}.$$

11.15. Applying eqn. (11.6) to the spherical surface gives

$$T_s - T_a = \frac{Q}{4\pi R^2 h}.$$

Therefore

$$T_s = 20 + \frac{60}{4\pi \times (0 \cdot 025)^2 \times 120} = 20 + 63 \cdot 7 = 83 \cdot 7^{\mathrm{o}}\mathrm{C}.$$

11.16.   Using formulae similar to those used in the solution to Example 11.6,

$$T_w - T_1 = \frac{Q}{2\pi r_1 h_1} = \frac{Q}{2\pi \times 0\cdot 08 \times 600} = 0\cdot 00332\,Q,$$

$$T_1 - T_2 = \frac{Q}{2\pi k_1} \log_e \frac{r_2}{r_1} = \frac{Q}{2\pi \times 30} \log_e \frac{0\cdot 10}{0\cdot 08} = 0\cdot 00118\,Q,$$

$$T_2 - T_3 = \frac{Q}{2\pi k_2} \log_e \frac{r_3}{r_2} = \frac{Q}{2\pi \times 0\cdot 25} \log_e \frac{0\cdot 20}{0\cdot 10} = 0\cdot 4413\,Q,$$

$$T_3 - T_a = \frac{Q}{2\pi r_3 h_2} = \frac{Q}{2\pi \times 0\cdot 20 \times 6\cdot 0} = 0\cdot 1326\,Q.$$

Therefore
$$Q = \frac{100 - 25}{0\cdot 5784} = 130 \text{ W/m}$$

and
$$T_1 = 100 - 0\cdot 43 = 99\cdot 57\,^{\circ}\text{C},$$

$$T_2 = 100 - 0\cdot 43 - 0\cdot 15 = 99\cdot 42\,^{\circ}\text{C},$$

$$T_3 = 25 + 17\cdot 3 = 42\cdot 3\,^{\circ}\text{C},$$

and, as a check,
$$T_2 = 42\cdot 3 + 57\cdot 2 = 99\cdot 5\,^{\circ}\text{C},$$

an accuracy of better than $0\cdot 1$ per cent.

11.17.   Using formulae similar to those used in the solution to Example 11.8,

$$\text{current density in the bar} = \frac{50 \times 10^3}{\pi \times (0\cdot 1)^2} = 1\cdot 59 \times 10^6 \text{ A/m}^2,$$

$$\text{rate of heat generation} = \rho = \frac{(1\cdot 59)^2 \times 10^{12}}{5\cdot 0 \times 10^7} = 5.07 \times 10^4 \text{ W/m}^3,$$

$$T_{\max} - T_s = \frac{\rho R^2}{4k},$$

$$T_s = T_a + \frac{R\rho}{2h} = 30 + \frac{0\cdot 1 \times 5\cdot 07 \times 10^4}{2 \times 20} = 156\cdot 6\,^{\circ}\text{C},$$

$$T_{\max} = 156 + \frac{5\cdot 07 \times 10^4 \times 0\cdot 01}{4 \times 400} = 156\cdot 6 + 0\cdot 32 = 156\cdot 9\,^{\circ}\text{C}.$$

11.18.   Using the formula derived in the solution to Problem 11.7 and eqn. (11.6),

$$T_1 - T_2 = -\frac{Q}{2\pi k} \log_e \frac{r_2}{r_1} + \frac{\rho}{2k}\left\{ \tfrac{1}{2}r_2^2 - r_1^2\left(\tfrac{1}{2} + \log_e \frac{r_2}{r_1}\right)\right\}.$$

$$T_w - T_1 = -\frac{Q}{2\pi r_1 h_1},$$

$$T_2 - T_a = \frac{(\text{total heat} - Q)}{2\pi r_2 h_2}.$$

Current density is given by

$$J = \frac{I}{\pi(r_2^2 - r_1^2)} = \frac{50 \times 10^3}{\pi(0\cdot01 - 0\cdot0001)} = 1\cdot61 \times 10^6 \text{ A/m}^3.$$

Therefore

$$\rho = \frac{(1\cdot61)^2 \times 10^{12}}{5\cdot0 \times 10^7} = 5\cdot18 \times 10^4 \text{ W/m}^3,$$

$$\text{total heat} = 5\cdot18 \times 10^4 \times \pi \times 0\cdot0099 = 1610 \text{ W}.$$

Therefore

$$T_w - T_a = 40 - 30 = -0\cdot0159Q - 0\cdot0009Q + 0\cdot306 + 128\cdot1 - 0\cdot0796Q.$$

Therefore
$$Q = \frac{118\cdot4}{0\cdot0964} = 1228 \text{ W}.$$

Therefore
$$T_1 = 40 + 0\cdot0159 \times 1228 = 59\cdot5^{O}C,$$

$$T_2 = 30 + 128\cdot1 - 0\cdot0796 \times 1228 = 60\cdot3^{O}C.$$

To check,
$$T_2 - T_1 = 0\cdot0009 \times 1228 - 0\cdot306 = 0\cdot8^{O}C.$$

11.19.   Again, using the formulae derived in the solution to Problem 11.7,

$$\rho = \frac{Q}{A} = \frac{8000}{\pi(0\cdot090 - 0\cdot0225)} = 3\cdot77 \times 10^4 \text{ W/m}^3,$$

$$T_1 - T_2 = 0 = -\frac{Q}{2\pi \times 1\cdot0} \log_e \frac{30}{15} + \frac{3\cdot77 \times 10^4}{2 \times 1\cdot0}\left\{\frac{0\cdot090}{2} - 0\cdot0225\left(\tfrac{1}{2} + \log_e \frac{30}{15}\right)\right\}.$$

Therefore
$$Q = \frac{342\cdot2}{0\cdot1103} = 3102 \text{ W/m}.$$

The position of the maximum temperature is given by

$$r_3^2 = r_1^2 + \frac{Q}{\pi\rho} = 0\cdot0225 + 0\cdot0262 = 0\cdot0487 \text{ m}^2.$$

Therefore
$$r_3 = 0\cdot2207 \text{ m},$$

$$T_{\text{max}} = T_1 + \frac{Q}{2\pi k} \log_e \frac{r_3}{r_1} - \frac{\rho}{2k}\left\{\tfrac{1}{2}r_3^2 - r_1^2\left(\tfrac{1}{2} + \log_e \frac{r_3}{r_1}\right)\right\}$$

$$= 30 + 190\cdot6 - 82\cdot7 = 137\cdot9^{O}C.$$

11.20.   Using formulae derived in the solution to Examples 11.2 and 11.7 and eqn. (11.6),

$$T_{max} - T_1 = \frac{\rho r_1^2}{6k_1},$$

$$T_1 - T_2 = \frac{Q}{4\pi k_2}\left(\frac{1}{r_1} - \frac{1}{r_2}\right),$$

$$T_2 - T_a = \frac{Q}{4\pi r_2^2 h},$$

$$Q = \frac{4}{3}\pi r_1^3 \rho = \frac{4}{3}\pi \times (0{\cdot}010)^3 \times 30 \times 10^6 = 40\pi \text{ W.}$$

Therefore

$$T_2 = 25 + \frac{40\pi}{4\pi \times (0{\cdot}060)^2 \times 50} = 80{\cdot}6\,^{\circ}\text{C},$$

$$T_1 = 80{\cdot}6 + \frac{40\pi}{4\pi \times 35}\left(\frac{1}{0{\cdot}010} - \frac{1}{0{\cdot}060}\right) = 104{\cdot}4\,^{\circ}\text{C},$$

$$T_{max} = 104{\cdot}4 + \frac{30 \times 10^6 \times (0{\cdot}010)^2}{6 \times 30} = 121{\cdot}1\,^{\circ}\text{C}.$$

CHAPTER 12

Potential Fluid Flow

THEORY

The relationship defining the potential in fluid flow through permeable media is given in eqn. (10.11)(page 199). For ideal fluid flow there is no resistance to flow and therefore no physical quantity equivalent to potential or potential gradient. However, the concept of potential in the fluid flow field is useful, and the fluid potential is defined by

$$\mathbf{v} = - \operatorname{grad} \phi. \tag{12.1}$$

The constant equivalent to the permittivity or permeability is dimensionless and equal to unity. Therefore in the ideal fluid flow field,

$$\text{flux density} = \text{field intensity} = \mathbf{v} .$$

The fluid potential function is defined by

$$\phi_b - \phi_a = \int_a^b - \mathbf{v} \cdot d\mathbf{l} , \tag{12.2}$$

and is called the *velocity potential function*. It is important to note that the velocity potential is *not* the same as the potential energy of the fluid or the pressure head of the fluid. The potential function can only exist in a conservative field, i.e. if

$$\oint \mathbf{v} \cdot d\mathbf{l} = 0. \tag{12.3}$$

However, under certain conditions for both real and ideal fluids, eqn. (12.3) is not satisfied, and integration of the expression in eqn. (12.3) around a closed path gives a quantity other than zero. It is called the *circulation* in the fluid and is defined by

$$k_c = \int_C \mathbf{v} \cdot d\mathbf{l} , \tag{12.4}$$

which is the circulation enclosed by the path C. *Vorticity* is defined as the circulation per unit area enclosed by the path as that area tends to zero. It is a vector quantity whose direction is related to the direction of integration around the path by the right-hand screw rule, that is, clockwise rotation in the plane of the paper moves along a vector normal to the paper away from the observer.

$$\text{Vorticity} = \mathbf{k},$$

$$|\mathbf{k}| = \lim_{\delta A \to 0} \left(\frac{\oint \mathbf{v} \cdot d\mathbf{l}}{\delta A} \right). \tag{12.5}$$

For a general two-dimensional field,

$$k = \frac{\partial v_y}{\partial x} - \frac{\partial v_x}{\partial y}. \tag{12.6}$$

A free vortex is a form of vortex flow that can occur in an ideal fluid. True velocity potential flow occurs everywhere in the fluid except at the origin of the vortex. In the free vortex there is no radial flow, and the circumferential velocity is given by

$$v_\theta = \frac{k}{2\pi r}, \tag{12.7}$$

where k is the strength of the *line vortex* causing the vortex flow. The potential function and flux function of a line vortex of strength k are given by

$$\phi = - \frac{k\theta}{2\pi}, \tag{12.8}$$

$$\psi = \frac{k}{2\pi} \log_e \frac{a}{r}. \tag{12.9}$$

EXAMPLES

12.1.   A pipe which terminates abruptly in the middle of a volume of fluid might be said to approximate to a point source of an ideal fluid. Obtain an expression for the velocity potential function of such a source having a discharge of $2 \cdot 0 \times 10^{-3}$ m³/s.

Answer. First an expression for the velocity potential function is derived entirely in symbols. Let the discharge from the point source be Q m³/s. Then the velocity at any distance r is given by

$$v = \frac{Q}{4\pi r^2}.$$

Substituting into eqn. (12.2) between the radii r_1 and r_2 gives

$$\phi_2 - \phi_1 = \int_{r_1}^{r_2} - \frac{Q}{4\pi r^2} \, dr = \frac{Q}{4\pi} \left(\frac{1}{r_2} - \frac{1}{r_1} \right).$$

If the potential function is zero when $r_1 = \infty$, $\phi_1 = 0$, and the absolute potential function for the fluid point source is given by

$$\phi = \frac{Q}{4\pi r} = \frac{1 \cdot 59 \times 10^{-4}}{r} \; \text{m}^2/\text{s}.$$

12.2.   A forced vortex occurs when a container of fluid rotates about its axis. Assuming that the whole of the fluid rotates as a body with a constant angular velocity, obtain an expression for the vorticity at any point in the fluid.

     Answer. Consider a small element in a rotating fluid as shown in Fig. 12.1, where the whole body of the fluid is rotating with a constant angular

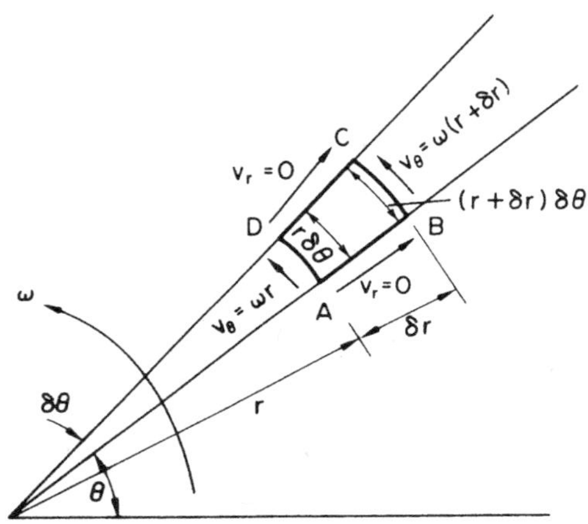

Fig. 12.1.   Velocities of a small element in a rotating fluid
which is a forced vortex.

velocity ω. Calculating the circulation of the element by integrating in a counter-clockwise direction around the element gives

$$k_c = 0 + \omega(r + \delta r)(r + \delta r)\delta\theta + 0 - \omega r r\,\delta\theta = \omega(2r + \delta r)\delta r\,\delta\theta.$$

If the element is taken to be a trapezium, its area is given by

$$\delta A = \tfrac{1}{2}(2r + \delta r)\delta r\,\delta\theta.$$

The vorticity is the circulation divided by the area, therefore

$$k = \frac{k_c}{\delta A} = 2\omega.$$

The vorticity at any point in the field of the fluid is constant and equal to twice the angular velocity of the fluid. However a separate calculation is necessary to verify what happens at the origin; if the path of integration is taken to be a circle about the centre of rotation radius r, then

$$k_c = (\omega r)(2\pi r) = 2\pi r^2 \omega$$

and the area is πr^2, therefore the vorticity is the same as before.

12.3.   A container of heavy viscous oil is placed on a gramophone turntable rotating at 33 r.p.m. Assume that after some time the oil is rotating as a solid body and calculate the vorticity in the oil.

Answer. The oil is in a forced vortex of constant angular velocity about the centre of rotation which due to its high viscosity can be assumed to be the same as the velocity of the container. The vorticity of a forced vortex of this type has been derived in the solution to the last example. Thirty-three r.p.m. is equal to 3·46 rad/s. Therefore the vorticity is given by

$$k = 6·9 \text{ rad/s}.$$

12.4.   Prove that eqn. (12.7) gives the velocity in a free vortex with true velocity potential flow.

Answer. In free vortex flow it is assumed that there is no radial velocity in the fluid but that there is a circular motion about some axis which is the origin of a polar coordinate system. By symmetry, the velocity will be constant at any radius, but the variation of this circumferential velocity with radius is unknown. An element of the fluid, together with its velocities, is shown in Fig. 12.2. Calculating the circulation of the element

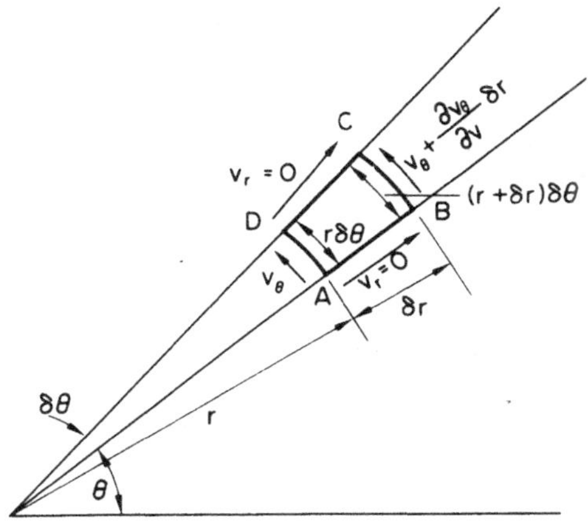

Fig. 12.2. Velocities of a small element of a free vortex.

by integration in a counter-clockwise direction around the element gives

$$k_c = 0 + \left(v_\theta + \frac{\partial v_\theta}{\partial r}\delta r\right)(r + \delta r)\delta\theta + 0 - v_\theta r\,\delta\theta.$$

If the second-order small term in $(\delta r)^2$ is neglected, the circulation becomes

$$k_c = \left(r\frac{\partial v_\theta}{\partial r} + v_\theta\right)\delta r\,\delta\theta.$$

For there to be true velocity potential flow, the circulation must be zero everywhere in the body of the fluid. Therefore k_c is zero and the velocity of the fluid is given by

$$r\frac{\partial v_\theta}{\partial r} + v_\theta = 0.$$

This expression may be integrated by the separation of variables. Therefore

$$\frac{\partial v_\theta}{v_\theta} = -\frac{\partial r}{r}$$

which when integrated becomes

$$\log_e v_\theta = -\log_e r + \log_e c = \log_e \frac{c}{r},$$

where c is the constant of integration. Therefore

$$v_\theta = \frac{c}{r}.$$

There is ideal fluid flow in any part of the field that does not include the origin of the vortex flow. Taking a circular path of integration at a constant radius from the origin gives

$$k_c = v_\theta\,2\pi r = \frac{c}{r}\,2\pi r = 2\pi c.$$

This is the value of the circulation for *any* path enclosing the origin. The strength of a line vortex is defined as the circulation for any path enclosing it, so that if k is the strength of the line vortex at the origin, $k = 2\pi c$, and the circumferential velocity in the vortex is given by

$$v_\theta = \frac{k}{2\pi r}.$$

12.5.   A tornado may be supposed to consist of a core of air rotating as a solid body surrounded by a region of free vortex flow. If the core diameter is 2·0 m and the velocity at a radius of 20 m is 1·0 m/s, find the maximum velocity occurring and the strength of the equivalent line vortex to give the same free vortex flow.

Answer. In the free vortex portion of the tornado, the velocity is given by

$$v_\theta = \frac{k}{2\pi r}.$$

Therefore

$$k = 2\pi r v_\theta = 2\pi \times 20 \times 1\cdot0 = 40\pi \text{ m}^2/\text{s}.$$

The maximum velocity will occur at the edge of the core, that is at a radius of $1\cdot0$ m, therefore

$$\text{maximum velocity} = \frac{40\pi}{2\pi \times 1\cdot0} = 20 \text{ m/s};$$

k is the strength of the equivalent line vortex, therefore

$$k = 126 \text{ m}^2/\text{s}.$$

12.6.   Derive expressions for the potential function and flux function of a line vortex of strength k m²/s.

Answer. The velocity in the field due to a line vortex of strength k at the origin is given by eqn. (12.7). The fluid potential function is obtained from the velocity in accordance with eqn. (12.2). The problem is that no path of integration must enclose the line of the vortex. The problem is solved by specifying an excluded region in the field which includes the line of the vortex. The excluded region consists of a cut in the field taken from infinity to the origin and enclosing the line of the vortex and negligible space elsewhere. Consider Fig. 12.3. The cut in the field is shown along the positive horizontal axis. Potential function calculations may be made so long as no path of integration crosses this positive horizontal axis. Since the velocity is entirely circumferential about the line of the vortex, the path of integration is taken along the circumference of a circle. Consider

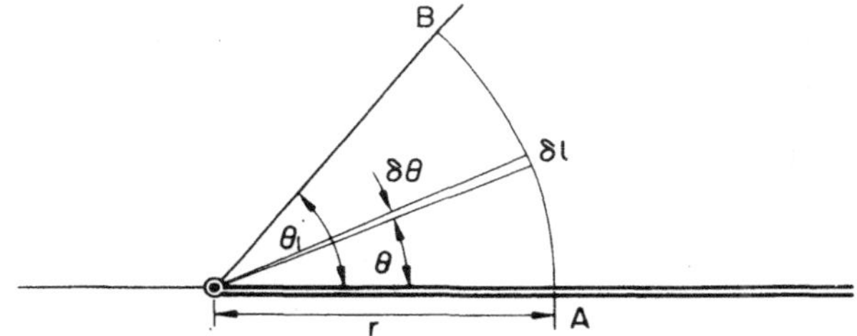

Fig. 12.3. Illustrating the terminology used in calculating the values of the potential function of a line vortex.

the path AB in Fig. 12.3. Then the potential function is given by

$$\phi_B - \phi_A = \int_r^B - \mathbf{v} \cdot d\mathbf{l} = \int_0^{\theta_1} - v_\theta r d\theta = -\int^{\theta_1} \frac{k}{2\pi r} r d\theta = \frac{k\theta_1}{2\pi}.$$

In more general terminology this equation becomes the same as eqn. (12.8).

The flux function lines will be parallel to the direction of flow. The value of flux function is obtained from the velocity in accordance with eqn. (6.2)(page 96). Referring again to Fig. 12.3, and considering unit depth, then the small element of area is a rectangular section, unit side by δr. Therefore the flux function is given by

$$\psi_r - \psi_a = \int_a^r - v_\theta dr = \int_a^r - \frac{k}{2\pi r} dr = -\frac{k}{2\pi} \log_e \frac{r}{a} = \frac{k}{2\pi} \log_e \frac{a}{r},$$

which is the same as eqn. (12.9).

12.7.   A doublet may be formed by the combination of two equal and opposite line vortices an infinitely small distance apart. If the strength of the vortices is k m^2/s and they are d m apart, then the strength of the doublet is given by $c = kd$ m^3/s. Find the potential function of such a line-vortex doublet.

Answer. The calculation of the potential function of a line-vortex doublet is similar to the calculation of the flux function of a line-source line-sink doublet as considered in the solution to Example 6.5 (page 99). The combination of two vortices a distance d apart is similar to the situation depicted in Fig. 3.4 (page 38). Then the potential function of the combination is given by

$$\phi = \phi_1 + \phi_2 = \frac{k}{2\pi} (- \theta_1 + \theta_2) = \frac{k\alpha}{2\pi}.$$

For the condition of the doublet where the distance d is infinitely small, the geometry is illustrated in Fig. 6.3 (page 100), and

$$\alpha = \frac{d \sin \theta}{r}.$$

Then the expression for the potential function becomes

$$\phi = \frac{k}{2\pi} \frac{d \sin \theta}{r} = \frac{c \sin \theta}{2\pi r}.$$

12.8.   A cylinder rotating in a uniform stream can be simulated in fields terminology by using the combination of a doublet with the uniform stream to

give a circular stagnation flux line and with the addition of a line vortex
to give the effect of the circular motion imparted by the rotation of the
cylinder on the fluid. Discuss the appropriateness of the model and find
expressions for both the flux function and potential function of the model.

Answer. In the solution to Example 3.8 (page 44), it is shown that the
combination of a doublet and a uniform field give a circular flux function
line that can be the boundary of a circular obstruction in a uniform flow.
This is a true ideal fluid field where there is no viscosity and the boundary
has no effect on the velocity of flow except to direct the flow parallel to
the boundary. For the rotating cylinder to impart circulatory motion to the
fluid of the problem, there must be viscosity so that the model cannot be an
ideal fluid. In order to make some estimate of the effect of rotation on a
real fluid, we make an arbitrary assumption that rotary motion is imparted to
the ideal fluid, realizing that the motion cannot be imparted by rotation of
the surface. In the model the rotary motion is generated by a line vortex.
The flux function of the combination of the uniform field, doublet, and line
vortex is obtained by summing the individual flux functions given in eqns.
(3.7)(page 37), (3.12)(page 39), and (12.9).

$$\psi = Ur \sin \theta - \frac{c}{2\pi r} \sin \theta + \frac{k}{2\pi} \log_e \frac{a}{r}.$$

If the zero of flux function is taken to be on the circular flux function line
at the surface of the cylinder at the radius a, then the solution to Example
3.8 (page 44) shows that

$$a^2 = \frac{c}{2\pi U}$$

and the flux function becomes

$$\psi = \frac{U \sin \theta}{r} (r^2 - a^2) + \frac{k}{2\pi} \log_e \frac{a}{r}.$$

The potential function of the combination is given by a similar combination of
the potential function of the individual components. They are to be found in
the solution to Example 8.4 (page 145), and eqns. (9.4)(page 167) and (12.8).
Therefore

$$\phi = - Ur \cos \theta - \frac{c \cos \theta}{2\pi r} - \frac{k\theta}{2\pi}.$$

Substituting the radius of the cylinder a into this expression gives

$$\phi = - \frac{U \cos \theta}{r} (r^2 + a^2) - \frac{k\theta}{2\pi}.$$

The flux and potential function lines appropriate to these equations are drawn in Fig. 12.4.

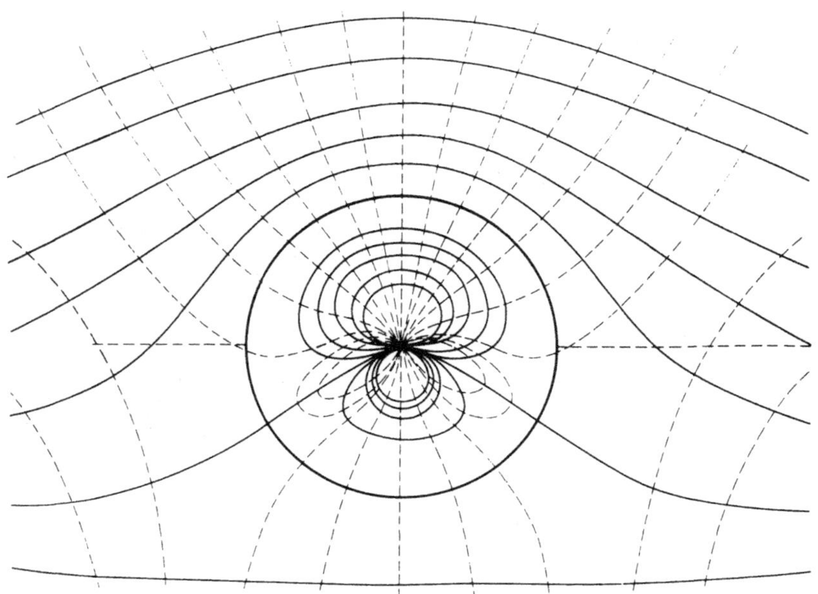

Fig. 12.4.   The flux and potential function lines for ideal fluid flow around a cylinder with circulation.   Potential function lines are shown dashed.

12.9.   A cylinder, 0·5 m diameter, is rotating at 10 rad/s and is mounted vertically in an infinitely wide stream of uniform flow of 2·0 m/s. Calculate the equivalent doublet strength to model the ideal fluid flow past the cylinder. Assume that the rotation of the cylinder imparts a free vortex flow to the fluid equivalent to the peripheral speed of the cylinder. Find the strength of the equivalent line vortex and the position of the stagnation points.

Answer.   The relationship between the radius of the cylinder and the strength of the equivalent doublet has been given in the solution to the last example.

$$a^2 = \frac{c}{2\pi U}.$$

Therefore, inserting numbers from the problem gives

$$c = 2\pi U a^2 = 2\pi \times 2 \cdot 0 \times \frac{0 \cdot 25}{4} = 0 \cdot 786 \ \text{m}^3/\text{s}.$$

For the line vortex, the fluid velocity is given by eqn. (12.7), therefore

$$k = 2\pi r v_\theta.$$

On the surface of the cylinder, the velocity due to rotation is given by

$$v_\theta = \omega r = 10 \times 0 \cdot 25 = 2 \cdot 5 \text{ m/s.}$$

Therefore
$$k = 2\pi \times 0 \cdot 25 \times 2 \cdot 5 = 3 \cdot 93 \text{ m}^2/\text{s.}$$

The stagnation points occur where the circumferential velocity is zero on the surface of the cylinder. The velocity can be obtained by differentiating the expression for the flux function which has been derived in the solution to the last example. Therefore

$$v_\theta = - \frac{\partial \psi}{\partial r} = - U \sin \theta \left(1 + \frac{a^2}{r^2}\right) + \frac{k}{2\pi r}.$$

On the surface of the cylinder, $r = a$ and the velocity becomes

$$v_\theta = - 2U \sin \theta + \frac{k}{2\pi a}.$$

At the stagnation point, $v_\theta = 0$, therefore

$$\sin \theta = \frac{k}{4\pi U a} = \frac{1 \cdot 25\pi}{4\pi \times 2 \cdot 0 \times 0 \cdot 25} = 0 \cdot 625.$$

Therefore
$$\theta = 38 \cdot 7^{\text{o}}.$$

12.10.   Calculate all the forces acting on the cylinder in the last example if the density of the liquid is 1000 kg/m³.

Answer. Since the model used in the solution to the last example is for an ideal fluid, and since the flow pattern is symmetrical in that the direction of flow cannot be determined from the flux function pattern, there is no net force acting on the cylinder in the direction of the fluid flow. However, there is a force acting perpendicular to the direction of flow, since the flow pattern is not symmetrical in this direction. The pressure acting on a small element of the surface of the cylinder is shown in Fig. 12.5. To calculate the force it is necessary to resolve the component of the pressure and then to integrate around the surface of the cylinder. Taking an element of surface of the cylinder of area $a\delta\theta$ for unit length of cylinder, the total force integrating around the surface of the cylinder is given by

$$F = \int_0^{2\pi} pa \sin \theta \, d\theta.$$

For horizontal flow the pressure is obtained from eqn. (6.5) (page 96) to be

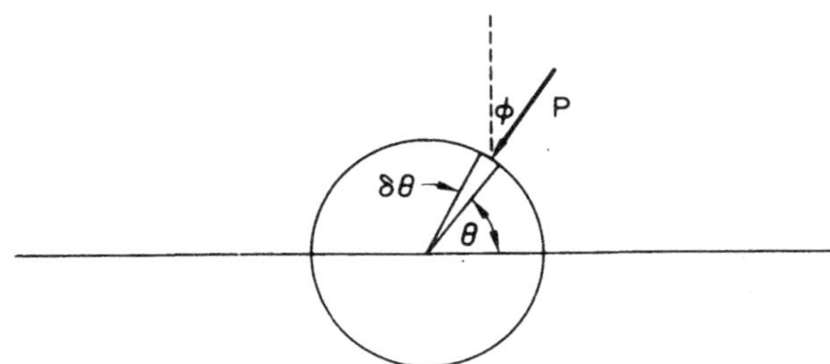

Fig. 12.5. Resolution of the pressure force on a small element of
the surface of a cylinder.

$$p = p_0 - \tfrac{1}{2}\rho v^2,$$

where p_0 is the pressure of the stationary fluid a long way away and ρ is the
density of the fluid. The velocity on the surface of the cylinder has been
derived in the solution to the last example. It is given by

$$v_\theta = -\,2U \sin \theta + \frac{k}{2\pi a}.$$

Therefore the pressure on the surface of the cylinder becomes

$$p = p_0 - \tfrac{1}{2}\rho\left(4U^2 \sin^2\theta - \frac{2Uk \sin \theta}{2\pi a} + \frac{k^2}{4\pi^2 a^2}\right).$$

The integration of odd powers of $\sin \theta$ from 0 to 2π is zero so that the
expression for the force simplifies to

$$F = \int_0^{2\pi} \frac{\rho Uk}{\pi} \sin^2 \theta \; d\theta = \rho Uk.$$

Substituting the numbers from the problem gives

$$F = 1000 \times 2{\cdot}0 \times 3{\cdot}93 = 7860 \text{ N/m}.$$

EXAMPLES

12.1.   Find expressions for the velocity potential function in both
rectangular and polar coordinates of a uniform stream of velocity U m/s.

12.2.   Find an expression for the velocity potential function of a line
source of strength q m²/m.

12.3.   In most field theory systems there is no physical equivalent to the line vortex; however, considering it purely as a mathematical exercise, find the **E** and **D** fields of an electrostatic line vortex of strength k C/m.

12.4.   Explain carefully the physical difference between *rotational* and *irrotational* fluid flow around a circular path. Which type of circulation is associated with an ideal fluid?

12.5.   By adding potential function values, find the equipotential line pattern of a combination of a uniform stream of 10 m/s and a line source of 2·0 m^2/s.

12.6.   A tornado may be supposed to consist of a core of air rotating as a solid body, surrounded by a region of free vortex flow. If the core radius is 2·0 m and the velocity at a radius of 100 m is 1·0 m/s, find the maximum velocity occurring and the strength of the equivalent line vortex.   (50 m/s; 628 m^2/s.)

12.7.   Find an expression for the circulation round an elemental area in a two-dimensional fluid flow field described in terms of polar coordinates. Hence find the way in which v_θ must vary with r if $v_r = 0$ for the flow to be irrotational.

12.8.   A long circular cylinder, 0·1 m in diameter lies with its axis normal to an infinitely wide stream of water flowing with a velocity of 3·0 m/s. It is rotating about its axis at 12·5 rad/s. Assume that the rotation of the cylinder causes a free vortex flow such that the water in contact with the cylinder moves at the same velocity as the solid surface in the absence of the uniform stream. Except for the impartation of vortex motion to the fluid by the cylinder, assume ideal fluid flow. If the density of the water is 1000 kg/m^3, find the force on the cylinder and the positions of the stagnation points.   (588 N/m; 6° off axis.)

12.9.   A line vortex of infinite length, circulation k, is located at a perpendicular distance x from an infinite plane wall. Determine the velocity induced by the line vortex at a general point on the wall, and using Bernoulli's equation, find the force exerted on the wall per unit length of vortex.   ($\rho k^2/4\pi x$.)

12.10.   The fluid speed at a distance of 1·0 m from the axis of a line vortex in an ideal fluid is 2·0 m/s. A similar vortex of the same strength is located at a distance of 3·0 m from a long wall with its axis parallel to the wall. Find the fluid speed at a point on the wall which is 7·0 m from the line of the vortex.   (0·245 m/s.)

SOLUTIONS

12.1.   A diagram of a general point in a uniform field is given in Fig. 8.4 (page 145). The value of the velocity potential function is obtained from eqn. (12.2). Referring to Fig. 8.4, if $\phi = 0$ at the point O, then the general velocity potential function in the field is given by

$$\phi = - U(OA) = - Ux = - Ur \sin \theta.$$

12.2.   A diagram of a general field of circular symmetry is shown in Fig. 8.8 (page 162). The value of the velocity potential function is again obtained from eqn. (12.2). For a line source, the flux density is given by a relationship similar to eqn. (2.15)(page 14),

$$v = \frac{q}{2\pi r} \text{ m/s}.$$

In terms of the dimensions shown in Fig. 8.8, the potential difference between two points in the field is given by

$$\phi_2 - \phi_1 = - \int_{r_1}^{r_2} v \, dr = - \int_{r_1}^{r_2} \frac{q}{2\pi r} \, dr = \frac{q}{2\pi} \log_e \frac{r_1}{r_2}$$

or, to give a positive value to the potential difference,

$$\phi_1 - \phi_2 = \frac{q}{2\pi} \log_e \frac{r_2}{r_1}.$$

It is impossible to specify an absolute value for the velocity potential due to a line source, but if $\phi = 0$ at $r = a$, then

$$\phi = \frac{q}{2\pi} \log_e \frac{a}{r}.$$

12.3.   By definition of the line vortex,

$$\frac{k}{\varepsilon_0} = \oint \mathbf{E} \cdot d\mathbf{l}.$$

k/ε_0 is used in order to maintain homogeneity of dimensions in the equation. At a fixed radius r from the line of the vortex, the field intensity will be constant, therefore

$$\frac{k}{\varepsilon_0} = \int_0^{2\pi} Er \, d\theta = 2\pi r E.$$

Therefore

$$E = \frac{k}{2\pi \varepsilon_0 r} \text{ V/m} \qquad \text{and} \qquad D = \frac{k}{2\pi r} \text{ C/m}^2,$$

where both these field quantities act tangentially to the circumference of
the circle of radius r.

12.4.   *Rotational flow* around a circular path in a fluid means that in
any small element of the fluid

$$\int \mathbf{v} \cdot d\mathbf{l} \neq 0.$$

Physically this means that any body suspended in or floating on the surface
of the fluid will tend to rotate on its own axis as well as moving in an
approximately circular orbit.

For *irrotational flow* around a circular path, although any small element
of fluid will be moving in a circular path, it is not rotating about its own
axis. In the small element of fluid,

$$\int \mathbf{v} \cdot d\mathbf{l} = 0$$

and any body suspended in or floating on the surface of the fluid will not
rotate about its own axis even though it will move in a circular path. The
body floating in the irrotational fluid will always have the same orientation
as it is moved around by the fluid.

12.5.   The equipotential line pattern of a uniform stream is equally
spaced straight lines perpendicular to the direction of flow. The
equipotential line pattern of a line source is circles centred on the line of
the source with logarithmic ratios between their radii. The completed line
pattern is shown in Fig. 12.6.

Fig. 12.6. The equipotential line pattern for a line source combined
with a uniform stream.   The solution to Problem 12.5.

12.6.   For free vortex flow, the tangential velocity is given by eqn. (12.7). Therefore the vorticity of the equivalent line vortex is given by

$$k = 2\pi r v_\theta = 2\pi \times 100 \times 1{\cdot}0 = 628 \text{ m}^2/\text{s}.$$

At the edge of the core,

$$v_\theta = \frac{628}{2\pi \times 2{\cdot}0} = 50 \text{ m/s},$$

which is the maximum velocity occurring. Inside the core, the velocity increases proportional to radius. The maximum velocity occurs at the edge of the core. Outside the core, the velocity decreases inversely proportional to radius.

12.7.   The velocities in a small element of a free vortex are shown in Fig. 12.2. In this problem it is also necessary to consider the existence of a radial component of velocity that is also a variable. Considering the trapezoidal element shown in Fig. 12.2, the circulation is given by

$$k_c = v_r \delta r + \left(v_\theta + \frac{\partial v_\theta}{\partial r}\delta r\right)(r + \delta r)\delta\theta - \left(v_r + \frac{1}{r}\frac{\partial v_r}{\partial\theta}r\delta\theta\right)\delta r - v_\theta r\delta\theta$$

$$= \left\{v_\theta + \frac{\partial v_\theta}{\partial r}(r + \delta r) - \frac{\partial v_r}{\partial\theta}\right\}\delta\theta\,\delta r.$$

For irrotational flow, $k_c = 0$; therefore

$$v_\theta + \frac{\partial v_\theta}{\partial r}(r + \delta r) - \frac{\partial v_r}{\partial\theta} = 0.$$

If $v_r = 0$,

$$v_\theta + \frac{\partial v_\theta}{\partial r}r = 0$$

neglecting the term in δr. Therefore

$$\frac{\partial v_\theta}{v_\theta} = -\frac{\partial r}{r}.$$

Integrating gives

$$\log_e v_\theta = -\log_e r + \log_e c = \log_e \frac{c}{r},$$

where c is the constant of integration. Therefore

$$v_\theta = \frac{c}{r}.$$

The velocity varies inversely with radius for the flow to be irrotational.

12.8.   All the equations relating to a cylinder rotating in a uniform stream were derived in the solutions to Examples 12.8 to 12.10. The strength of the equivalent line vortex is given by

$$k = 2\pi r v_\theta = 2\pi \times 0\cdot05 \times (0\cdot05 \times 12\cdot5) = 0\cdot196 \ \text{m}^2/\text{s}.$$

At the stagnation points,

$$\sin\theta = \frac{k}{4\pi Ua} = \frac{0\cdot196}{4\pi \times 3\cdot0 \times 0\cdot05} = 0\cdot104.$$

Therefore $\theta = 6^\circ$ or $(180 - 6)^\circ$.

The lift force $= \rho Uk = 1000 \times 3\cdot0 \times 0\cdot196 = 588$ N/m.

12.9.   The wall is a flux boundary. The problem can be solved by replacing the wall with an image vortex of opposite hand on the opposite side of the wall as shown in Fig. 12.7. At any point on the wall the contributions

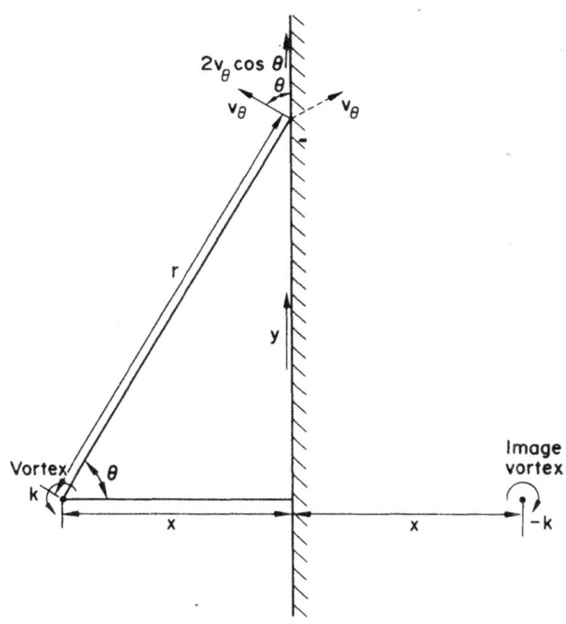

Fig. 12.7. Illustrating the geometry of Problem 12.9. A line vortex parallel to a plane flux boundary which is replaced by an image line vortex.

to the velocity parallel to the wall from each line vortex will be equal and will add together, whereas their contributions to the velocity perpendicular to the wall will be equal and opposite, so cancelling. The velocity due to one vortex is given by eqn. (12.7). From the geometry as shown in the figure,

$$r = \frac{x}{\cos\theta};$$

therefore, substituting into eqn. (12.7),

$$v_\theta = \frac{k\cos\theta}{2\pi x}.$$

On the wall, the component parallel to the wall from each vortex will add so that the total velocity along the wall is given by

$$v = 2v_\theta \cos\theta = \frac{k\cos^2\theta}{\pi x}.$$

From Bernoulli's equation, the pressure is given by

$$p = p_0 - \tfrac{1}{2}\rho v^2 = p_0 - \tfrac{1}{2}\rho\left(\frac{k\cos^2\theta}{\pi x}\right)^2.$$

The total force is given by

$$F = \int_\infty^\infty p\,dy - \int_{-\frac{1}{2}\pi}^{\frac{1}{2}\pi} \frac{pr}{\cos\theta}\,d\theta = \int_{-\frac{1}{2}\pi}^{\frac{1}{2}\pi} \frac{px}{\cos^2\theta}\,d\theta.$$

The contribution p_0 to the total pressure is the static pressure and is independent of the vortex, so in order to find the pressure exerted by the vortex it is necessary to make $p_0 = 0$. Therefore

$$F = -\tfrac{1}{2}\rho\,\frac{k^2}{\pi^2 x} \int_{-\frac{1}{2}\pi}^{\frac{1}{2}\pi} \cos^2\theta\,d\theta = -\rho\frac{k^2}{4\pi x}$$

12.10.   This problem is similar to Problem 12.9. To find the strength of the vortex,

$$k = 2\pi r v_\theta = 2\pi \times 1{\cdot}0 \times 2{\cdot}0 = 4\pi \ \mathrm{m^2/s}.$$

At the point on the wall, the angle θ is given by

$$\cos\theta = \frac{3}{7}.$$

Then the velocity on the wall is given by

$$v = \frac{k\cos^2\theta}{\pi x} = \frac{4\pi}{\pi \times 3{\cdot}0} \times \left(\frac{3}{7}\right)^2 = 0{\cdot}245 \ \mathrm{m/s}.$$

FURTHER PROBLEMS

12.11.   Obtain an expression for the potential function of a point source of fluid having a discharge of $0{\cdot}314 \ \mathrm{m^3/s}$ and hence or otherwise obtain an expression for the velocity anywhere due to this discharge.

12.12.   Find expressions for the potential function in both rectangular and polar coordinates of a uniform fluid flow of 10 m/s.

12.13.   Obtain an expression for the velocity potential function of a dipole of strength $0 \cdot 126$ m⁴/s.

12.14.   Obtain an expression for the vorticity in a general two-dimensional fluid field in rectangular coordinates.

12.15.   A centrifuge operating at 5000 r.p.m. is a circular chamber completely filled with liquid. If at speed, the liquid inside the centrifuge chamber rotates as a solid body, find an expression for the vorticity at any point in the liquid.   (1050 rad/s.)

12.16.   Liquid enters a circular container of 10 m diameter tangentially at the walls with a velocity of 10 m/s and leaves via a hole in the centre. If the liquid approximates to an ideal fluid, find the strength of the equivalent line vortex which could describe the free vortex flow in the container.   (314 m²/s.)

12.17.   Obtain an expression for the potential function of a line vortex of strength $6 \cdot 28$ m²/s.

12.18.   Draw the equipotential line pattern due to a line vortex of strength 16 m²/s which is parallel to and distant $4 \cdot 0$ m from an equal line vortex having the opposite hand of rotation.

12.19.   A long circular cylinder $0 \cdot 1$ m radius lies with its axis normal to a uniform flow of 10 m/s. The boundaries are at such a distance that their effect may be ignored. The cylinder is rotating about its own axis at 100 r.p.m. Assume that the rotation causes a free vortex flow such that the water in contact with the cylinder moves at the same velocity as the solid surface of the cylinder in the absence of the uniform stream. If the density of the water is 1000 kg/m³, find the strength of the doublet and uniform flow needed to model the circular boundary in an ideal fluid and the strength of the vortex needed to provide the vortex flow imparted to the fluid by the rotation of the cylinder. Find also the positions of the stagnation points and the force on the cylinder.   (0·628 m³/s; 10 m/s; 0·66 m²/s; 3°; 6·6 kN/m.)

12.20.   As an alternative to the conventional fluid flow field theory analogy used in this book, it would be possible to use mass flow for the flux rather than the volume discharge. Develop expressions for all the analogous quantities in this new field theory analogy keeping the velocity potential function.

FURTHER SOLUTIONS

12.11.   Following the solution to Example 12.1, the potential function is given by

$$\phi = \frac{Q}{4\pi r} = \frac{0\cdot314}{4\pi r} = \frac{2\cdot5 \times 10^{-2}}{r} \ \text{m}^2/\text{s},$$

$$v = -\ \text{grad}\ \phi = -\frac{\partial\phi}{\partial r} = \frac{2\cdot5 \times 10^{-2}}{r^2} \ \text{m/s}.$$

12.12   Following the solution to Problem 12.1, the potential function is given by

$$\phi = -\ Ux = -\ Ur\ \sin\phi = -\ 10x = -\ 10r\ \sin\theta \ \text{m}^2/\text{s}.$$

12.13.   The potential function of a dipole is given in eqn. (9.2) (page 167) and has been derived in the solution to Example 9.4 (page 170). Therefore

$$\phi = -\frac{m\ \cos\theta}{4\pi r^2} = -\frac{0\cdot126\ \cos\theta}{4\pi r^2} = -\ 1\cdot0 \times 10^{-2}\ \frac{\cos\theta}{r^2} \ \text{m}^2/\text{s}.$$

12.14.   A small element of area of the field is shown in Fig. 12.8 with

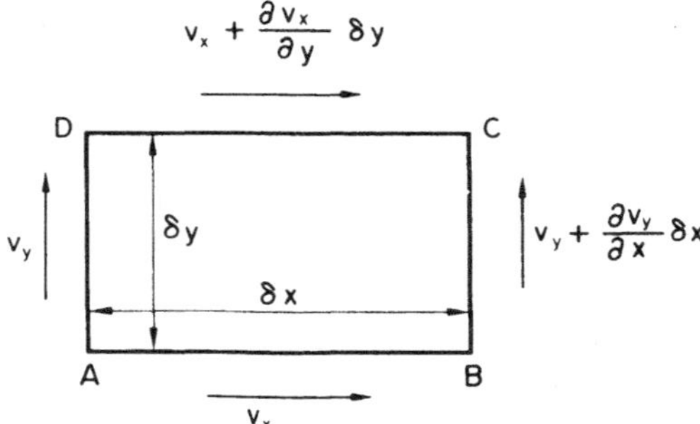

Fig. 12.8. Velocities in a general two-dimensional fluid field in rectangular coordinates.

the velocities shown in the diagram. The circulation around the rectangular contour *ABCD* is given by

$$k_c = \oint \mathbf{v} \cdot d\mathbf{l} = v_x\delta x + \left(v_y + \frac{\partial v_y}{\partial x}\delta x\right)\delta y - \left(v_x + \frac{\partial v_x}{\partial y}\delta y\right)\delta x - v_y\delta y$$

$$= \left(\frac{\partial v_y}{\partial x} - \frac{\partial v_x}{\partial y}\right)\delta x\ \delta y.$$

Therefore the vorticity is given by

$$k = \frac{k_c}{\text{area}} = \frac{\partial v_y}{\partial x} - \frac{\partial v_x}{\partial y}.$$

12.15.   The vorticity of a forced vortex is given in the solution to Example 12.2. The liquid in the centrifuge in this problem may be assumed to be rotating as a forced vortex, therefore the vorticity is given by

$$k = 2\omega = \frac{2 \times 2\pi \times 5000}{60} = 1047 \text{ rad/s.}$$

12.16.   The relationships for a free vortex flow are given in the solution to Example 12.4, therefore the strength of the equivalent line vortex is given by

$$k = 2\pi r v_\theta = 2\pi \times 5\cdot 0 \times 10 = 314 \text{ m}^2/\text{s.}$$

12.17.   The required relationship is given in eqn. (12.8) which has been derived in the solution to Example 12.6, therefore

$$\phi = -\frac{k\theta}{2\pi} = \frac{6\cdot 28\ \theta}{2\pi} = \theta \text{ m}^2/\text{s.}$$

12.18.   The equipotential line pattern for a line vortex is similar to the flux function line pattern of a line source as shown in Fig. 3.2 (page 37). Then for the vortex of strength 16 m²/s, the potential difference between each of the lines in Fig. 3.2 will be 1·0 m²/s. A vortex of the opposite hand of rotation is a vortex with a negative value for its strength, so that the combination of the equipotential line pattern for the two vortices is similar to that shown in Fig. 3.14 (page 54). The complete equipotential line pattern without the construction lines is the same as the flux line pattern shown in Fig. 3.3 (page 38).

12.19.   The solution to this problem uses results that have been derived in the solutions to Examples 3.8 (page 44), 12.8, 12.9, and 12.10. Therefore, the strength of the equivalent doublet is given by

$$c = 2\pi U a^2 = 2\pi \times 10 \times 0\cdot 01 = 0\cdot 628 \text{ m}^3/\text{s.}$$

The strength of the uniform stream is unchanged at 10 m/s.
The strength of the equivalent line vortex is given by

$$k = 2\pi a v_\theta = 2\pi \times 0\cdot 1 \times \frac{100 \times 2\pi \times 0\cdot 1}{60} = 0\cdot 66 \text{ m}^2/\text{s.}$$

The positions of the stagnation points is given by

$$\sin\theta = \frac{k}{4\pi a U} = \frac{0\cdot 66}{4\pi \times 0\cdot 1 \times 10} = 0\cdot 0524.$$

Therefore $\theta = 3\cdot 0^\circ.$

The lift force on the cylinder is given by

$$F = \rho U k = 1000 \times 10 \times 0{\cdot}66 = 6{\cdot}6 \times 10^3 \text{ N/m} = 6{\cdot}6 \text{ kN/m}.$$

12.20.   The analogous quantities in the proposed fluid field system are:

flux = mass flow, ρQ kg/s;

flux density = mass flow density, ρv kg/m^2 s;

field intensity = velocity, $\dot{v}$ m/s;

dimensional constant = density, ρ kg/m^3;

potential difference = velocity potential difference, Φ;

stored energy = $\frac{1}{2}\rho v^2$ J/m^3, kinetic energy per unit volume.

CHAPTER 13

Magnetic Potential

THEORY

In the absence of any electric currents, the magnetic field is a conservative field. For any magnetic field,

$$\oint \mathbf{H} \cdot d\mathbf{l} = 0 \tag{13.1}$$

provided that the path of integration does not enclose any electric current. Then a magnetic potential function can be defined by

$$\phi_B - \phi_A = \int_A^B - \mathbf{H} \cdot d\mathbf{l}. \tag{13.2}$$

It is necessary to remember that the magnetic potential function only exists in the absence of current-carrying wires. In the presence of any current, the line integral of the field intensity around a closed path enclosing the current is equal to the value of the current enclosed. In fields terminology, the total quantity of enclosed current gives rise to circulation in the magnetic field. Therefore

$$\oint \mathbf{H} \cdot d\mathbf{l} = \Sigma I = k_c. \tag{13.3}$$

This relationship between magnetic field intensity and electric current is called Ampere's law. Electric current in a single wire is equivalent to a magnetic line vortex.

$$\oint \mathbf{H} \cdot d\mathbf{l} = I = \frac{k}{\mu_o}. \tag{13.4}$$

A coil of wire carrying an electric current can be used to generate a magnetic flux; the effect of the current-carrying coil can be considered to be equivalent to a source of magnetic potential which causes the magnetic flux in the field. There is a useful analogy between the battery in the electric circuit carrying an electric current flux and the current-carrying coil around a magnetic path which makes a circuit for a magnetic flux. The magnetic potential of the coil is called the *magnetomotive force* (m.m.f.) of the coil. Therefore

$$\text{m.m.f.} = \Phi = \oint - \mathbf{H} \cdot d\mathbf{l} = -I. \tag{13.5}$$

The potential rise around any closed path enclosing some electric current is equal to the m.m.f. of that current.

It may be assumed that the magnetic flux flowing in an iron core is entirely confined to that core. Flux is considered to flow in any air gap in series with the magnetic core, but the flux in the air surrounding the core may be ignored. Therefore there is a direct analogy between electric current flow in a conducting medium and magnetic flux flow through an iron core. Magnetic circuit calculations may be performed using the following analogies:

$$\left.\begin{array}{l} \text{m.m.f.} \equiv \text{e.m.f.,} \\ \text{magnetic flux} \equiv \text{electric current,} \\ \text{magnetic reluctance} \equiv \text{electric resistance.} \end{array}\right\} \qquad (13.6)$$

The relationship for the magnetic reluctance is similar to eqn. (10.9),

$$R_m = \frac{\Phi}{\Psi}. \qquad (13.7)$$

For a homogeneous material of uniform cross-sectional area A and length d, the reluctance is given by

$$R_m = \frac{1}{\mu}\frac{d}{A}. \qquad (13.8)$$

Ampere's law is only applicable in a symmetrical field configuration. For a non-symmetrical field, the relationship equivalent to Ampere's law is given by the *Biot-Savart law*. The magnetic field δH due to the short element of electric current is given by

$$\delta H = \frac{I \,\delta l \, \sin\theta}{4\pi r^2}, \qquad (13.9)$$

where the dimensions and relative directions of the field and the current are defined in Fig. 13.1. The total field is determined by integrating along the

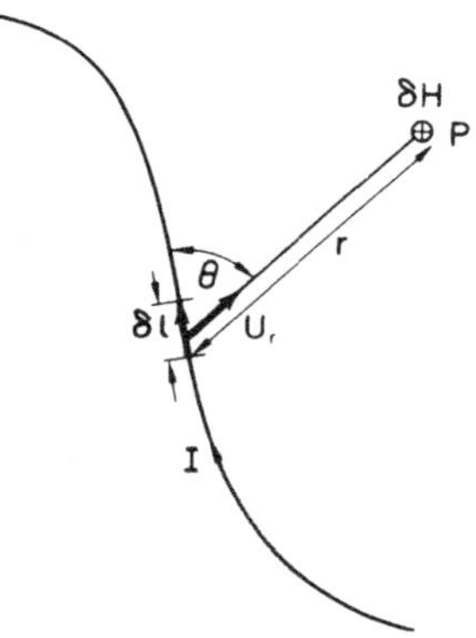

Fig. 13.1. Illustrating the Biot-Savart law.

length of the current path. As a general field theory relationship for the
field intensity due to a general vortex filament, using the vector notation
shown in Fig. 13.1,

$$\text{field intensity} = -\,\text{grad }\phi = \frac{k}{4\pi}\int_{\text{along length of vortex}} \frac{d\mathbf{l}\times\mathbf{U}_r}{r^2} \qquad (13.10)$$

where the vector $d\mathbf{l}$ is taken in the same direction as the vortex.

In permanent magnet design, the m.m.f. across the air gap is equated to
the negative m.m.f. which will reduce the flux density in the circuit to less
than the remanence.

EXAMPLES

13.1.   Obtain an expression for the flux density inside a toroidal coil
of N turns carrying a current I A, wound on a non-magnetic former.

Answer. Let the dimensions of the coil be as shown in the diagram of
Fig. 13.2. It is found that the flux density is uniform across the cross-
section of the coil provided that $D \gg d$ The total m.m.f. around the coil
is NI A. As the field is uniform around the toroid, then

$$Hl = H\pi D = NI \quad \text{and} \quad B = \mu_0 H = \frac{\mu_0 NI}{\pi D}.$$

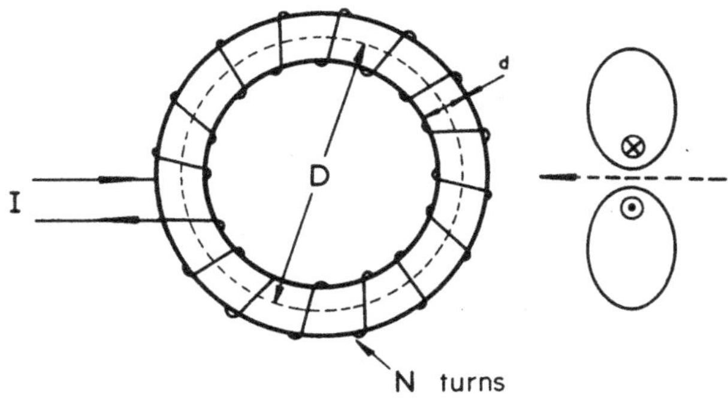

Fig. 13.2.   A toroidal coil and its magnetic field.

13.2.   Calculate the flux density and the total flux inside a closely
wound solenoidal coil of 7000 turns of fine copper wire in which a current of
1·0 A is flowing. The coil is of circular cross-section of 30 mm diameter and
is 150 mm long.

Answer. The calculation must assume that the solenoid approximates to a

section of infinitely long coil with the same number of turns per unit length,
i.e. it may be considered as a toroidal coil of infinite diameter. The
relationship for a uniform field may be expressed as

$$H = \frac{NI}{l} = \text{(ampere turns per unit length)}.$$

Therefore the flux density in the coil is given by

$$B = \mu_0 H = \frac{\mu_0 NI}{l} = \frac{4\pi \times 10^{-7} \times 7000 \times 1 \cdot 0}{0 \cdot 15} = 5 \cdot 86 \times 10^{-2} \text{ T} = 58 \cdot 6 \text{ mT}.$$

Assuming that the flux density is uniform across the cross-section of the coil,
the total flux is given by

$$\Psi = B \times \text{(area)} = 5 \cdot 86 \times 10^{-2} \times \pi \times (15)^2 \times 10^{-6} = 4 \cdot 15 \times 10^{-5} \text{ Wb} = 41 \cdot 5 \text{ } \mu\text{Wb}.$$

13.3.   Calculate the field intensity at a distance of 10 m from a long
straight conductor carrying a current of $1 \cdot 0$ A. Hence or otherwise find the
value of the m.m.f. due to the current.

Answer. The flux density due to a long straight current is given in
eqn. (7.5)(page 119). Therefore the field intensity is given by

$$H = \frac{B}{\mu_0} = \frac{I}{2\pi d} = \frac{1 \cdot 0}{2\pi \times 10} = 1 \cdot 59 \times 10^{-2} \text{ A/m}.$$

The m.m.f. is given by eqn. (13.5). Therefore

$$\Phi = \oint - \mathbf{H} \cdot d\mathbf{l} = - 1 \cdot 59 \times 10^{-2} \times 2\pi \times 10 = - 1 \cdot 0 \text{ A},$$

which is equal to the total current enclosed.

13.4.   Find expressions for the flux function and potential function in
the magnetic field set up by a long straight current filament I.

Answer. The flux function and potential function in the field due to a
long straight line vortex of strength k are given by eqns. (12.9) and (12.8)
(page 239) respectively. There is a direct equivalence between the line
vortex and the current flowing in a long straight conductor,

$$k = \mu_0 I.$$

There is one difference which does not allow direct substitution into the
equations of Chapter 12. In the fluid flow field the dimensions of the flux
density and the field intensity are the same. In the magnetic field, as with
all other fields, they are different so that in substituting into eqns. (12.8)
and (12.9), the permeability enters into the expression for the flux function
whereas it does not enter into the expression for the potential function. It

is necessary to check the dimensional consistency of each relationship.
Therefore

$$\psi = \frac{\mu_o I}{2\pi} \log_e \frac{a}{r},$$

$$\phi = -\frac{I\theta}{2\pi}.$$

13.5.   How is it possible to specify the potential function of the
magnetic field due to a current-carrying wire? If it is possible, how does
the m.m.f. enter the situation?

   Answer. In the consideration of the potential function of the line
vortex in the solution to Example 12.6 (page 243), it was necessary to take
a cut in the field and to specify an infinitely small excluded region
consisting of a cut in the field extending from infinity to the line of the
vortex and enclosing the vortex and then returning to infinity along the line
of the cut. The cut is shown in Fig. 12.3. No potential function is possible
in any region which includes an electric current because there is not a
conservative field everywhere in the space and it is possible to find paths
of integration which enclose electric current. The only satisfactory way of
specifying a magnetic potential function is to restrict the field to a region
which does not include the electric current. The cut in the field shown in
Fig. 12.3 does just this. The cut excludes a space of negligible area but it
does ensure that no path of integration can encircle the line of the vortex
or current. By finding a cut or other limit to the extent of the field, it
is possible to exclude any electric current from the region in which it is
desired to specify the magnetic potential function and there is a conservative
field in the region under consideration. The m.m.f. is equal to the abrupt
change of potential function that occurs between the two sides of the cut.
Considering a line at a constant radius from a long straight current-carrying
wire, the potential difference between the two sides of the cut causes the
magnetic flux to flow in the field. Therefore the m.m.f. across the cut is
equivalent to a magnetic battery causing the magnetic flux to flow round the
magnetic circuit.

13.6.   Find the m.m.f. generated by a coil of 1000 turns carrying a
current of 100 mA.

   Answer. The m.m.f. is given by eqn. (13.5). The path of integration
is chosen so that it threads the coil; it encloses a total current of 100 mA
flowing in 1000 parallel wires giving a total current of 100 A. Therefore
the m.m.f. of the coil is 100 A.

13.7.   Find the magnetic reluctance of an iron bar, 300 mm long of uniform cross-section of area $0 \cdot 01$ m^2, which may be assumed to have a constant relative permeability of 1000. The bar is to be used under uniform magnetization along its length.

Answer. Let there be a uniform magnetic field inside the bar with a flux density and field intensity of B and H respectively. Then the total flux in the bar is given by

$$\Psi = B \times \text{(area)} = 0 \cdot 01 B.$$

The total potential difference between the ends of the bar is given by

$$\Phi = H \times \text{(length)} = 0 \cdot 3 H.$$

The reluctance is given by eqn. (13.8), therefore

$$R_m = \frac{\Phi}{\Psi} = \frac{0 \cdot 3 H}{0 \cdot 01 B} = \frac{30}{\mu_r \mu_o} = \frac{30 \times 10^7}{4\pi \times 1000} = 2 \cdot 38 \times 10^4 \ \text{H}^{-1}.$$

13.8.   Find the number of ampere turns needed to provide a flux density of $1 \cdot 5$ T in the air gap of the magnet shown in Fig. 13.3, when l_1 = 200 mm, l_2 = 160 mm, l_3 = 10 mm, A_1 = $0 \cdot 0016$ m^2, A_2 = $0 \cdot 0008$ m^2. The steel used for the magnetic core is that whose magnetization curve is given in Fig. 7.9 and repeated in Fig. 13.4.

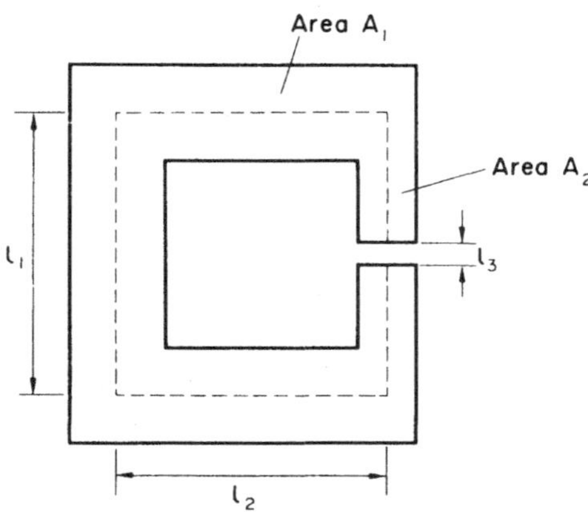

Fig. 13.3 Magnet of Example 13.8.

Answer. It will be assumed that the magnetic flux density is uniform across any cross-section of the core or the air gap. Then the reluctance of each section of the core is given by an expression similar to that used in the solution to the last example. The total reluctance of the circuit is the sum

of the reluctance of each section of the core in exactly the same way as the
total resistance of an electric circuit is the sum of all the resistances in
series in the circuit. Because the core of the circuit changes in cross-
section, the flux density will be different in different parts of the circuit
and because the magnetization curve is a non-linear relationship, the
permeability will also be different in the different parts of the core having
different cross-sections. The flux density in the arms of smaller cross-
section is the same as that in the air gap. The total flux in the circuit is
the product of the flux density and the cross-sectional area:

$$\Psi = 1 \cdot 5 \times 0 \cdot 0008 = 0 \cdot 0012 \text{ Wb.}$$

Therefore the flux density in the arms of larger cross-section is given by

$$B = \frac{0 \cdot 0012}{0 \cdot 0016} = 0 \cdot 75 \text{ T.}$$

It is necessary to turn to the magnetization curve for the steel in order to
find the relative permeability of the steel at these flux densities. The
magnetization curve is repeated in Fig. 13.4. From the curve it is seen that

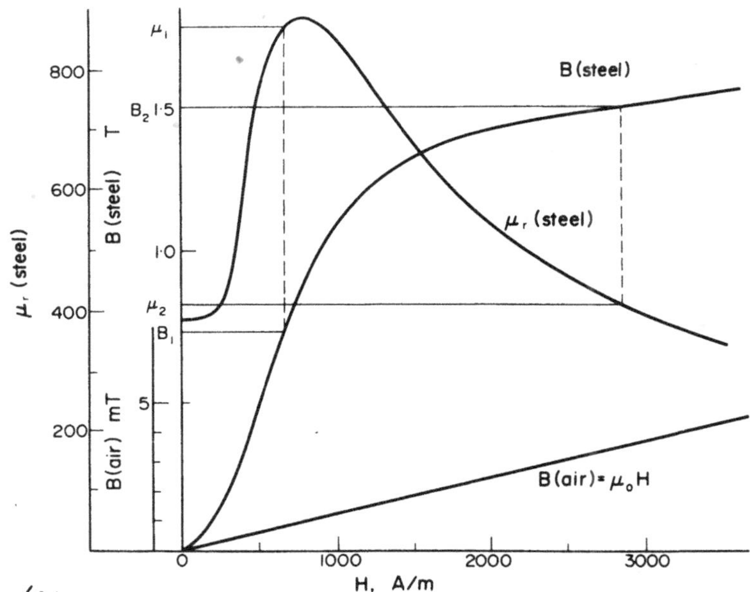

Fig. 13.4. Magnetization curve of the steel used for the magnet
discussed in Example 13.8 together with the construction lines
used to find the relative permeability at various flux densities.

the two permeabilities are given by

$$\mu_1 = 870\ \mu_0, \qquad\qquad \mu_2 = 410\ \mu_0.$$

It will be noticed that the higher value of permeability is associated with the larger area of cross-section and smaller flux density. Therefore the total reluctance of the core and air gap is given by

$$R_m = \frac{(l_1 + 2l_2)}{\mu_1 A_1} + \frac{(l_1 - l_3)}{\mu_2 A_2} + \frac{l_3}{\mu_0 A_2}$$

$$= \left(\frac{(200 + 320)}{870 \times 1 \cdot 6} + \frac{(200 - 10)}{410 \times 0 \cdot 8} + \frac{10}{0 \cdot 8}\right)\frac{1}{\mu_0} = \frac{13 \cdot 5}{\mu_0}.$$

The m.m.f. is given by

$$\text{m.m.f.} = NI = \Psi R_m = \frac{0 \cdot 0012 \times 13 \cdot 5}{4\pi \times 10^{-7}} = 12\ 900\ \text{Ampere turns.}$$

13.9.   Derive an expression for the magnetic field intensity due to a long straight current filament I both by integrating from the Biot-Savart law and by using the fields relationships of a line vortex.

Answer. The geometry of the problem is shown in Fig. 13.5. In order to

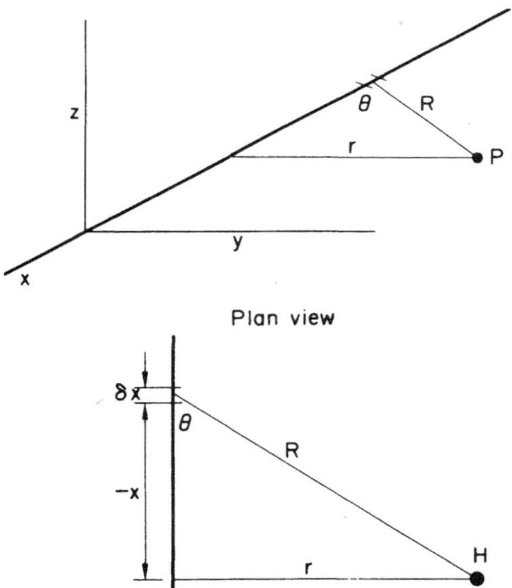

Fig. 13.5.   The geometry of a point near to a long straight wire, illustrating the solution to Example 13.9.

apply the Biot-Savart law it is necessary to take a small element of the line vortex and then to integrate along the length of the vortex. Taking the geometry and dimensions from the diagram and applying eqn. (13.9) gives

$$\delta H = \frac{I\delta x \, \sin\theta}{4\pi R^2}.$$

In order to perform the integration, it is necessary to express the two variables R and x in terms of θ. Using the substitutions

$$R = \frac{r}{\sin\theta} \qquad \text{and} \qquad x = r\cot\theta,$$

and by differentiation

$$dx = \frac{r}{\sin^2\theta}\, d\theta,$$

the integral becomes

$$H = \frac{I}{4\pi} \int_{-\infty}^{\infty} \frac{\sin\theta}{R^2}\, dx = \frac{I}{4\pi r} \int_{0}^{\pi} \sin\theta \, d\theta = \frac{I}{2\pi r}.$$

For a line vortex of strength I, the potential function is given by eqn. (12.8) (page 239) to be

$$\phi = -\frac{I\theta}{2\pi}.$$

The field intensity is obtained from the potential function by a relationship similar to eqn. (8.6) (page 142),

$$\mathbf{H} = -\operatorname{grad}\phi.$$

Therefore

$$H_\theta = -\frac{1}{r}\frac{\partial\phi}{\partial\theta} = \frac{I}{2\pi r}.$$

The radial component of the magnetic field is zero.

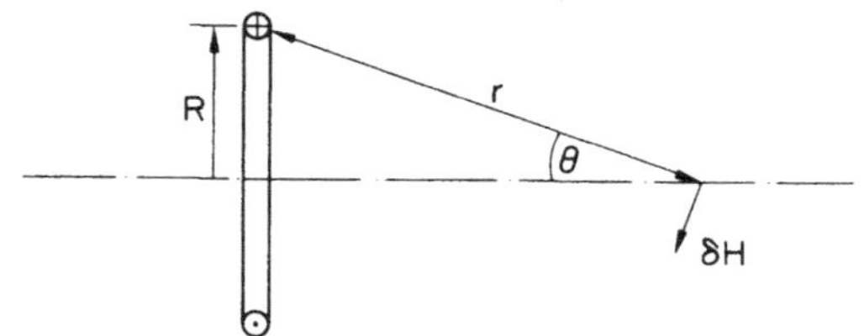

Fig. 13.6.   Illustrating Example 13.10.   The field on the axis of
a coil.

13.10.   Find an expression for the magnetic field intensity on the axis
of a current loop of radius 25 mm carrying a current of 1·0 A.

Answer. The geometry of the problem is shown in Fig. 13.6. The problem
will be solved in terms of symbols first and then the numbers will be inserted
at the end in order to give the required answer. The magnetic field intensity
due to one small element of the current in the loop is shown in the diagram,
which ·also gives all the dimensions used in the derivation. Substituting into
the Biot-Savart law, eqn. (13.9), gives

$$\delta H = \frac{I \delta l}{4\pi r^2}$$

because the angle between the distance r and the small element of current is
always 90°. By symmetry, the radial components of the magnetic field will be
zero on integrating around the loop of current; the resultant field will be
axial, along the axis of the current loop. Taking the axial component of δH
and integrating δl around the circumference of the current loop gives an
expression for the total magnetic field intensity,

$$H = \frac{IR \sin \theta}{2r^2}.$$

This expression includes two dimensions which are functions of the position
at which the field is to be derived. $R = r \sin \theta$ and the expression for the
field intensity becomes

$$H = \frac{IR^2}{2r^3} = \frac{I}{2R} \sin^3\theta.$$

Inserting numbers from the problem gives

$$H = \frac{1 \cdot 0}{2 \times 25 \times 10^{-3}} \sin^3\theta = 20 \sin^3\theta \ \text{A/m}.$$

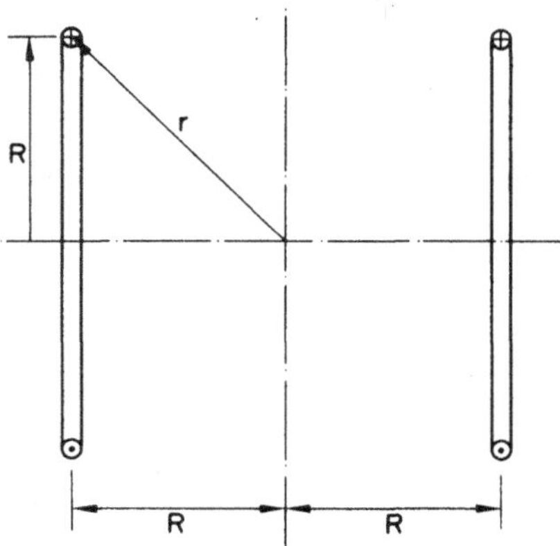

Fig. 13.7. The geometry of the two coils of Example 13.11.

13.11.   A particular coil consists of two identical circular loops of
wire mounted coaxially spaced apart by a distance equal to their diameter.
Find an expression for the magnetic field intensity at the centre of the two
coils in terms of the current in the wire and the diameter of the loops.

Answer. The magnetic field intensity on the axis of a single coil is
given in the solution to the last example. The geometry of the two coils in
this problem is shown in Fig. 13.7. Then the magnetic field at the centre is
given by

$$H = 2\,\frac{IR^2}{2r^3} = \frac{IR^2}{2\sqrt{(2)}R^3} = \frac{I}{2\sqrt{(2)}R}.$$

13.12.   A permanent magnet has the shape and size shown in Fig. 13.8 and is made of a magnetic steel whose hysteresis curve is shown in Fig. 13.9. Find the maximum flux density in the air gap.

Answer. Assume that there is a uniform flux density B in the air gap from the principles of magnetic circuit calculations. Then B will also be the flux density in the magnet. The m.m.f. in the air gap is given by

$$\text{m.m.f.} = H_a l_a = \frac{B}{\mu_o} l_a = \frac{0\cdot01 B}{4\pi \times 10^{-7}} = \frac{10^5 B}{4\pi}.$$

If the magnetic steel is in a state characterized by the upper half of the hysteresis curve, when the magnetic flux density in the magnet is less than remanence, there must be a negative field intensity in the magnet of $- H_m$. Therefore the m.m.f. in the magnet is given by

$$\text{m.m.f.} = - H_m l_m = - 0\cdot1\pi H_m.$$

Equating these two expressions for m.m.f. gives

$$H_m = - \frac{10^6}{4\pi^2} B = - \frac{10^6}{39\cdot5} B.$$

A straight line may be drawn on the hysteresis curve of the magnetic material for this equation, and the intersection of the two curves gives the required value of flux density. The line has been added to Fig. 13.9 and gives a result of

$$B = 1\cdot28 \text{ T.}$$

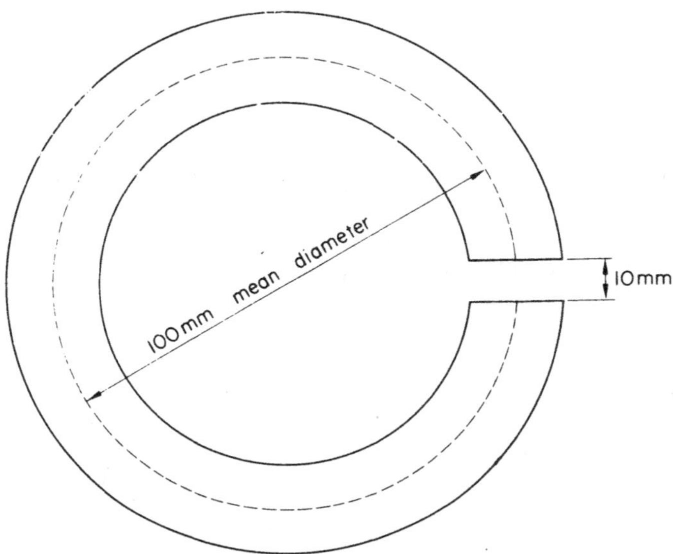

Fig. 13.8.   The shape and size of the permanent magnet of Example 13.12.

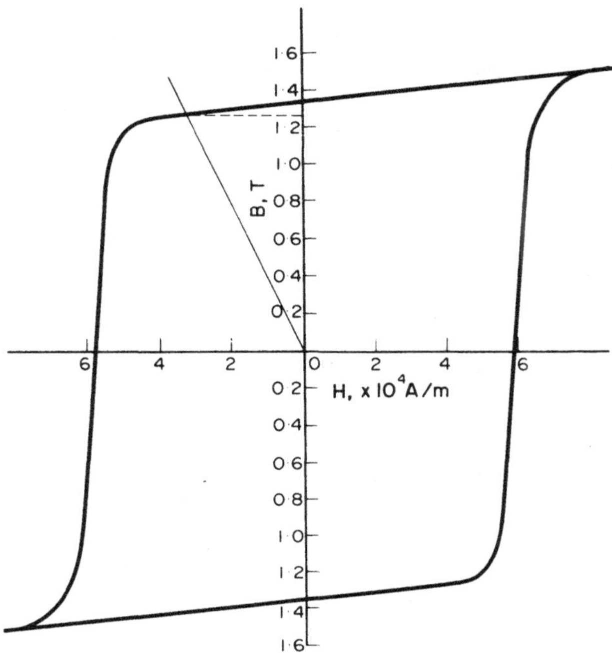

Fig. 13.9.   Hysteresis curve of a permanent magnet steel.

PROBLEMS

13.1.   A toroidal coil of 1000 turns is evenly wound on a wooden former of 0·1 m mean diameter. It is carrying a current of 1·0 A. Calculate the flux density in the coil and the force on a wire carrying 25 mA situated in a narrow slot in the coil former. $(4.0 \times 10^{-3}$ T; 10^{-4} N/m.)

13.2.   A toroidal coil of 500 turns is wound on a steel ring 0·5 m mean diameter. An excitation of 4000 A/m produces a flux density of 1·0 T. If a 10 mm long gap is cut in the ring, find the current required to maintain the flux density at 1·0 T. Neglect all leakage and fringing. (28·5 A.)

13.3.   Calculate the rise in magnetic potential in moving a distance of 10 mm against the direction of a uniform magnetic flux density of 1·0 T in air.   (7960 A.)

13.4.   Calculate the flux density at a distance of 1·0 m from a long straight wire carrying a current of 5·0 A in air.   (1·0 μT.)

13.5.   Determine the m.m.f. generated by a multilayer coil of 1000 turns of fine wire which is 10 mm long and 5·0 mm thickness of winding carrying a current of 1·0 mA.   (1·0 A.)

13.6.   Calculate the ampere turns required to maintain a flux of 1·5 T in the gap of the magnet shown in Fig. 13.10. The *B–H* relationship for the steel is given in Fig. 13.4.   (6350 A.)

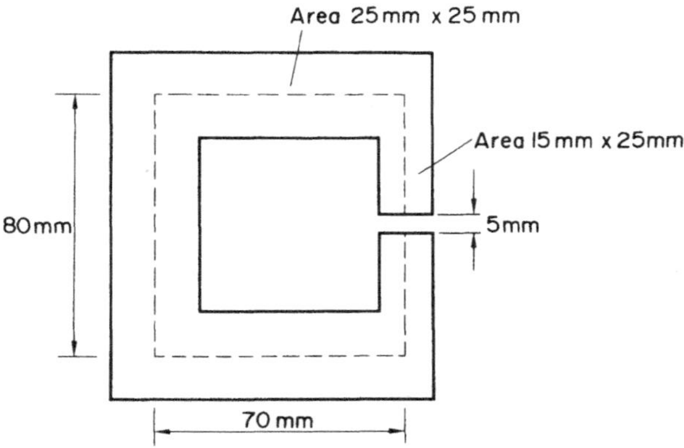

Fig. 13.10. The magnet of Problem 13.6.

13.7.   A steel circular lifting magnet has an outside diameter of 1·3 m.
The diameter of the inner circular pole is 0·4 m and the outer annular pole
has the same area as the inner one. The mean length of the magnetic path in
the magnet is 1·5 m. The magnetization curve of the steel is given in Fig.
13.4. The magnet lifts steel plates of negligible reluctance with an
effective air gap of 6·0 mm at each pole. Find the m.m.f. required to produce
a flux density in the air gaps of 1·0 T.   (10 900 A.)

13.8.   Derive an expression for the magnetic flux density on the axis
of a closely wound solenoid of radius r and length l in terms of the angle
subtended by the ends of the coil. $\left(\mu_0 \dfrac{NI}{2l} (\cos \theta_1 - \cos \theta_2). \right)$

13.9.   Find an expression for the flux density at a point a perpendicular
distance R away from the centre of a long straight wire length $2l$ carrying a
current I. What is the minimum length of straight wire which could be used to
produce a flux density at the point differing by not more than two per cent
from that for a wire of infinite length? $\left(H = \dfrac{Il}{2\pi R \sqrt{(l^2 + R^2)}}; \; l = 4 \cdot 9R. \right)$

13.10.   Three closely wrapped coils, each having the same m.m.f., have
circular, hexagonal, and square shapes respectively, the distance between
opposite sides of the last two being equal to the diameter of the first. Use
the Biot-Savart law to show that the magnetic intensities at the centres of
these coils have relative magnitudes of 1·00:0·955:0·901 respectively.

SOLUTIONS

13.1.   The flux density of a toroidal coil is given in the solution to Example 13.1, therefore

$$B = \frac{\mu_0 NI}{\pi D} = \frac{4\pi \times 10^{-7} \times 1000 \times 1 \cdot 0}{\pi \times 0 \cdot 1} = 4 \cdot 0 \times 10^{-3} \text{ T.}$$

The force is given by eqn. (7.6), therefore

$$f = 4 \cdot 0 \times 10^{-3} \times 25 \times 10^{-3} = 10^{-4} \text{ N/m.}$$

13.2.   The excitation is given by

$$\frac{NI}{l} = H = 4000 \text{ A/m.}$$

Therefore the current in the coil is given by

$$I = \frac{H\pi D}{N} = \frac{4000 \times \pi \times 0 \cdot 5}{500} = 4\pi \text{ A.}$$

If the flux density is maintained at $1 \cdot 0$ T, the field intensity in the air gap is given by

$$H = \frac{B}{\mu_0} = \frac{1 \cdot 0}{4\pi \times 10^{-7}} = 7 \cdot 958 \times 10^5 \text{ A/m.}$$

The reluctance of the steel ring may be assumed to be unchanged, therefore the total m.m.f. is given by

$$NI = \int \mathbf{H} \cdot d\mathbf{l} = 4000 \times \pi \times 0 \cdot 5 + 7 \cdot 958 \times 10^5 \times 10^{-2} = 14 \cdot 24 \times 10^3 \text{ A.}$$

Therefore
$$I = \frac{14 \cdot 24 \times 10^3}{500} = 28 \cdot 48 \text{ A.}$$

13.3.   The magnetic potential difference is given by eqn. (13.2). The field intensity is given by

$$H = \frac{B}{\mu_0} = \frac{1 \cdot 0}{4\pi \times 10^{-7}} = 7 \cdot 96 \times 10^5 \text{ A/m.}$$

Therefore, moving 10 mm against a uniform field gives

$$\Phi = Hl = 7 \cdot 96 \times 10^5 \times 10^{-2} = 7960 \text{ A.}$$

13.4.   The flux density due to a long straight wire is given by eqn. (7.5). Therefore

$$B = \frac{\mu_0 I}{2\pi r} = \frac{4\pi \times 10^{-7} \times 5 \cdot 0}{2\pi \times 1 \cdot 0} = 1 \cdot 0 \times 10^{-6} \text{ T} = 1 \cdot 0 \text{ } \mu\text{T.}$$

13.5.   The m.m.f. of a coil is given by the product, $\Phi = NI$, and it is independent of the dimensions of the coil. Therefore

$$\text{m.m.f.} = NI = 10^3 \times 10^{-3} = 1 \cdot 0 \text{ A.}$$

13.6.   Assume that there is a uniform flux distribution in each arm of the magnet. The flux density in the gap of the magnet is 1·5 T. Therefore, assuming that there is no leakage of flux outside the magnetic circuit, the flux density in the larger arms of the magnet is given by

$$B = 1.5 \times \frac{15}{25} = 0.9 \text{ T.}$$

From the B–H relationship in the steel given in Fig. 13.4, the field intensities in the two parts of the magnet are given by

$$H_1 = 2950 \text{ A/m}; \quad H_2 = 800 \text{ A/m}; \quad \text{and } H_a = \frac{1.5}{4\pi \times 10^{-7}} = 1.19 \times 10^6 \text{ A/m.}$$

Ampere turns $= NI =$ m.m.f. $= \int \mathbf{H} \cdot d\mathbf{l}$

$$= 75 \times 10^{-3} \times 2950 + 220 \times 10^{-3} \times 800 + 1.19 \times 10^6 \times 5.0 \times 10^{-3}$$

$$= 6350 \text{ A.}$$

13.7.   The flux density in the air gaps is assumed to be the same as that in the magnet. Then, from Fig. 13.4 the field intensity in the steel is given by

$$H = 900 \text{ A/m.}$$

In the air gap, it is

$$H_a = \frac{1.0}{4\pi \times 10^{-7}} = 7.96 \times 10^5 \text{ A/m.}$$

Then the m.m.f. is given by

$$\Phi = \int \mathbf{H} \cdot d\mathbf{l} = 900 \times 1.5 + 7.96 \times 10^5 \times 2 \times 6.0 \times 10^{-3} = 1.09 \times 10^4 \text{ A.}$$

13.8.   This problem makes use of the field on the axis of a single coil of wire as derived in the solution to Example 13.10 and illustrated in Fig. 13.6:

$$H = \frac{I}{2R} \sin^3\theta.$$

The solenoid may be considered to consist of a large number of small elemental coils of thickness δx, as shown in Fig. 13.11. Then the current in each elemental coil is given by

$$i = \frac{NI\delta x}{l}$$

and the field due to one element of the coil at any point on the axis is given by

$$\delta H = \frac{NI\delta x}{2rl} \sin^3\theta.$$

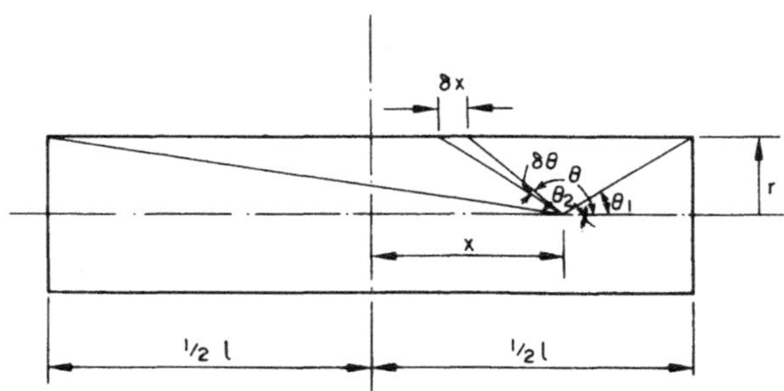

Fig. 13.11. A solenoid of finite length, illustrating the solution
to Problem 13.8.

By geometry,
$$\delta x = \frac{r\delta\theta}{\sin^2\theta}.$$

Therefore the total field intensity at any point on the axis of the coil is
given by

$$H = \int_{\theta_1}^{\theta_2} \frac{NIr\,\sin^3\theta}{2lr\,\sin^2\theta}\,d\theta = \frac{NI}{2l}\int_{\theta_1}^{\theta_2} \sin\theta\,d\theta = \frac{NI}{2l}(\cos\theta_1 - \cos\theta_2).$$

Therefore the flux density is given by

$$B = \frac{\mu_0 NI}{2l}(\cos\theta_1 - \cos\theta_2).$$

13.9.   The application of the Biot-Savart law to a long straight current
filament has been done in the solution to Example 13.9 and is illustrated in
Fig. 13.5. Using the results derived in that solution, the field due to a
current filament of limited length in a plane perpendicular to the filament
and at its centre, is given by

$$H = \frac{I}{4\pi R}\int_{\theta_1}^{(\pi - \theta_1)} \sin\theta\,d\theta = \frac{I}{2\pi R}\cos\theta_1,$$

where $\tan\theta_1 = \dfrac{R}{l}$.

Therefore
$$H = \frac{Il}{2\pi R\sqrt{(l^2 + R^2)}}.$$

When l is infinitely large, this reduces to the result for a long wire,

$$H_0 = \frac{I}{2\pi R}.$$

In this problem, we are given that $H = 0.98\, H_o$, therefore

$$\frac{l}{\sqrt{(l^2 + R^2)}} = 0.98.$$

Therefore, $l = 4.9R$.

13.10.   By direct application of the Biot-Savart law, the field at the centre of a circular coil can be shown to be given by

$$H = \frac{I}{2R},$$

where R is the radius of the coil. For the straight-sided coils, the field due to each side is given by the result obtained in the solution to the last problem. Therefore, for the hexagonal coil, $l = R/\sqrt{3}$, and

$$H = \frac{6I}{2\pi R}\, \frac{R}{\sqrt{3}}\, \frac{\sqrt{3}}{\sqrt{(R^2 + 3R^2)}} = 0.955\,\frac{I}{2R}.$$

For the square coil, $l = R$, therefore

$$H = \frac{4I}{2\pi R}\, \frac{R}{\sqrt{(R^2 + R^2)}} = 0.901\,\frac{I}{2R}.$$

FURTHER PROBLEMS

13.11.   A toroidal coil of 700 turns is uniformly wound on a non-magnetic former of 200 mm mean diameter and having a circular cross-section of 10 mm diameter. Find the magnetic field intensity inside the coil former when there is an electric current of 250 mA flowing in the coil.   (278 A/m.)

13.12.   A circular ring of steel has a mean diameter of 150 mm and is of square cross-section with a side of 20 mm. It is wound with 150 turns of fine wire carrying an electric current of 10 mA. The magnetizing curve for the steel is given in Fig. 13.4. Find the flux density in the steel.   (1·6 mT.)

13.13.   In order to measure the flux density in the steel ring of Problem 13.12, a 1·0 mm wide air gap is cut in the steel ring. Find the value of the flux density that will be measured in the air gap.   (0·86 mT.)

13.14.   Determine the m.m.f. of a coil of 10 000 turns carrying a current of 100 mA.   (1000 A.)

13.15.   A long straight copper wire is embedded 10 mm below the surface of a large block of magnetic steel. The wire is electrically insulated from

the steel and carries a current of 100 mA. Obtain an expression for the
magnetic field intensity just inside the surface of the steel and its value
at the point, still on the surface of the steel, nearest to the wire. It may
be assumed that the wire is negligibly thin and that all other surfaces of
the steel are an infinite distance away.   (3·18 A/m.)

13.16.   Find the magnetic reluctance between the ends of an iron bar,
1·0 m long of cross-section 25 mm by 100 mm. It may be assumed to have a
constant permeability of 500.   (6·37 × 10⁵ H⁻¹.)

13.17.   Plot a graph of the relationship between the current in the coil
of 1000 turns and the flux density in the air gap of 2·0 mm for the magnetic
circuit shown in Fig. 13.12. Use the magnetization curve of Fig. 13.4 for
the steel.

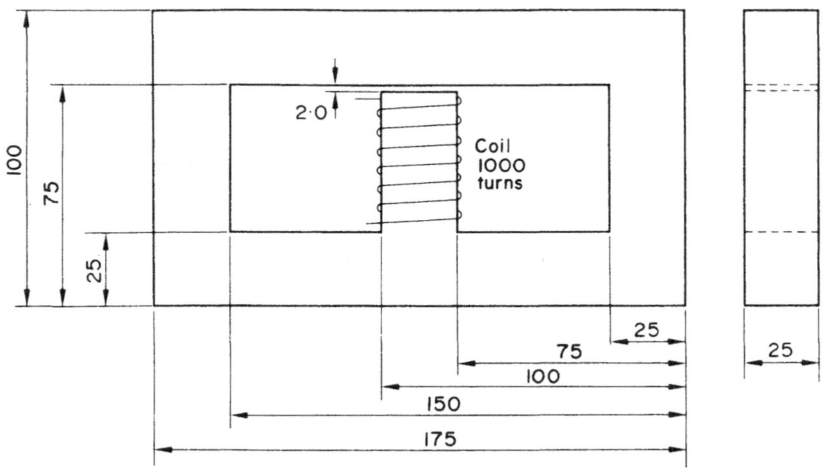

Fig. 13.12. The magnet of Problem 13.17. Dimensions in millimetres.

13.18.   A solenoid consists of a single layer winding of 1000 turns of
fine wire carrying a current of 5·0 mA. It has a circular cross-section of
100 mm diameter and it is 250 mm long. Calculate the field intensity at
eleven equally spaced points along the axis inside the solenoid including one
point at each end.   (9·8, 14·2, 16·8, 17·9, 18·4, 18·6, 18·4, 17·9, 16·8,
14·2, 9·8 A/m.)

13.19.   Find the maximum magnetic field intensity at a distance of 20 mm
from a straight wire 100 mm long carrying a current of 10 mA.   (0·074 A/m.)

13.20.   A magnet is constructed of a block of magnetic steel and two mild

steel pole pieces as shown in Fig. 13.13. The relative permeability of the
mild steel may be assumed to be 1000 and the hysteresis curve for the magnetic
steel is given in Fig. 13.9. Estimate the maximum flux density in the air
gap.   (1·56 T.)

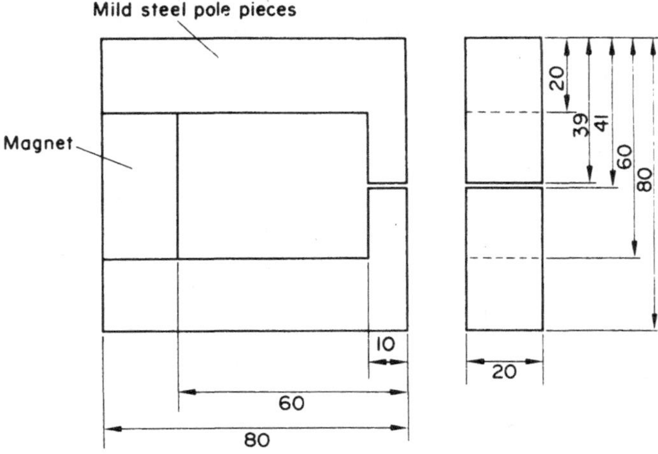

Fig. 13.13. The permanent magnet of Problem 13.20. Dimensions
in millimetres.

FURTHER SOLUTIONS

13.11.   This problem is similar to Example 13.1. Using the result derived in the solution to that example,

$$H = \frac{NI}{\pi D} = \frac{700 \times 0\cdot 25}{\pi \times 0\cdot 2} = 278 \text{ A/m}.$$

13.12.   This is also a toroidal coil; therefore the field intensity is given by the result used in the solution to the last problem:

$$H = \frac{NI}{\pi D} = \frac{150 \times 0\cdot 01}{\pi \times 0\cdot 15} = 3\cdot 18 \text{ A/m}.$$

The value of field intensity is too small to read the flux density directly from the magnetization curve, but the permeability is constant for small values of field. It is called the initial permeability. From the curve, we have μ_r = 390. Therefore

$$B = \mu_0 \mu_r H = 4\pi \times 10^{-7} \times 390 \times 3\cdot 18 = 1\cdot 6 \times 10^{-3} \text{ T} = 1\cdot 6 \text{ mT}.$$

13.13.   The m.m.f. is given by

$$\text{m.m.f.} = \int \mathbf{H} \cdot d\mathbf{l}.$$

Therefore
$$150 \times 0\cdot 01 = \frac{B}{390\,\mu_0} \times \pi \times 0\cdot 15 + \frac{B}{\mu_0} \times 0\cdot 001.$$

Therefore
$$B = \frac{150 \times 0\cdot 01 \times 4\pi \times 10^{-7}}{0\cdot 00218} = 0\cdot 86 \times 10^{-3} \text{ T} = 0\cdot 86 \text{ mT}.$$

13.14.   This problem is similar to Example 13.6. Therefore

$$\text{m.m.f.} = NI = 10\ 000 \times 0\cdot 1 = 1000 \text{ A}.$$

13.15.   This is a problem involving a line vortex near to a flux boundary. Any magnetic flux external to the steel has to be ignored. Then the problem is the same in fields terms as Problem 12.9. The field intensity on the boundary is parallel to the boundary and has been derived in the solution to Problem 12.9 (page 253). The geometry is illustrated in Fig. 12.7. Therefore

$$H = \frac{I \cos^2\theta}{\pi x} = \frac{0\cdot 1 \cos^2\theta}{\pi \times 0\cdot 01} = 3\cdot 18 \cos^2\theta \text{ A/m}.$$

The maximum value of field intensity occurs at the point on the surface nearest to the wire. Therefore

$$H_{\text{max}} = 3\cdot 18 \text{ A/m}.$$

13.16.   This problem is similar to Example 13.7. Using the result derived in the solution to that example,

$$R = \frac{l}{\mu_0 \mu_r A} = \frac{1 \cdot 0}{4\pi \times 10^{-7} \times 500 \times 25 \times 100 \times 10^{-6}} = 6 \cdot 37 \times 10^5 \text{ H}^{-1}.$$

13.17.   This is a magnetic circuit problem similar to Example 13.8. It must be remembered that the two side arms of the magnetic circuit are in parallel so that each will have half of the flux flowing through the centre arm and the air gap. It must be assumed that the magnetic flux is uniform in each section of the steel. Taking the path length to be along the centre line of each section of the core, the reluctance is given by

$$R_m = \frac{225}{\mu_0 \mu_1 \times 2 \times 0 \cdot 625} + \frac{73}{\mu_0 \mu_2 \times 0 \cdot 625} + \frac{2 \cdot 0}{\mu_0 \times 0 \cdot 625}$$

Therefore the m.m.f. equation is

$$1000 I = \frac{B \times 625 \times 10^{-6}}{\mu_0} \left(\frac{225}{\mu_1 \times 2 \times 0 \cdot 625} + \frac{73}{\mu_2 \times 0 \cdot 625} + \frac{2 \cdot 0}{0 \cdot 625} \right).$$

In order to plot the relationship between I and B it is necessary to obtain values for μ_1 and μ_2 from Fig. 13.4. These values can only be obtained when B is known so that the values of I required to give a certain B will be calculated. Simplifying the equation gives

$$I = \frac{B \times 10^{-6}}{1000 \times 10^{-3} \times 4\pi \times 10^{-7}} \left(\frac{112 \cdot 5}{\mu_1} + \frac{73}{\mu_2} + 2 \cdot 00 \right).$$

Or, simplifying, gives

$$I = 0 \cdot 796 \times B \times S.$$

The calculation of some results from the last two equations are given in Table 13.1. The results are plotted in Fig. 13.14. These show the dominant effect of the small air gap on the magnetic circuit and the way in which it has produced a linear relationship for the electromagnet despite a grossly non-linear relationship for the steel.

13.18.   This problem is similar to Problem 13.8. Using the result derived in the solution to that problem gives

$$H = \frac{NI}{2l} (\cos \theta_1 - \cos \theta_2).$$

The geometry of the problem is illustrated in Fig. 13.11 where all the dimensional symbols are defined. Then the angles are given by

TABLE 13.1. SOLUTION TO PROBLEM 13.17

B (T)	μ_1	μ_2	$\dfrac{112\cdot5}{\mu_1}$	$\dfrac{73}{\mu_2}$	S	I (A)
0·25	410	530	0·275	0·138	2·413	0·48
0·50	530	780	0·212	0·094	2·306	0·92
0·75	670	880	0·168	0·083	2·251	1·34
1·00	780	880	0·144	0·083	2·227	1·77
1·25	840	760	0·134	0·096	2·230	2·22
1·50	880	410	0·128	0·178	2·306	2·76

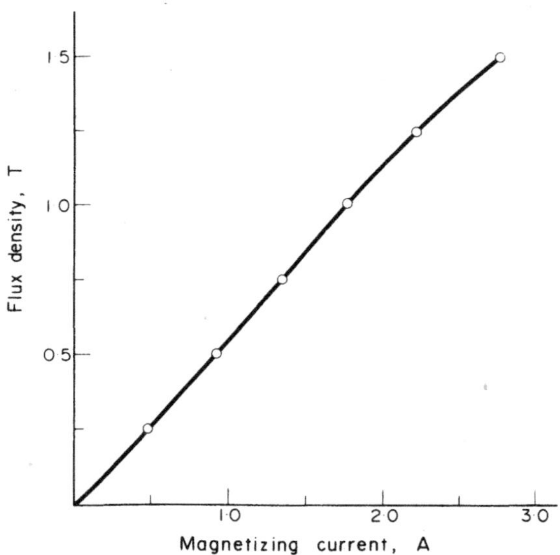

Fig. 13.14. The solution to Problem 13.17. A graph of the relationship between the flux density in the air gap of the magnet and the current flowing in the coil of a thousand turns.

$$\tan\theta_1 = \frac{r}{\frac{1}{2}l - x}, \qquad \tan\theta_2 = -\frac{r}{\frac{1}{2}l + x}.$$

The required results are given in Table 13.2.

TABLE 13.2. SOLUTION TO PROBLEM 13.18

(Dimensions are in millimetres)

x	$\tfrac{1}{2}l+x$	$-\tan\theta_2$	$180°-\theta_2$	$\tfrac{1}{2}l-x$	$\tan\theta_1$	θ_1	$-\cos\theta_2$	$\cos\theta_1$	diff.	H (A/m)
0	125	0·400	21·8°	125	0·400	21·8°	0·928	0·928	1·856	18·6
25	150	0·333	18·4°	100	0·500	26·6°	0·949	0·894	1·843	18·4
50	175	0·285	15·9°	75	0·667	33·7°	0·962	0·832	1·794	17·9
75	200	0·250	14·0°	50	1·000	45·0°	0·970	0·707	1·677	16·8
100	225	0·222	12·5°	25	2·000	63·5°	0·976	0·446	1·422	14·2
125	250	0·200	11·3°	0		90·0°	0·980	0	0·980	9·8

13.19.   The solution for the magnetic field intensity in the region of a
straight current-carrying wire of finite length has been derived in the
solution to Problem 13.9. The maximum field intensity occurs in a plane
perpendicular to the line of the wire passing through the centre point of the
length of the wire. Therefore for the dimensions given in this problem,

$$H = \frac{Il}{2\pi R\sqrt{(l^2 + R^2)}} = \frac{10^{-2} \times 50}{2\pi \times 0\cdot02 \times \sqrt{(2500 + 400)}} = 0\cdot074 \text{ A/m}.$$

13.20.   This problem is similar to Example 13.12. The m.m.f. of the air
gap and the mild steel pole pieces is given by

$$\text{m.m.f.} = H_a l_a + H_s l_s = \frac{2B}{\mu_0} l_a + \frac{2B}{\mu_0 \mu_r} \left(l_1 + \frac{2l_2}{2}\right).$$

The lengths l_1 and l_2 are the total lengths taken along the centre line of
each limb of the magnet for the sections of the two cross-sectional areas.
Therefore

$$\text{m.m.f.} = \frac{2B \times 10^{-3}}{\mu_0} \left(2 + \frac{58 + 75}{1000}\right) = \frac{2B}{\mu_0} \times 10^{-3} \times 2\cdot133.$$

It is seen that the path length through the steel contributes a small amount
to the total m.m.f.; any error in estimating the path length contributes a
negligible error to the total m.m.f. The m.m.f. in the magnet is given by

$$\text{m.m.f.} = -H_m l_m = -H_m \times 40 \times 10^{-3}.$$

Equating these two expressions gives

$$-H_m = \frac{2B \times 10^{-3} \times 2\cdot133}{4\pi \times 10^{-7} \times 40 \times 10^{-3}} = \frac{10^5 B}{1\cdot178}.$$

A straight line may be drawn on the hysteresis curve of the magnetic material
for this equation and the intersection of the two curves gives the required
value of flux density. The result is

$$B = 0 \cdot 778 \text{ T.}$$

The flux density in the air gap is $2B$. Therefore

$$\text{flux density in air gap} = 1 \cdot 56 \text{ T.}$$

CHAPTER 14

Electromagnetic Induction

THEORY

If a closed electric conducting circuit encloses a changing
magnetic field, an electric potential difference is developed around the
circuit which is equal to the rate of change of the magnetic field. This is
called *Faraday's law*. Mathematically it is given by

$$\int \mathbf{E} \cdot d\mathbf{l} = -\frac{\partial \Psi}{\partial t}. \tag{14.1}$$

This potential difference is often called the back e.m.f. of the circuit. It
is generated either by a change in the geometry of the system giving a change
of the total flux threading the circuit, or by a change of flux density for
an unchanging geometry, or both.

A coil of wire carrying an electric current generates a magnetic flux.
If the current in the coil of wire is changing, the magnetic flux threading
the coil is also changing which in its turn generates a back e.m.f. in the
coil to oppose the change in current. The constant of proportionality between
the back e.m.f. and the rate of change of current is called the *inductance*.
The inductance is defined by the equation

$$V = L\,\frac{\partial I}{\partial t}, \tag{14.2}$$

where L is the inductance. It is equal to the permeability multiplied by a
constant which is determined entirely by the geometry of the coil comprising
the inductor. If the coil of the inductor is wound on to a magnetic core, it
is extremely difficult to calculate the inductance of the coil since the
relative permeability is dependent on the flux density in the core which is
dependent on the current in the coil. For an iron-cored coil the inductance
varies as the instantaneous current in the coil varies.

An electric charge q moving with a velocity v in an electric and
magnetic field experiences a force according to the relationship

$$\mathbf{f} = q(\mathbf{E} + \mathbf{v} \times \mathbf{B}), \tag{14.3}$$

which is called the *Lorentz equation*.

If an electric current flows through a conductor in the presence of a
magnetic field, a potential difference is generated between two points

285

perpendicular to both the direction of current flow and of the magnetic field.
It is called the *Hall voltage*. It is given by

$$\mathbf{E}_h = R\,\mathbf{J} \times \mathbf{B}\,, \tag{14.4}$$

where R is the Hall constant, which is a property of the material.

The energy stored in a magnetic field is the integral under the $H{-}B$
relationship for the magnetic material used to support the magnetic field.
The energy stored per unit volume is given by

$$w = \int_0^{B_1} H\,dB \;\; \text{J/m}^3, \tag{14.5}$$

where B_1 is the maximum flux density in the material. For a non-magnetic
material, this becomes

$$W = \int_{\text{volume}} \tfrac{1}{2}\, BH\,dv \;\; \text{J}. \tag{14.6}$$

The force on the pole piece of a magnet is given by

$$f = \tfrac{1}{2}BH \;\; \text{N/m}^2. \tag{14.7}$$

To summarize, some electromagnetic relationships:

Gauss's law:

$$\iint_{\text{area}} \mathbf{D} \cdot d\mathbf{A} \;=\; \iiint_{\text{volume}} \rho\,dv,$$

$$\iint_{\text{area}} \mathbf{B} \cdot d\mathbf{A} = 0.$$

Ampere's law:

$$\oint \mathbf{H} \cdot d\mathbf{l} = \iint_{\text{area}} \mathbf{J} \cdot d\mathbf{A}\,.$$

Faraday's law:

$$\oint \mathbf{E} \cdot d\mathbf{l} = -\frac{\partial}{\partial t} \iint_{\text{area}} \mathbf{B} \cdot d\mathbf{A}\,.$$

Total electromagnetic energy

$$w = \tfrac{1}{2}DE + \tfrac{1}{2}BH + \int JE\,dt \;\; \text{J/m}^3.$$

A summary of many of the relvant field theory formulae are given in Appendix 2.

EXAMPLES

14.1.   Starting with the forces acting on a rectangular coil rotating
in a uniform magnetic field about one of its axes of symmetry in the plane of
the coil, derive an expression for the e.m.f. developed in the coil and show
that it is the same as that given by the application of Faraday's law.

Answer. A section through the coil giving dimensions and showing the
direction of the forces is shown in Fig. 14.1. If the length of the side of

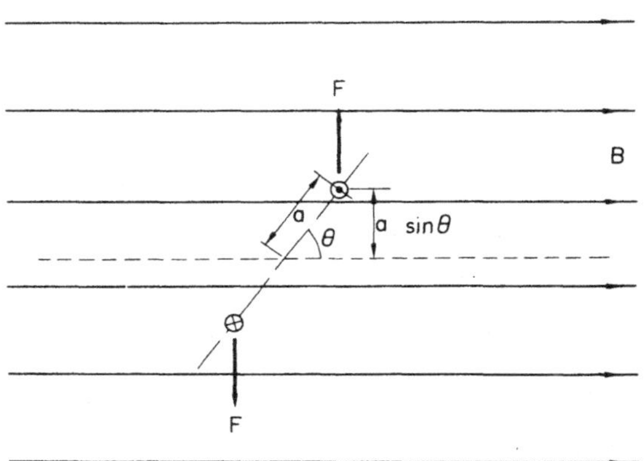

Fig. 14.1. A rectangular coil set at an angle to a uniform
magnetic field.

the coil perpendicular to the plane of the diagram is l, the force is given by
application of eqn. (7.6) (page 119). Therefore

$$F = IBl.$$

It is assumed that the current in the coil is supplied by an external circuit
which is used to maintain the current constant. The force exerted on the
other parts of the circuit and on the other two sides of the coil act parallel
to the axis of rotation and are symmetrical so that no work is done by these
forces during rotation. If the coil starts from the position $\theta = 0$, the work
done by the force during movement of the coil is given by

$$W = 2Fa \sin \theta = 2IBla \sin \theta.$$

The only source of energy is the e.m.f. driving the current around the coil.
The total electrical energy in the time T of the rotation of the coil is
given by

$$W = \int_0^T VI\,dt.$$

As the work is the same in each instance and as the current I is kept constant,

$$\int_{0}^{T} VI\,dt = 2IBla \sin \theta,$$

or, differentiating,

$$V = 2Bla \cos \theta \, \frac{d\theta}{dt}.$$

If ω is the speed of rotation of the coil, the potential difference becomes

$$V = 2B\omega la \cos \theta.$$

From Faraday's law the potential difference is given by

$$V = \frac{\partial \Psi}{\partial t}.$$

The total flux enclosed by the coil when it is at an angle θ is given by

$$\Psi = 2laB \sin \theta.$$

Therefore, by application of Faraday's law, the potential difference becomes

$$V = 2Bla \cos \theta \, \frac{\partial \theta}{\partial t},$$

which is the same result as that derived above.

14.2.   A coil of variable size consists of a wire around three sides of a rectangular factory building. The fourth side is attached to a mobile crane in the roof of the factory having contacts at each end of the roof with two opposite sides of the wire of the coil. Assuming that there is a uniform magnetic field threading the coil, find the e.m.f. generated per unit length of the mobile side of the coil when the crane is moving with a uniform speed of 1·0 m/s.

Answer. If the straight wire is of length l, the rate of increase of flux enclosed by the loop will be equal to

$$Blv = \text{total e.m.f. generated in the loop.}$$

It is assumed that the e.m.f. is generated uniformly along the moving wire and that the stationary sides of the coil do not contribute to the e.m.f. Then the e.m.f. per unit length of moving wire is equal to the total e.m.f. generated in the loop divided by the length of the wire. Therefore

$$\text{e.m.f. per unit length} = Bv = B \times 1{\cdot}0 = B \text{ V/m},$$

where B is that component of the uniform flux density threading the coil which is perpendicular to the direction of motion of the wire.

14.3.   A solenoid consists of 500 turns of fine wire uniformly wound on a non-magnetic coil former, 100 mm long and 50 mm diameter. A search coil, 10 mm in diameter, with 20 closely wound turns is placed centrally and coaxially in the solenoid. Determine the e.m.f. induced in the search coil when the solenoid current changes uniformly from 0 to $2 \cdot 0$ A in 15 ms.

Answer. The field on the axis of a short solenoid has been derived in the solution to Problem 13.8 (page 275) to be

$$B = \frac{\mu_0 NI}{2l} (\cos \theta_1 - \cos \theta_2).$$

Inserting the dimensions of the coil, for the field at the centre, gives

$$B = \frac{4\pi \times 10^{-7} \times 500 I}{2 \times 0 \cdot 1} \left(2 \times \frac{2}{\sqrt{5}}\right) = 5 \cdot 62 \times 10^{-3} I \text{ T}.$$

The total flux threading the search coil is given by

$$\Psi = B \times (\text{area}) = 25\pi B \times 10^{-6}.$$

The rate of change of current is given by

$$\frac{dI}{dt} = \frac{2 \cdot 0}{15 \times 10^{-3}} = 133 \text{ A/s}.$$

Therefore the e.m.f. is given by

$$\Phi = 20 \times 25\pi \times 10^{-6} \times 5 \cdot 62 \times 10^{-3} \times 133 = 1 \cdot 17 \times 10^{-3} \text{ V} = 1 \cdot 17 \text{ mV},$$

where the e.m.f. generated in one turn has been multiplied by the number of turns.

14.4.   Derive an expression for the inductance of a coil of N turns uniformly wound on a non-magnetic toroidal former having a cross-sectional area A and a mean magnetic path length l.

Answer. The flux density inside a toroidal coil has been derived in the solution to Example 13.1 (page 261). Therefore, if the coil is carrying a current I, the flux density is given by

$$B = \mu_0 \frac{NI}{l}.$$

The total flux threading the circuit is given by

$$\Psi = BA = \mu_0 \frac{ANI}{l}.$$

The back e.m.f. for one turn is given by Faraday's law, eqn. (14.1); for N turns, the total back e.m.f. is N times as large, therefore back e.m.f. is given by

$$V = N \frac{\partial \Psi}{\partial t} = \mu_o \frac{AN^2}{l} \frac{\partial I}{\partial t}.$$

The inductance is given by eqn. (14.2). By equating this with the above relationship, gives

$$L = \mu_o \frac{AN^2}{l} \text{ H}.$$

14.5.   If the toroidal coil of Example 14.4 has a further winding of n turns added on top of the existing winding and coaxially with it, find the mutual inductance between the new winding and the original winding on the toroid.

Answer. Mutual inductance is defined, similarly to (self)-inductance, from the e.m.f. generated in one electric circuit due to the rate of change of electric current in a magnetically linked circuit, by the equation

$$V = M \frac{\partial I}{\partial t},$$

where M is the mutual inductance between the two coils. From the solution to the previous example, the total flux threading the magnetic circuit, which is also the total flux threading the new winding, is given by

$$\Psi = \mu_o \frac{ANI}{l}.$$

Therefore the back e.m.f. in the n turns of the new coil due to a change in current, is given by

$$V = \mu_o \frac{nAN}{l} \frac{\partial I}{\partial t}.$$

Therefore, the mutual inductance is given by

$$M = \mu_o \frac{AnN}{l}.$$

14.6.   In a cathode-ray tube, the electron beam is accelerated by a potential difference of 12·0 kV. Find the final velocity of the electrons.

Answer. From mechanical energy considerations, the energy gained by the electron is given by the kinetic energy,

$$W = \tfrac{1}{2}mv^2,$$

where m is the mass of the electron and v is its velocity. In moving through the electrical field, the electron loses electrical energy which may be directly derived from the definition of electrical potential difference, eqn. (8.1) (page 141), therefore

$$W = e\Phi,$$

where e is the charge on the electron. Equating these two energies gives

$$\tfrac{1}{2}mv^2 = e\Phi.$$

The sizes of the charge and mass of the electron are given in Appendix 1, therefore

$$v^2 = \frac{2e\Phi}{m} = \frac{2 \times 1\cdot602 \times 10^{-19} \times 12\cdot0 \times 10^3}{9\cdot109 \times 10^{-31}} = 4\cdot23 \times 10^{15}.$$

Therefore

$$v = 6\cdot5 \times 10^7 \text{ m/s.}$$

14.7.   An electron travelling at a speed of 10^6 m/s encounters a uniform magnetic field of $1\cdot0$ mT acting perpendicularly to its direction of motion. If the electron is travelling initially in a straight line, describe the electron path in the magnetic field.

Answer. If the moving electron is considered to be equivalent to an electric current, the force on the electron due to the magnetic field will be perpendicular to the direction of the magnetic field and perpendicular to the direction of motion of the electron. The electron will move in a circular orbit. Then the force due to the magnetic field acting perpendicularly to the direction of motion will be equal to the centripetal force of the circular motion. The equivalent current effect of the electron movement is given by

$$I = ev$$

and the magnetic force is given by substituting into eqn. (7.6) (page 119),

$$f = IB = evB.$$

For a circular orbit of radius r, the centripetal force is given by

$$f = \frac{mv^2}{r}.$$

Equating these two forces gives

$$evB = \frac{mv^2}{r}.$$

Therefore

$$r = \frac{mv}{eB} = \frac{9\cdot109 \times 10^{-31} \times 10^6}{1\cdot602 \times 10^{-19} \times 10^{-3}} = 5\cdot68 \times 10^{-3} \text{ m} = 5\cdot68 \text{ mm.}$$

The electron performs a circular path of radius 5·68 mm.

14.8.   Derive an expression for the Hall voltage in terms of the various quantities depicted in Fig. 14.2.

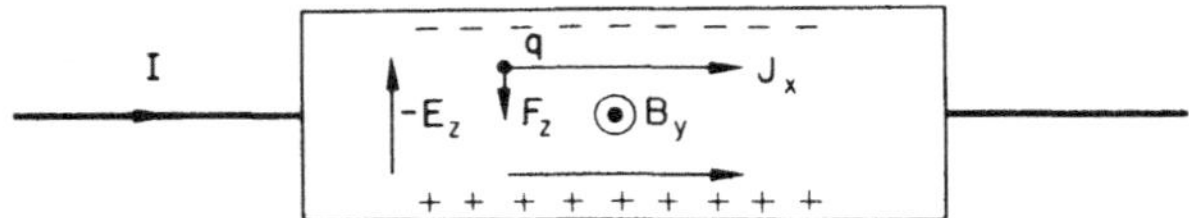

Fig. 14.2. Hall effect in a conductor.

Answer. The current is shown in the diagram to be moving through the
conductor in a direction perpendicular to the magnetic flux density.
Therefore the individual charged particles in the conductor will experience
a force given by

$$F_z = J_x B_y,$$

where the subscripts give the directions in which the various quantities act
in a rectangular coordinate system. There will be an accumulation of charge
on one surface of the conductor and a negative charge on the opposite surface
as shown in the figure. The surface charge will give rise to an electric
field intensity E_z such that the electrostatic forces on the charge conductors
exactly equal the magnetic forces. It ought to be noticed that the charged
particles are not necessarily electrons and are given a positive sign in this
derivation. Then, equating these forces,

$$qE_z = J_x B_y = qv_x B_y.$$

The field intensity is the Hall voltage as defined in eqn. (14.4), therefore
the Hall voltage is given by

$$E_z = v_x B_y.$$

The actual potential difference measured between the two edges of the
conducting slab is this field intensity multiplied by the thickness in the
direction of the electric field intensity.

14.9.   Derive, from first principles, the energy stored in a magnetic
field inside a magnetic material.

Answer. Energy is expended in setting up a magnetic field. This energy
is then stored in the field. It is possible to obtain a measure of the energy
stored if an electrical circuit is used to set up the magnetic field. A
continuous electric current is required to maintain the magnetic field constant
but there is no impedance to the flow of the electric current from the magnetic

field unless the current and the magnetic field change strength. When the
magnetic field is changing strength, a back e.m.f. is developed which opposes
any change in the electric current. The electrical energy expended in
overcoming the back e.m.f. is equal to the energy stored in the magnetic
field. For a coil of N turns, the back e.m.f. is given by

$$\text{back e.m.f.} = V = N \frac{\partial \Psi}{\partial t}.$$

The power expended in any small time δt by the electrical circuit is given by

$$\delta W = VI\delta t = NI \frac{\partial \Psi}{\partial t} \delta t = NI\delta \Psi.$$

But NI is the m.m.f., therefore

$$W = (\text{m.m.f.})\delta \Psi = \Phi \delta \Psi.$$

Please note that in this answer V is used for the electrical potential
difference and Φ is used for the magnetic potential difference or m.m.f.
Since for any particular magnet there is only a geometrical relationship
between the flux and the flux density and between the field intensity and the
m.m.f., the B–H relationship for any particular magnetic material becomes the
Ψ-m.m.f. relationship to a different scale for a magnet made of that material.
Then the total energy stored in any magnet is given by

$$W = \int_{0}^{\Psi_1} \Phi d\Psi \ \text{J}.$$

For a uniform field in a rectangular shape of magnetic material of length l
and cross-sectional area A, the total flux and potential difference are
given by

$$\Psi = BA, \qquad \Phi = Hl.$$

Therefore the incremental change of energy is given by

$$\delta W = \Phi \delta \Psi = Hl\delta(BA);$$

but A is a constant, therefore

$$\delta W = H\delta B(lA) = H\delta B(\text{volume}).$$

Therefore the change in energy per unit volume is given by

$$\delta w = H\delta B \ \text{J/m}^3$$

and the total energy stored in the magnetic field is given by

$$w = \int_{0}^{B_1} HdB \ \text{J/m}^3.$$

This is the area above the $B\text{-}H$ curve. For a non-magnetic material, there is
a linear relationship between B and H given by eqn. (7.9)(page 119),
therefore the energy per unit volume becomes

$$w = \frac{B^2}{2\mu_0} = \tfrac{1}{2}\mu_0 H^2 = \tfrac{1}{2}BH \ \ \text{J/m}^3 .$$

14.10.   Estimate the electrical energy needed to magnetize a small magnet
to saturation. The magnet is 50 mm long by 20 mm by 10 mm and the hysteresis
curve of the magnetic material is given in Fig. 14.3. Assume that the magnet
is initially demagnetized.

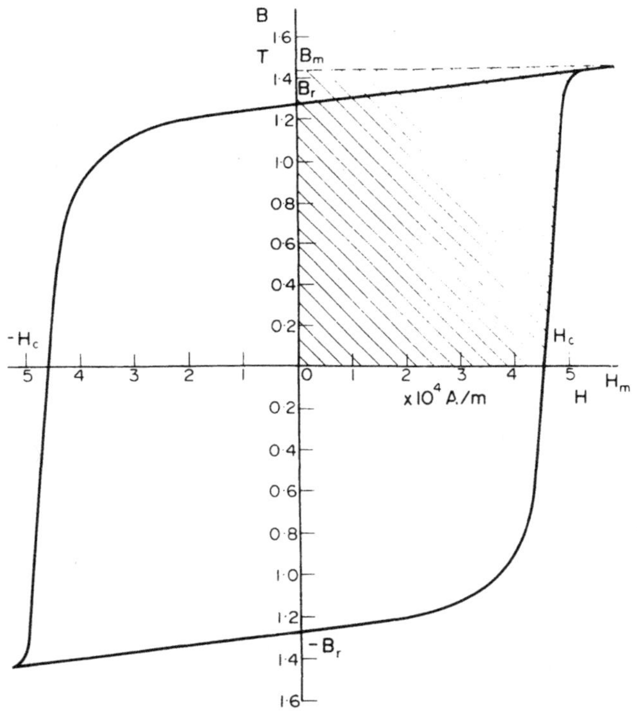

Fig. 14.3. Hysteresis curve of a permanent magnet steel used in
Example 14.10.

Answer. Saturation is when the magnet is magnetized to its
maximum flux density. This is shown at B_m on Fig. 14.3. From the hysteresis
curve it is only possible to estimate the $B\text{-}H$ relationship when starting from
the demagnetized condition. So as not to underestimate the energy required,
it will be assumed that negligible flux is created in the magnet until the

intensity exceeds the coercivity. Then the energy per unit area is given by
the area above the $B-H$ curve up to the maximum value of flux density required.
This energy is given by the size of the shaded area on Fig. 14.3; measuring
the area gives

$$w = 6 \cdot 9 \times 10^4 \ \mathrm{J/m^3}.$$

The volume of the magnet is given by

$$v = 50 \times 10 \times 20 \times 10^{-9} = 10^{-5} \ \mathrm{m^3}.$$

Therefore the total energy is given by

$$W = wv = 0 \cdot 69 \ \mathrm{J}.$$

14.11.   By the principle of virtual work, derive an expression for the
force on the pole piece of a magnet.

   Answer. Assume that there is a uniform field in the gap between two
pole pieces of a magnet as shown in Fig. 14.4. If the pole pieces are moved

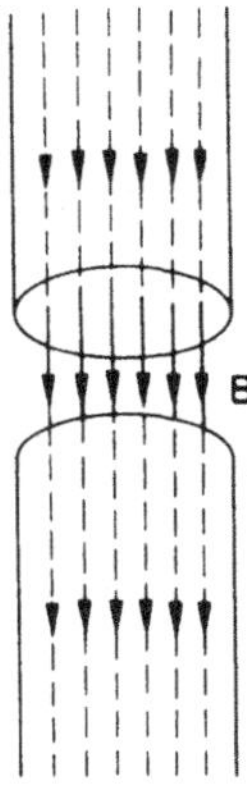

Fig. 14.4. The uniform field in the gap between two pole pieces
of a magnet.

a small distance apart, the increase in energy of the magnetic field will be
equal to the energy expended in moving the pole pieces. It is necessary to
assume also that the movement is so small that there is no change in the total
flux threading the magnetic circuit and no change in the total reluctance of
the magnetic circuit, and hence no change in the stored magnetic energy
elsewhere in the magnet. As the magnetic flux remains constant, no energy is
extracted from the electrical circuit which is maintaining the magnetic flux
constant, therefore all the mechanical work done in moving the pole pieces

will appear as an increase in energy stored in the field. Let there be a force of attraction F between the two pole pieces. If the pole pieces are of cross-sectional area A and are moved a distance δx, the mechanical work is $F\delta x$ and the magnetic energy is given by eqn. (14.6). Therefore

$$F\delta x = \tfrac{1}{2}BHA\delta x.$$

Therefore $F = \tfrac{1}{2}BHA$ N.

Therefore the force per unit area is given by eqn. (14.7).

14.12.   The magnetic flux density between the poles of a magnet is measured to be 1·6 T. If the pole face is a square of 50 mm side, find the force between the poles of the magnet.

  Answer. The force between the pole pieces of a magnet is given by eqn. (14.7). Therefore substituting into that equation gives

$$F = \tfrac{1}{2}BH(\text{area}) = \frac{B^2}{2\mu_0}\,(\text{area}) = \frac{1{\cdot}6^2 \times 25 \times 10^{-4}}{2 \times 4\pi \times 10^{-7}} = 2{\cdot}55 \times 10^3 \text{ N.}$$

PROBLEMS

14.1.   The surface of the stator of a linear electric motor consists of a number of permanent magnet pole pieces of alternating polarity such that the magnetic flux perpendicular to the surface is given by

$$B_y = B_0 \sin ax \text{ T,}$$

where x is a dimension parallel to the surface. The moving portion consists of a number of straight conductors lying parallel to the surface and perpendicular to the x-direction, each carrying a current I A, which are constrained to move in the x-direction. Find the force on one conductor and, if the conductor is moving with a velocity v, the back e.m.f. generated.

14.2.   A solenoid, 0·2 m long, consists of 200 turns uniformly wound on a 60 mm diameter cardboard former. A search coil, 10 mm in diameter, with 20 closely wound turns is placed centrally and coaxially in the solenoid. Determine the e.m.f. induced in the search coil when the solenoid current changes uniformly from 0 to 10 A in 10 ms.   (1·89 mV.)

14.3.   A toroidal coil of 500 turns is wound on a steel ring of 0·5 m mean diameter and $2{\cdot}0 \times 10^{-3}$ m² cross-sectional area. An excitation of 4000 A/m produces a flux density of 1·0 T. Find the inductance of the coil. If a 10 mm long gap is cut in the ring, find the inductance under these new conditions assuming that the excitation is increased to maintain the flux density at 1·0 T. Neglect all leakage and fringing. (See Problem 13.2.) (80 mH;   35 mH.)

14.4.   Find the mutual inductance of the two coupled coils of Problem 14.2. (1·89 μH.)

14.5.   By assuming that the flux density is uniform across any cross-section of the coil, calculate the inductance of a coil of circular cross-section of diameter 10 mm of 1000 turns uniformly distributed along a length of 20 mm. (*Hint*: calculate the flux density at twenty equally spaced points along the length of the coil.) (3·85 mH.)

14.6.   In a cathode-ray tube, the electron beam is accelerated by a potential difference of 8·0 kV. Find the final velocity of the electrons. ($5\cdot3 \times 10^7$ m/s.)

14.7.   An electron in a uniform magnetic field is found to travel in a circular path. Prove that the frequency of rotation of the electron in this circular path is independent of the radius of the path and find the frequency of rotation in a uniform field of 10 mT.   ($2\cdot8 \times 10^8$ Hz.)

14.8.   Calculate the maximum Hall voltage which occurs across a 1·0 mm thick slice of germanium for a total current flow of 0·1 A and a magnetic flux density of 1·2 T. The Hall coefficient for germanium is $R = -\ 10^{-3}$ m^3/C. (120 mV.)

14.9.   A magnetically operated electrical contactor consists of a magnetic pole piece which moves a distance of 5·0 mm to close an air gap in the magnetic circuit of an electromagnet when it is excited to a flux density of 1·0 T. Calculate the initial closing force and the mechanical energy dissipated in the impact of closure if the area of the pole piece is 10^{-3} m^2 and if the flux density can be assumed to be uniform in the air gap. (400 N; 2·0 J.)

14.10.   Calculate the force of attraction between the magnet and plates of Problem 13.7 (page 273).   (10^5 N.)

SOLUTIONS

14.1.   The flux density is given in the question. The conductors lie in the z-direction and may move in the x-direction. As the current flow is perpendicular to the magnetic flux density, the force is given by application of eqn. (7.6)(page 119). Therefore

$$f_x = I_z B_y = IB_0 \sin ax \text{ N/m}.$$

If the force is moving with the velocity v, the power expended by the force is given by

$$P = f_x v = IB_0 v \sin ax \text{ W/m}.$$

If V is the back e.m.f. per unit length of moving conductor, then this power is the same as the electrical power used to keep the current flowing in the circuit, therefore

$$P = VI \text{ W/m} \qquad \text{and} \qquad V = B_0 v \sin ax \text{ V/m}.$$

14.2.   From the solution to Problem 13.8 (page 275), the flux density on the axis of a short solenoid is given by

$$B = \frac{\mu_0 NI}{2l} (\cos \theta_1 - \cos \theta_2).$$

At the centre of the solenoid,

$$\cos \theta_1 = - \cos \theta_2 = \frac{l}{\sqrt{(l^2 + 4r^2)}} = \frac{200}{\sqrt{(40\ 000 + 3600)}} = 0{\cdot}958.$$

Therefore the flux density is given by

$$B = \frac{\mu_0 NI}{2l} \times 1{\cdot}916 = \frac{4\pi \times 10^{-7} \times 200 \times 1{\cdot}916 I}{2 \times 0{\cdot}2} = 1{\cdot}203 \times 10^{-3} I \text{ T}.$$

The total flux in the search coil is given by

$$\Psi = BA = 1{\cdot}203 \times 10^{-3} I \times \pi \times 0{\cdot}005^2 = 9{\cdot}45 \times 10^{-8} I \text{ Wb}.$$

Therefore the e.m.f. in the search coil is given by

$$V = N \frac{\partial \Psi}{\partial t} = 20 \times 9{\cdot}45 \times 10^{-8} \frac{\partial I}{\partial t} = 1{\cdot}89 \times 10^{-6} \times \frac{10}{10^{-2}} = 1{\cdot}89 \times 10^{-3} \text{ V} = 1{\cdot}89 \text{ mV}.$$

14.3.   From the solution to Problem 13.2 (page 274), there is needed a current of 4π A to provide the flux density of $1{\cdot}0$ T. There is a constant of proportionality between the current and the flux such that

$$\Psi = KI.$$

At the flux levels of this problem,

$$K = \frac{BA}{I} = \frac{1 \cdot 0 \times 2 \cdot 0 \times 10^{-3}}{4\pi} = \frac{10^{-3}}{2\pi} \text{ H.}$$

The back e.m.f. is given by

$$V = N \frac{\partial \Psi}{\partial t} = NK \frac{\partial I}{\partial t}.$$

Therefore the inductance is given by

$$L = NK = \frac{500 \times 10^{-3}}{2\pi} = 7 \cdot 96 \times 10^{-2} \text{ H} = 79 \cdot 6 \text{ mH.}$$

If the slot is cut in the ring, a greater current is required to provide the same flux density and K will have a different value. Using the results from the solution to Problem 13.2,

$$K = \frac{2 \cdot 0 \times 10^{-3}}{28 \cdot 4} \text{ H.}$$

Therefore

$$L = NK = \frac{500 \times 2 \cdot 0 \times 10^{-3}}{28 \cdot 4} = 3 \cdot 53 \times 10^{-2} \text{ H} = 35 \text{ mH.}$$

14.4.   The definition of mutual inductance is given in the solution to Example 14.5. From the solution to Problem 14.2,

$$V = 1 \cdot 89 \times 10^{-6} \frac{\partial I}{\partial t}.$$

Therefore the inductance is $1 \cdot 89 \ \mu\text{H}$.

14.5.   The flux density inside the coil is given by the formula derived in the solution to Problem 13.8 (page 275). It is assumed that this flux density is uniform across the cross-section of the coil.

$$B = \frac{\mu_0 NI}{2l} (\cos \theta_1 - \cos \theta_2).$$

As the flux pattern in the coil is symmetrical, it is possible to use only ten calculations to find the flux density at twenty equally spaced points. The calculation is similar to that used in the solution to Problem 13.18 (page 278). The calculation of the angles is given in Table 14.1. In this case we have calculated the field at twenty-one points along the coil. At each point, the flux can be assumed to link with fifty turns at that point except for the ends where the linkage is only with twenty-five turns. At any point, the flux in the coil is given by

$$\Psi_n = \frac{\mu_0 NI}{2l} S_n(\text{area}).$$

TABLE 14.1.   SOLUTION TO PROBLEM 14.5

n	x	$\frac{1}{2}l - x$	$\cos\theta_1$	$-\cos\theta_2$	S_n
0	0	0·020	0	0·970	0·970
1	0·001	0·019	0·196	0·968	1·164
2	0·002	0·018	0·372	0·964	1·336
3	0·003	0·017	0.515	0·960	1·475
4	0·004	0·016	0·625	0.955	1·580
5	0·005	0·015	0·707	0.949	1·656
6	0·006	0·014	0·769	0·942	1·711
7	0·007	0·013	0·814	0·934	1·748
8	0·008	0·012	0·849	0·924	1·773
9	0·009	0·011	0·875	0·911	1·786
10	0·010	0·010	0·895	0·895	1·790

The back e.m.f. generated in fifty turns at that point is given by

$$\Delta V = \frac{\mu_0 N}{2l}\, S_n(\text{area}) \times 50\, \frac{\partial I}{\partial t}.$$

The total back e.m.f. is the sum of all these twenty e.m.f.s , therefore

$$V = \sum \Delta V = \frac{\mu_0 N}{2l}\left(\sum S_n\right)(\text{area}) \times 50\, \frac{\partial I}{\partial t}.$$

The sum is given by

$$\sum S_n = S_0 + S_{10} + \sum_{n=1}^{9} 2S_n = 31\cdot218,$$

with values taken from Table 14.1.   Therefore the inductance is given by

$$L = \frac{\mu_0 N}{2l}\left(\sum S_n\right)(\text{area})\ 50 = \frac{4\pi \times 10^{-7} \times 10^3 \times 31\cdot218 \times \pi \times (0\cdot005)^2 \times 50}{2 \times 0\cdot02}$$

$$= 3\cdot85 \times 10^{-3}\ \text{H} = 3\cdot85\ \text{mH}.$$

A simple approximation to the required solution may be used by applying the formula for the inductance of a toroidal coil to a short solenoid.   The formula has been derived in the solution to Example 14.4.   Therefore the estimate of inductance gives

$$L = \frac{\mu_0 A N^2}{l} = \frac{4\pi \times 10^{-7} \times \pi \times (0\cdot005)^2 \times 10^6}{0\cdot02} = 4\cdot93 \times 10^3\ \text{H}.$$

Assuming that the earlier result is correct, and even in that calculation there have been some approximations, the estimate of inductance from the formula for a toroid gives a result that is 28 per cent too large.

14.6.   This problem is similar to Example 14.6. Using the formula derived in the solution to that example,

$$v^2 = \frac{2e\Phi}{m} = \frac{2 \times 1 \cdot 602 \times 10^{-19} \times 8 \cdot 0 \times 10^3}{9 \cdot 109 \times 10^{-31}} = 2 \cdot 81 \times 10^{15}.$$

Therefore $v = 5 \cdot 3 \times 10^7$ m/s.

14.7.   This problem is similar to Example 14.7. In the presence of a magnetic field, the electron travels in a circular path of radius r and at a velocity v.   Then its equation of motion is given by

$$eB = \frac{mv}{r}.$$

Therefore the angular velocity of rotation of the electron around its orbit is given by

$$\omega = \frac{v}{r} = \frac{eB}{m}.$$

The frequency of rotation of the electron is given by

$$f = \frac{\omega}{2\pi} = \frac{eB}{2\pi m}.$$

This expression for the frequency is independent of the radius. To find the frequency for a particular magnetic flux density,

$$f = \frac{1 \cdot 602 \times 10^{-19} \times 10^{-2}}{2\pi \times 9 \cdot 109 \times 10^{-31}} = 2 \cdot 8 \times 10^8 \text{ Hz.}$$

14.8.   The Hall effect voltage is given by eqn. (14.4). Let the width of the slice of germanium, that is the dimension across which the Hall voltage is measured, be d. Then the cross-sectional area across which the current flows will be $10^{-3}d$.   Therefore substituting into eqn. (14.4) and using positive values of field intensity, gives

$$E_h = 10^{-3} \times \frac{0 \cdot 1}{10^{-3}d} \times 1 \cdot 2 = \frac{0 \cdot 12}{d}.$$

The measured voltage is given by

$$V = E_h d = 0 \cdot 12 \text{ V} = 120 \text{ mV.}$$

14.9.   If the flux density in the air gap remains constant as the air gap closes, the force will remain constant and is given by eqn. (14.7), therefore

$$F = \tfrac{1}{2}BH(\text{area}) = \frac{B^2}{2\mu_o}\,(\text{area}) = \frac{1\cdot0 \times 10^{-3}}{2 \times 4\pi \times 10^{-7}} = 398 \text{ N.}$$

The mechanical energy dissipated in the impact of closure will be equal to the magnetic energy stored in the air gap. The magnetic energy is calculated assuming that the flux density is uniform in the air gap, therefore, from eqn. (14.6),

$$W = \tfrac{1}{2}BH(\text{volume}) = \frac{B^2}{2\mu_o}(\text{area})(\text{distance}) = 398 \times 0\cdot005 = 1\cdot99 \text{ J.}$$

14.10.   The force of attraction is given by eqn. (14.7), therefore taking the values calculated in the solution to Problem 13.7 (page 275), and remembering that there are two pole faces,

$$F = \frac{2 \times 1\cdot0 \times \pi \times (0\cdot2)^2}{2 \times 4\pi \times 10^{-7}} = 10^5 \text{ N.}$$

FURTHER PROBLEMS

14.11.   A coil of 200 turns is rotating at a uniform speed of 2·0 rad/s about one diameter in a uniform flux density of 0·1 T. If the cross-sectional area of the coil is 10^{-3} m², find the e.m.f. developed across the winding. (40 cos θ mV.)

14.12.   A toroidal coil of 120 turns is uniformly wound on a non-magnetic former of cross-sectional area $2\cdot0 \times 10^{-4}$ m² and mean diameter 0·1 m. Find the inductance of the coil.   (11·5 μH.)

14.13.   In a three-phase induction motor, the magnetic flux density generated by the stator can be represented at the position of the rotor conductors by

$$B = B_o \sin (\omega t - 80),$$

where ω = 100π rad/s for a supply frequency of 50 Hz, and θ is an angle of rotation measured about the axis of rotation of the motor. The magnetic field represented by B is rotating about the axis of the motor at one-eighth of the supply frequency and has eight cycles of variation per revolution. The squirrel-cage rotor is 1·5 m in diameter and consists of a number of straight conductors, 1·5 m long, parallel to the axis of the motor, on the surface of the rotor. If the rotor speed is 366 r.p.m., find the e.m.f. developed in each conductor in terms of the maximum magnetic flux developed by the stator. If the effective resistance of the circuit connected to each conductor is 1·0 Ω, find the torque developed by the force on each conductor. $(4 \cdot 8B_o^2 \text{ N m.})$

14.14.   A solenoid of 50 mm diameter and 0·3 m long has 2000 turns. Find its self-inductance. A small search coil of 10 mm diameter having twenty closely wound turns is placed at the centre of the solenoid coaxially with the solenoid. If the solenoid is connected to a d.c. electrical supply of 10 V, find the e.m.f. generated in the search coil assuming that the resistance of the wire of the solenoid is negligible. Hence, or otherwise, find the mutual inductance between the coils.   (33 mH; 4·0 mV; 13·1 μH.)

14.15.   Consider a toroidal coil on a non-magnetic former and prove that the stored energy in the magnetic field is given by $\frac{1}{2}LI^2$, where L is the inductance and I is the electric current.

14.16.   Prove the relationship given in Problem 7.2 (page 129) by the principle of virtual work and stored energy.

14.17.   An electron travelling at 10^7 m/s in a vacuum tube is deflected by an electric field intensity of 10 kV/m for a distance of 10 mm. Find the angle through which the electron is deflected. $(10^{\circ}.)$

14.18.   An electron is emitted from a flat metal plate which forms the source of a uniform electric field intensity of 100 kV/m. There is also a uniform electric flux density of 1·6 T which is perpendicular to the electric field intensity and parallel to the surface of the plate. Prove that an electron emitted from the plate will return to the plate a distance of 1·4 μm from where it was emitted and after a time of 22 ps.

14.19.   A magnetic door catch consists of a permanent magnet with mild steel pole pieces each of which presents an area of 20×10^{-6} m² to a mild steel striker plate on the door. If, when the door is closed, the flux density in the pole pieces is uniform at 0·2 T, find the force exerted by the magnet to keep the door closed.   (0·64 N.)

14.20.   Find the mechanical force between the pole pieces of the magnet of Problem 13.20 (page 278).   (421 N.)

FURTHER SOLUTIONS

14.11.   It may be helpful to understand the mechanics of this problem by making reference to Fig. 14.1. The total flux linked by the coil is given by

$$\Psi = BA \sin\theta.$$

Then from Faraday's law, eqn. (14.1),

$$\text{e.m.f.} = NBA \cos\theta\, \frac{\partial\theta}{\partial t} = \omega NBA \cos\theta.$$

Inserting the numbers from the problem gives

$$\text{e.m.f.} = 2{\cdot}0 \times 200 \times 0{\cdot}1 \times 10^{-3} \cos\theta = 0{\cdot}04 \cos\theta \text{ V} = 40 \cos\theta \text{ mV}.$$

14.12.   Using the relationship derived in the solution to Example 14.4, gives

$$L = \mu_0 \frac{AN^2}{l} = \frac{4\pi \times 10^{-7} \times 2{\cdot}0 \times 10^{-4} \times (120)^2}{\pi \times 0{\cdot}1} = 1{\cdot}15 \times 10^{-5} \text{ H} = 11{\cdot}5 \text{ }\mu\text{H}.$$

14.13.   The magnetic field is rotating at a speed of

$$f_1 = \frac{50}{8} = 6{\cdot}25 \text{ rev/s.}$$

The rotor is rotating at a slower speed of

$$f_2 = \frac{366}{60} = 6{\cdot}10 \text{ rev/s.}$$

The rotor is *slipping* behind the rotating magnetic field. Therefore the rotor conductors are moving through the field at a speed equivalent to a frequency of

$$f_3 = (f_1 - f_2)(\text{number of field cycles per revolution}) = 0{\cdot}15 \times 8 = 1{\cdot}2 \text{ Hz.}$$

Therefore the e.m.f. is given by

$$\Phi = Bvl \sin 2\pi f_3 t = B_0 \times \tfrac{1}{2} \times 1{\cdot}5 \times 2\pi \times 1{\cdot}2 \times 1{\cdot}5 \sin 2\pi \times 1{\cdot}2t$$

$$= 8{\cdot}48 B_0 \sin 7{\cdot}55t \text{ V.}$$

The current in each conductor is given by

$$I = \frac{\Phi}{R} = 8{\cdot}48 B_0 \sin 7{\cdot}55t \text{ A.}$$

Assuming that the magnetic field and the direction of the electric current are perpendicular, as they are in an electric motor, the force on one conductor is given by

$$F = IBl = 8{\cdot}48 B_0^2 \times 1{\cdot}5 \sin^2 7{\cdot}55t \text{ N.}$$

Therefore the torque is given by

$$\text{torque} = Fr = 8{\cdot}48 \times 1{\cdot}5 \times 0{\cdot}75 B_0^2 \sin^2 7{\cdot}55t = 9{\cdot}55 B_0^2 \sin^2 7{\cdot}55t \text{ Nm.}$$

Therefore the average value of torque contributed by each conductor is

$$\tfrac{1}{2} \times 9 \cdot 55 B_o^2 = 4 \cdot 8 B_o^2 \text{ Nm.}$$

14.14.   The inductance of a toroidal coil is given in the solution to Example 14.4. Assuming that the long solenoid approximates to a toroidal coil, the inductance is given by

$$L = \mu_o \frac{AN^2}{l} = \frac{4\pi \times 10^{-7} \times \pi \times (25)^2 \times 10^{-6} \times (2000)^2}{0 \cdot 3} = 3 \cdot 29 \times 10^{-2} \text{ H} = 32 \cdot 9 \text{ mH.}$$

A more accurate estimate of the inductance of such a coil could be found using the method suggested in Problem 14.5 and used in its solution. However, there is much less error in using the long solenoid approximation to calculate the magnetic field at the centre of the solenoid. Therefore the calculation of the e.m.f. in the coupled coil in terms of the rate of change of the current in the solenoid will be more accurate. The field at the centre of the coil is given by

$$B = \mu_o \frac{NI}{l}.$$

Therefore the e.m.f. is given by

$$\Phi = \mu_o \frac{Nna}{l} \frac{\partial I}{\partial t}.$$

But the rate of change of current is given by

$$V = L \frac{\partial I}{\partial t}, \qquad \text{therefore} \qquad \frac{\partial I}{\partial t} = \frac{V}{L},$$

and the e.m.f. is given by

$$\Phi = \mu_o \frac{NnaV}{lL} = \frac{4\pi \times 10^{-7} \times 2000 \times \pi \times 10^{-4} \times 20 \times 10}{4 \times 0 \cdot 3 \times 3 \cdot 29 \times 10^{-2}} = 4 \cdot 00 \times 10^{-3} \text{ V} = 4 \cdot 0 \text{ mV.}$$

Mutual inductance is defined in the solution to Example 14.5. Therefore the e.m.f. is given by

$$\Phi = M \frac{\partial I}{\partial t} = \mu_o \frac{Nna}{l} \frac{\partial I}{\partial t}.$$

Therefore

$$M = \mu_o \frac{Nna}{l} = \frac{4\pi \times 10^{-7} \times \pi \times 10^{-4} \times 20 \times 2000}{4 \times 0 \cdot 3} = 1 \cdot 315 \times 10^{-5} \text{ H} = 13 \cdot 1 \text{ μH.}$$

Application of the formula derived in the solution to Problem 13.8 (page 275), shows that there is an error of $1 \cdot 4$ per cent in using the assumption of an infinitely long solenoid in the calculation of this mutual inductance.

14.15.   The stored energy is given by eqn. (14.6). If the toroidal coil is assumed to have N turns, a cross-sectional area A, a mean flux path length

l, and to be carrying an electric current I, then the magnetic flux and field intensity inside the coil is given by

$$H = \frac{NI}{l} \qquad \text{and} \qquad B = \mu_o\frac{NI}{l}.$$

Therefore the stored energy is given by

$$W = \tfrac{1}{2}BH(\text{volume}) = \mu_o\,\frac{N^2 I^2}{2l^2}\,(lA) = \mu_o\,\frac{N^2 A}{2l}\,I^2.$$

But the inductance is given by

$$L = \mu_o\,\frac{N^2 A}{l}.$$

Therefore $\qquad\qquad\qquad\qquad W = \tfrac{1}{2}LI^2.$

14.16.   In Problem 7.2 (page 129) we are asked to prove that the bursting pressure on the sheath of a coaxial cable carrying equal and opposite currents I is given by

$$p = \frac{\mu_o I^2}{8\pi^2 r^2}\ \text{N/m}^2.$$

In the solution to that problem, use was made of the force on a current-carrying wire in a magnetic field. However, the careful student may have observed that the required result was obtained only by making the assumption that the magnetic flux density exactly at the position of the current in the sheath was half the value of the flux density just inside the sheath. The difficulty occurs because the current contributes towards the generation of the flux density. A more rigorous solution may be obtained working from the principle of virtual work and the change in stored energy.

Assume that there is an inward acting pressure p at the outer conductor radius r. Then the total force acting around the circumference of the cable is given by

$$F = 2\pi rp.$$

Expand the outer conductor by increasing the radius an infinitesimal amount δr. Then the change in energy due to moving against the force F a distance δr is given by

$$\delta W_p = 2\pi rp\delta r.$$

The increase in stored magnetic energy is given by

$$\delta W_m = \tfrac{1}{2}BH\,2\pi r\delta r \qquad \text{and} \qquad B = \mu_o H = \frac{\mu_o I}{2\pi r}.$$

Therefore

$$\delta W_m = \tfrac{1}{2}\mu_o\left(\frac{I}{2\pi r}\right)^2 2\pi r\delta r = \frac{\mu_o I^2}{4\pi r}\,\delta r.$$

There will also be some electrical energy contributed from the supply due to the change in inductance of the cable. The back e.m.f. due to the change in size is given by

$$V = \frac{\partial \Psi}{\partial t} = B\,\frac{\partial r}{\partial t} = \left(\frac{\mu_0 I}{2\pi r}\right)\frac{\partial r}{\partial t}.$$

The electrical energy expended in overcoming this change is given by

$$\delta W_e = IV\delta t = I\left(\frac{\mu_0 I}{2\pi r}\right)\frac{\partial r}{\partial t}\,\delta t = \frac{\mu_0 I^2}{2\pi r}\,\delta r.$$

The energy from the electrical supply is used to increase the stored magnetic energy and to do mechanical work on the surroundings. Therefore

$$\delta W_p + \delta W_m = \delta W_e.$$

Therefore

$$2\pi r p \delta r + \frac{\mu_0 I^2}{4\pi r}\,\delta r = \frac{\mu_0 I^2}{2\pi r}\,\delta r.$$

Therefore the internal pressure is bursting and is given by

$$p = \frac{\mu_0 I^2}{8\pi^2 r^2}\ \text{N/m}^2.$$

14.17.   The time that the electron is under the influence of the field is given by

$$t = \frac{l}{v} = \frac{10^{-2}}{10^7} = 10^{-9}\ \text{s} = 1\!\cdot\!0\ \text{ns}.$$

The force acting on the electron is given by

$$F = eE$$

and the acceleration of the electron in the direction of the force is given by

$$\frac{dv}{dt} = \frac{eE}{m}.$$

Therefore the final speed in that direction is given by

$$v = \frac{eEt}{m} = \frac{1\!\cdot\!602 \times 10^{-19} \times 10^4 \times 10^{-9}}{9\!\cdot\!109 \times 10^{-31}} = 1\!\cdot\!76 \times 10^6\ \text{m/s}.$$

Therefore the angle of deflection is given by the ratio of the two perpendicular velocities,

$$\text{angle} = \tan^{-1}\frac{1\!\cdot\!76}{10} = 10^0.$$

14.18.   Let the electric field intensity E be acting in the x-direction and the magnetic flux density B be acting in the z-direction. It is assumed that the electron is emitted from the plate with zero velocity. Then substituting into eqn. (14.3) for motion in the x- and y-directions,

$$m \frac{d^2x}{dt^2} = - e \left(E + \frac{dy}{dt} B \right) \quad \text{and} \quad m \frac{d^2y}{dt^2} = - e \left(- \frac{dx}{dt} B \right).$$

Integrating the second equation gives

$$m \frac{dy}{dt} = eBx + C.$$

If $x = 0$ when $\frac{dy}{dt} = 0$, then $C = 0$. Therefore substituting into the first equation gives

$$m \frac{d^2x}{dt^2} + \frac{e^2 B^2 x}{m} = -eE.$$

Provided that $x = 0$ and $\frac{dx}{dt} = 0$ when $t = 0$, this differential equation has the solution

$$x = - \frac{Em}{eB^2} \left\{ 1 - \cos\left(\frac{eBt}{m} \right) \right\}.$$

Substituting and integrating to find y gives

$$y = - \frac{E}{B} \left\{ t - \frac{m}{eB} \sin\left(\frac{eBt}{m} \right) \right\}.$$

The electron will move in a direction perpendicular to the magnetic flux density and will return to the plate from which it started. Inserting numbers from the problem will give the required results. The time to return to the plate is given by

$$\frac{eBt}{m} = 2\pi.$$

Therefore $\qquad t = \dfrac{2\pi m}{eB} = \dfrac{2\pi \times 9 \cdot 109 \times 10^{-31}}{1 \cdot 602 \times 10^{-19} \times 1 \cdot 6} = 2 \cdot 23 \times 10^{-11}$ s.

The distance moved is given by substituting into the equation for y,

$$y = \frac{10^5}{1 \cdot 6} \left(2 \cdot 23 \times 10^{-11} - \frac{m}{eB} \sin 2\pi \right) = 1 \cdot 40 \times 10^{-6} \text{ m} = 1 \cdot 40 \text{ } \mu\text{m}.$$

14.19.   The force is given by eqn. (14.7), therefore for the two poles attracting the striker plate,

$$\text{force} = \frac{B^2}{2\mu_0}(\text{area}) = \frac{0 \cdot 2 \times 0 \cdot 2 \times 2 \times 20 \times 10^{-6}}{2 \times 4\pi \times 10^{-7}} = 0 \cdot 637 \text{ N}.$$

14.20.   The force is given by eqn. (14.7). Therefore

$$\text{force} = \frac{B^2}{2\mu_0}(\text{area}) = \frac{2 \cdot 3 \times 2 \cdot 3 \times 10 \times 20 \times 10^{-6}}{2 \times 4\pi \times 10^{-7}} = 421 \text{ N}.$$

Physical Constants

Gravitational constant $\qquad G = 6{\cdot}67 \times 10^{-11}$ m³/kg s²

Gravitational permittivity $\qquad k = -1.19 \times 10^{9}$ kg s²/m³

Permeability constant $\qquad \mu_0 = 4\pi \times 10^{-7}$ H/m

Permittivity constant $\qquad \varepsilon_0 = \dfrac{1}{36\pi \times 10^{9}}$ F/m

Charge on the electron $\qquad e = 1{\cdot}602 \times 10^{-19}$ C

Mass of the electron $\qquad m = 9{\cdot}109 \times 10^{-31}$ kg

APPENDIX 2

Field Theory: Comparative Summary of Relevant Formulae

	Electrostatics	Magnetism	Gravitation
FLUX $\Psi =$	Charge Q coulombs	Magnetic charge, webers	Mass m kg
	Q C	Ψ Wb	m kg
FLUX DENSITY	D C/m^2	B T	D_g kg/m^2
FIELD INTENSITY	E N/C or V/m	H N/Wb or A/m	g N/kg or m/s^2
	$D = \varepsilon E$	$B = \mu H$	$D_g = -\dfrac{1}{4\pi G}\, g$
DIMENSIONAL CONSTANT	Permittivity ε F/m	Permeability μ H/m	$-\dfrac{1}{4\pi G}$ kg s^2/m^3
POTENTIAL DIFFERENCE Φ	V e.m.f. Volts	Φ m.m.f. amperes	gravitational potential
STORED ENERGY per unit volume	$\tfrac{1}{2}DE$ J/m^3	$\tfrac{1}{2}BH$ J/m^3	$\tfrac{1}{2}D_g g$ J/m^3
FLUX FUNCTION $\psi =$	$\iint \mathbf{D} \cdot d\mathbf{A}$	$\iint \mathbf{B} \cdot d\mathbf{A}$	$\iint \mathbf{D}_g \cdot d\mathbf{A}$
GAUSS'S LAW (integral over a closed surface)	$\iint \mathbf{D} \cdot d\mathbf{A} = \Sigma Q$	$\iint \mathbf{B} \cdot d\mathbf{A} = 0$	$\iint \mathbf{D}_g \cdot d\mathbf{A} = \Sigma m$
POTENTIAL FUNCTION $\phi_B - \phi_A =$	$-\int_A^B \mathbf{E} \cdot d\mathbf{l} = -\dfrac{1}{\varepsilon}\int_A^B \mathbf{D} \cdot d\mathbf{l}$	$-\int_A^B \mathbf{H} \cdot d\mathbf{l} = -\dfrac{1}{\mu}\int_A^B \mathbf{B} \cdot d\mathbf{l}$	$-\int_A^B \mathbf{g} \cdot d\mathbf{l}$
POTENTIAL GRADIENT	$\mathbf{E} = -\operatorname{grad} V$	$\mathbf{H} = -\operatorname{grad} \phi$	$\mathbf{g} = -\operatorname{grad} \phi$
AMPERE'S LAW	m.m.f. $= \int -\mathbf{H} \cdot d\mathbf{l} = \iint -\mathbf{J} \cdot d\mathbf{A}$		
FARADAY'S LAW	e.m.f. $= \int -\mathbf{E} \cdot d\mathbf{l} = \dfrac{\partial}{\partial t} \iint \mathbf{B} \cdot d\mathbf{A}$		

Electrical conduction	Conductive heat transfer	Fluid flow through permeable media	Ideal fluid flow
Current I amperes I A J A/m^2 E V/m	Heat flow Q watts Q W q W/m^2 Thermal gradient K/m	Discharge Q m^3/s Q m^3/s Velocity v m/s Pressure gradient N/m^3	Discharge Q m^3/s Q m^3/s Velocity v m/s Velocity v m/s
$J = \sigma E$ Conductivity σ S/m V e.m.f. volts IE W/m^3 (power dissipated)	$q = -k \operatorname{grad} T$ Thermal conductivity k W/m K T temperature K	$v = -k \operatorname{grad} p$ Permeability k m^3 s/kg p pressure N/m^2 $v \operatorname{grad} p$ W/m^3 (power dissipated)	$v = v$ l Φ velocity potential $\frac{1}{2}v^2$ J/kg (kinetic energy per unit mass)
$\iint \mathbf{J} \cdot d\mathbf{A}$ $\iint \mathbf{J} \cdot d\mathbf{A} = \Sigma I$	$\iint \mathbf{q} \cdot d\mathbf{A}$ $\iint \mathbf{q} \cdot d\mathbf{A} = \Sigma Q$	$\iint \mathbf{v} \cdot d\mathbf{A}$ $\iint \mathbf{v} \cdot d\mathbf{A} = \Sigma Q$	$\iint \mathbf{v} \cdot d\mathbf{A}$ $\iint \mathbf{v} \cdot d\mathbf{A} = \Sigma Q$
$-\int_A^B \mathbf{E} \cdot d\mathbf{l} = -\dfrac{1}{\sigma} \int_A^B \mathbf{J} \cdot d\mathbf{l}$ $\mathbf{E} = -\operatorname{grad} V$	$-\dfrac{1}{k} \int_A^B \mathbf{q} \cdot d\mathbf{l}$ $q = -k \operatorname{grad} T$	$-\dfrac{1}{k} \int_A^B \mathbf{v} \cdot d\mathbf{l}$ $v = -k \operatorname{grad} p$	$-\int_A^B \mathbf{v} \cdot d\mathbf{l}$ $\mathbf{v} = -\operatorname{grad} \phi$

APPENDIX 3

Notation

$\mathbf{A}, A$	arbitrary vector
$\mathbf{A}, A$	area
a	distance
a	radius of a circle
$\mathbf{B}, B$	arbitrary vector
$\mathbf{B}, B$	magnetic flux density
b	boundary value
$\mathbf{C}, C$	arbitrary vector
C	capacitance
c	constant
c	strength of a doublet
$\mathbf{D}, D$	electric flux density
$\mathbf{D}_g, D_g$	gravitational flux density
D	diameter
d	distance
d	differential coefficient
$\mathbf{E}, E$	electric field intensity
e	exponential coefficient
e	electronic charge
$\mathbf{F}, F, \mathbf{f}, f$	force
f	arbitrary function
G	gravitational constant
$\mathbf{g}, g$	gravitational field intensity, acceleration due to gravity
$\mathbf{H}, H$	magnetic field intensity
h	distance, height
h	surface heat transfer coefficient
I	moment of inertia
I, i	current
$\mathbf{J}, J$	current density
K, k	constant
k	gravitational permittivity
k	thermal conductivity

k	permeability (of a permeable medium)
$\mathbf{k},k$	vorticity
k	line vortex strength
L	inductance
l	length
$\mathbf{M},M$	magnetization
M	mutual inductance
m	mass
m	dipole strength, dipole moment
N,n	number of turns
N,n	an integer
P	power
P,p	pressure
Q	strength of source
Q	rate of heat flow
Q	charge
q	point charge
q	linear charge density
$\mathbf{q},q$	thermal flux density
R	radial distance
R	electrical resistance, thermal resistance
R	Hall constant
R_m	reluctance
r	radial coordinate in polar coordinate system
S,s	surface
T	temperature
t	time
$\mathbf{U}$	unit vector
U	uniform velocity
V	electrical potential
$\mathbf{v},v$	velocity
v	volume
W	work, energy
w	energy density
x,y,z	dimensional coordinate in rectangular coordinate system
z	height above some datum
α,β	angle

δ	differential coefficient
ε	permittivity
ε_0	permittivity constant
ε_r	relative permittivity
θ	angle
θ	angular coordinate in the polar coordinate system
μ	permeability
μ_0	permeability constant
μ_r	relative permeability
ρ	mass density
ρ	charge density
ρ	volume rate of heat generation
ρ	resistivity
σ	conductivity
σ	surface charge density
σ	surface tension
Φ	potential difference
ϕ	potential function
Ψ	flux
ψ	flux function
ω	angular velocity
∇	differential coefficient (nabla)
∂	partial differentiation coefficient

Index

Printed in France by Amazon
Brétigny-sur-Orge, FR

48269746R00188